Springer Complexity

Springer Complexity is an interdisciplinary program publishing the best research and academic-level teaching on both fundamental and applied aspects of complex systems – cutting across all traditional disciplines of the natural and life sciences, engineering, economics, medicine, neuroscience, social and computer science.

Complex Systems are systems that comprise many interacting parts with the ability to generate a new quality of macroscopic collective behavior the manifestations of which are the spontaneous formation of distinctive temporal, spatial or functional structures. Models of such systems can be successfully mapped onto quite diverse "real-life" situations like the climate, the coherent emission of light from lasers, chemical reaction–diffusion systems, biological cellular networks, the dynamics of stock markets and of the Internet, earthquake statistics and prediction, freeway traffic, the human brain, or the formation of opinions in social systems, to name just some of the popular applications.

Although their scope and methodologies overlap somewhat, one can distinguish the following main concepts and tools: self-organization, nonlinear dynamics, synergetics, turbulence, dynamical systems, catastrophes, instabilities, stochastic processes, chaos, graphs and networks, cellular automata, adaptive systems, genetic algorithms and computational intelligence.

The two major book publication platforms of the Springer Complexity program are the monograph series "Understanding Complex Systems" focusing on the various applications of complexity, and the "Springer Series in Synergetics", which is devoted to the quantitative theoretical and methodological foundations. In addition to the books in these two core series, the program also incorporates individual titles ranging from textbooks to major reference works.

Understanding Complex Systems

Founding Editor: J.A. Scott Kelso

Future scientific and technological developments in many fields will necessarily depend upon coming to grips with complex systems. Such systems are complex in both their composition – typically many different kinds of components interacting simultaneously and nonlinearly with each other and their environments on multiple levels – and in the rich diversity of behavior of which they are capable.

The Springer Series in Understanding Complex Systems series (UCS) promotes new strategies and paradigms for understanding and realizing applications of complex systems research in a wide variety of fields and endeavors. UCS is explicitly transdisciplinary. It has three main goals: First, to elaborate the concepts, methods and tools of complex systems at all levels of description and in all scientific fields, especially newly emerging areas within the life, social, behavioral, economic, neuro- and cognitive sciences (and derivatives thereof); second, to encourage novel applications of these ideas in various fields of engineering and computation such as robotics, nano-technology and informatics; third, to provide a single forum within which commonalities and differences in the workings of complex systems may be discerned, hence leading to deeper insight and understanding.

UCS will publish monographs, lecture notes and selected edited contributions aimed at communicating new findings to a large multidisciplinary audience.

A. Fuchs · V. K. Jirsa (Eds.)

Coordination: Neural, Behavioral and Social Dynamics

With 123 Figures

 Springer

Armin Fuchs
Viktor K. Jirsa
Center for Complex Systems &
Brain Sciences
Florida Atlantic University
777 Glades Road
Boca Raton, FL 33431-0991, USA
fuchs@ccs.fau.edu
jirsa@ccs.fau.edu

ISBN: 978-3-540-74476-4   e-ISBN: 978-3-540-74479-5

Understanding Complex Systems ISSN: 1860-0832

Library of Congress Control Number: 2007937246

Cover Design: WMX Design GmbH

Printed on acid-free paper

9 8 7 6 5 4 3 2 1

springer.com

To our mentor, good friend, and a great scientist, Scott Kelso

Preface

In February 2007 an international conference was held in Boca Raton, Florida, in honor of J.A. Scott Kelso's 60th birthday. The theme of this meeting "Coordination: Neural, Behavioral and Social Dynamics" reflects the richness of a scientific field that has emerged during the last quarter century. This book contains invited contributions by leading scientists who have played a major role in pushing this field from its infancy to a state where it plays an established role among the more senior disciplines in the life sciences.

One of the most striking features of Coordination Dynamics is its interdisciplinary character. The problems we aim to solve in this field range from understanding of behavioral phenomena in interlimb and sensorimotor coordination, perception–action cycles through neural pattern dynamics to applications in clinical and social domains. It is not surprising that close collaborations among scientists from different fields such as psychology, kinesiology, neurology, and even physics are imperative when attempting to understand a system as complex as the human brain.

The chapters in this volume are not simply summaries of the lectures given by the experts at the meeting but are written with the mindset to provide sufficient introductory information to be comprehensible and useful for all interested scientists and students.

We wish to thank all those who were involved in the making of this book, especially to the experts who contributed the most essential parts of any book, its content. The editorial help we received from students at the Center for Complex Systems & Brain Sciences at Florida Atlantic University is much appreciated. Last but not least we are most grateful to Editor Dr. Thomas Ditzinger at Springer Verlag for his advice and guidance throughout the editorial process.

Boca Raton, Florida   *Armin Fuchs*
June 2007   *Viktor K. Jirsa*

List of Contributors

Danielle S. Bassett
Unit for Systems Neuroscience in
Psychiatry, and Neuroimaging Core
Facility, Genes, Cognition and
Psychosis Program, National
Institute for Mental Health, GCAP,
NIMH,DHHS, NIH, 10-3C103, 9000
Rockville Pike, Bethesda, MD
20892-1365, USA
bassettda@mail.nih.gov

Niels Birbaumer
Institute of Medical Psychology
and Behavioral Neurobiology,
University of Tübingen,
Gartenstrasse 29, D-72074 Tübingen,
Germany
niels.birbaumer@uni-tuebingen.de

Douglas Cheyne
Program in Neurosciences and
Mental Health, Hospital for Sick
Children Research Institute, 555
University Avenue, Toronto, Ontario
M5G 1X8, Canada and Department
of Medical Imaging, University of
Toronto, Toronto, Ontario, Canada
douglas.cheyne@utoronto.ca

Brett R. Fajen
Department of Cognitive Science,
Rensselaer Polytechnic Institute,
Troy, NY 12180-3590

Carol A. Fowler
Haskins Laboratories and University
of Connecticut, 300 George
Street, Storrs, CT 06511, USA
carol.fowler@haskins.yale.edu

Armin Fuchs
Department of Physics, Center for
Complex Systems and Brain
Sciences, Florida Atlantic University,
777 Glades Road, Boca Raton,
FL 33431, USA
fuchs@ccs.fau.edu

Klaus Haagen
Department of Economy,
University of Trento, Via Inama 5,
I-38100 Trento, Italy
klaus.haagen@unitn.it

Raoul Huys
Theoretical Neuroscience Group,
UMR 6152 Institut Mouvement et

Perception, Université de la
Méditerranée and CNRS, 163
av. de Luminy, CP910, F-13288
Marseille Cedex 09, France
raoul.huys@univmed.fr

Kelly J. Jantzen
Department of Psychology, Western
Washington University, 516 High
Street, Bellingham, WA 98225, USA
kelly.jantzen@wwu.edu

Viktor K. Jirsa
Theoretical Neuroscience Group,
UMR 6152 Institut Mouvement et
Perception, Université de la
Méditerranée and CNRS,
Marseille, Cedex 09, France and
Center for Complex Systems and
Brain Sciences, Florida Atlantic
University, 777 Glades Road,
Boca Raton, FL 33431, USA
jirsa@ccs.fau.edu

James R. Lackner
Ashton Graybiel Spatial Orientation
Laboratory, Brandeis University,
Waltham, MA 02254-9110, USA

Yeou-Teh Liu
Graduate Institute of Exercise and
Sport Science, National Taiwan
Normal University, 88 Ting-Zhou
Road Section 4, 116, Taipei, Taiwan
yeouth@ntnu.edu.tw

Kerry L. Marsh
University of Connecticut, UConn
Hartford Campus, 85 Lawler Road,
West Hartfod, CT 06117, USA
Kerry.L.Marsh@uconn.edu

Marcello Massimini
Department of Psychiatry, University
of Wisconsin, 6001 Research
Park Blvd, Madison, WI 53719, USA
mmassimini@wisc.edu

Gottfried Mayer-Kress
Department of Kinesiology, The
Pennsylvania State University, 276
Recreation Building, University
Park, PA 16802, USA
gxm21@psu.edu

Anthony R. McIntosh
Rotman Research
Institute of Baycrest, Baycrest
University of Toronto, 3560 Bathurst
St, Toronto, Ontario, Canada,
M6A 2E1
rmcintosh@rotman-baycrest.on.ca

Andreas Meyer-Lindenberg
Central Institute of Mental Health,
J5. D-68159 Mannheim, Germany
A.Meyer-Lindenberg@zi-mannheim.de

Karl M. Newell
Department of Kinesiology, The
Pennsylvania State University, 210
Henderson, University Park, PA
16802, USA
kmn1@psu.edu

Olivier Oullier
Laboratoire de Neurobiologie Huimaine (UMR 6149),
Aix-Marseille Université,
3 place Victor Hugo, F-13331
Marseille Cedex 3, France
olivier@oullier.fr

Hyeongsaeng Park
Department of Psychology, Seoul
National University, Shilim-dong,
Kwanak-gu, Seoul, 151-742, Korea
and Department of Psychology,
Center for Ecological Study of
Perception and Action, University of
Connecticut, Storrs, CT 06269-
1020, USA
parkie@snu.ac.kr

Nicole Rheaume
Health and Kinesiology, Purdue
University, 800 W. Stadium Blvd.,
West Lafayette, IN 47907, USA
nrheaume@purdue.edu

Michael J. Richardson
Department of Psychology, Colby
College, Mayflower Hill Drive,
Waterville, ME 04901-8885, USA
and Department of Psychology,
College of the Holy Cross and
University of Connecticut,
Waterville, ME 04901-8885, USA
mjrichar@colby.edu

Richard C. Schmidt
Department of Psychology, College
of the Holy Cross and University
of Connecticut, PO Box 176A,
1 College Street, Worcester,
MA 01610, USA
rschmidt@holycross.edu

Kevin D. Shockley
University of Cincinnati, 229-E Dyer
Hall, PO Box 210376,
Cincinnati, OH 45221-3124, USA
kevin.shockley@uc.edu

Surjo R. Soedakar
Department for Psychiatry and
Psychotherapy, Neurostimulation
Unit, University of
Tübingen, Osianderstrasse 24,
D-72076 Tübingen, Germany

Dagmar Sternad
Departments of Kinesiology and
Integrative Biosciences, Pennsylvania
State University, 266 Rec Hall,
University Park, PA 16802, USA
dxs48@psu.edu

Brenna E. Studenka
Health and Kinesiology, Purdue
University, West Lafayette, IN
47907, USA

Emmanuelle Tognoli
Center for Complex Systems and
Brain Sciences, Florida Atlantic
University, 777 Glades Road, Boca
Raton, FL 33431, USA
tognoli@ccs.fau.edu

Giulio Tononi
Department of Psychiatry, University
of Wisconsin, 6001 Research
Park Blvd., Madison, WI 53719, USA
gtononi@wisc.edu

Betty Tuller
Center for Complex Systems and
Brain Sciences, Florida Atlantic
University, Boca Raton, FL
33431, USA

Michael T. Turvey
Department of Psychology,
Center for the Ecological Study of
Perception
and Action, University of
Connecticut, U 20,
406 Babbidge Road,
Unit 1020, Storrs,
CT 06269-1020, USA
michael.turvey@uconn.edu

William H. Warren
Department of Cognitive and Lin-
guistic Sciences, Brown University,
Providence, RI 02912, USA

Howard N. Zelaznik
Health and Kinesiology, Purdue
University, 800 W. Stadium Blvd.,
West Lafayette, IN 47907, USA
hnzelaz@purdue.edu

Contents

Part I

Behavioral Dynamics

Imperfect Symmetry and the Elementary Coordination Law

Hyeongsaeng Park[1,2] and Michael T. Turvey[2]

[1] Department of Psychology, Seoul National University, Seoul 151–742, Korea
[2] Department of Psychology, Center for the Ecological Study of Perception and Action, University of Connecticut, Storrs, CT 06269-1020, USA

1 Introduction

The ability to synchronize rhythmically moving limbs and limb segments is one of the most fundamental abilities of vertebrate and invertebrate movement systems. As Kelso [18] has underscored, the ability is a primary expression of how movements (a) are organized in space and time, (b) resolve issues of efficiency, and (c) meet the competing challenges of stability and flexibility. In broad theoretical terms, 1:1 frequency locking of two or more limb segments is one of biology's original models for collective behavior – the organizing of multiple interactions among neural, muscular, metabolic, and mechanical processes under task-specific intentional constraints. Given the complexity, the task of formulating and validating quantitative mathematical models of monofrequency rhythmic coordination based on physicochemical principles, neurobiological facts, and assumptions about intentionality has proven to be extremely difficult and may well be intractable. An alternative approach, the one adopted more than two decades ago by Haken, Kelso, and Bunz [15], attempts to develop a qualitative dynamical model that incorporates, in broad strokes, the essential features of synchrony between and among the components of, in principle, any biological movement system. The model-independent approach taken by Kelso and his colleagues accords with elementary lessons from the study of complexity [11].

- *Lesson 1*: Even in simple situations, Nature produces complex structures and even in complex situations Nature obeys simple laws.
- *Lesson 2*: Revealing large-scale structure requires a description that is phenomenological and aggregated and directed specifically at the higher level. A modeling strategy that includes very many processes and parameters obscures (qualitative) understanding.

As argued by Stewart and Golubitsky [37], a model-independent approach explores level-independent regularities of phenomena and strives to optimize those lawful properties into principles. The latter is an endeavor aptly described as the extracting of invariants or the discovering of symmetries. The present chapter takes this endeavor as its focus.

2 The Elementary Coordination Law: Bistability and Haken–Kelso–Bunz (HKB) Dynamics

In 1984, Kelso [17] reported a simple experiment showing that a person moving the hands rhythmically at a common frequency tends to do so primarily in two stable patterns of coordination, in-phase and antiphase. Furthermore, with an increase in the common frequency, Kelso [17] observed a tendency for antiphase to switch spontaneously to in-phase but not vice versa. This bistable 1:1 frequency locking of limbs can be characterized by relative phase ϕ, with the observed interlimb patterns mapped onto point attractors at $\phi = 0$ and $\phi = \pi$. The simplest dynamics of ϕ satisfying the aforementioned behavior for two limbs or limb segments of the same type (e.g., left and right index fingers) are given by

$$\dot{\phi} = -\frac{\mathrm{d}V}{\mathrm{d}\phi}, \tag{1}$$

where V is the potential function

$$V(\phi) = -a \cos \phi - b \cos 2\phi. \tag{2}$$

The form of V is depicted in Fig. 1. The valleys (at 0 and $\pm\pi$) are attractors, the hilltops (at $\pm\pi/2$ and $\pm3\pi/2$) repellers. Of significance to present concerns is the fact that Fig. 1 expresses two symmetries, namely reflectional symmetry, $V(\phi) = V(-\phi)$, and time-translational symmetry, $V(\phi) = V(2\pi + \phi)$. It identifies 1:1 frequency-locked rhythmic coordination of two limbs as a dynamic that is invariant over reflectional and temporal transformations. Given (2), (1) becomes

$$\dot{\phi} = -a \sin \phi - 2b \sin 2\phi. \tag{3}$$

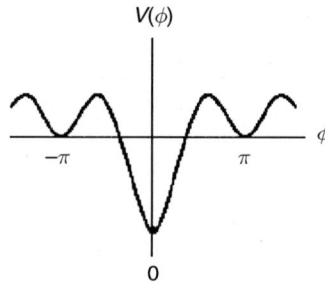

Fig. 1. The potential function $V(\phi)$

Equation (3) expresses the elementary coordination law identified by Haken, Kelso, and Bunz (hereafter HKB) [15]. As is now well known, the law captures the observed transition between the patterns. For $a < 4b$, stable states exist at $\phi = 0$ and $\phi = \pm\pi$; for $a \geq 4b$, only $\phi = 0$ is retained (for details see Kelso [19]). Formally, a change of the foregoing kind is a bifurcation: a change in the number and/or type of fixed points. Specifically, it is a subcritical pitchfork bifurcation. It is depicted in Fig. 2A. The 'collision' of unstable fixed points at $\phi = \pi/2$ and $3\pi/2$ with the stable fixed point at $\phi = \pi$ changes the fixed point at π from an attractor to a repeller. The bifurcation arises spontaneously when b/a of (3) reaches a critical value. The parameter b/a is a means of controlling the passage of the coordinated system through its qualitatively distinct organizations.

3 Unfolding the Elementary Coordination Law

The guiding thesis of this chapter is that the status of (3) (alias the HKB equation or HKB dynamics) as the elementary coordination law is based on the assumption that all possible patterns in the 1:1 coordination of biological components are derivable from it by the adding of a finite number of small parameters. In terms of dynamical systems theory, (3) is the organizing center of monofrequency coordination dynamics and the added parameters are imperfection parameters. They render the perfect symmetry of (3) imperfect.

As with laws in general, the HKB equation defines what is possible. The addendum of circumstances determines what actually occurs [5]. That is, paralleling the general case of 'laws + circumstances = actuality', the nature of 1:1 rhythmic coordination is 'elementary coordination law + the circumstances of coordination = the actual pattern of coordination'. The importance of the latter formulation is that one expects to see the footprint of HKB dynamics in each and every manifestation of monofrequency rhythmic behavior – most notably, the feature of reflectional symmetry and the distinction between the stability of coordination at (or in the vicinity of) 0 rad and that at (or in the vicinity of) π rad.

At issue in unfolding (3) is identifying the number and type of imperfection parameters to be added. A working assumption is that the number is not only finite but also very small [18]. For the number to be very small, it would have to be the case that the imperfections arising from the multiple circumstances in which HKB dynamics are embedded differ only superficially as far as the biology instantiating those dynamics is concerned. That is, it would have to be the case that the circumstances rendering the symmetry of (3) imperfect comprise relatively few natural kinds.

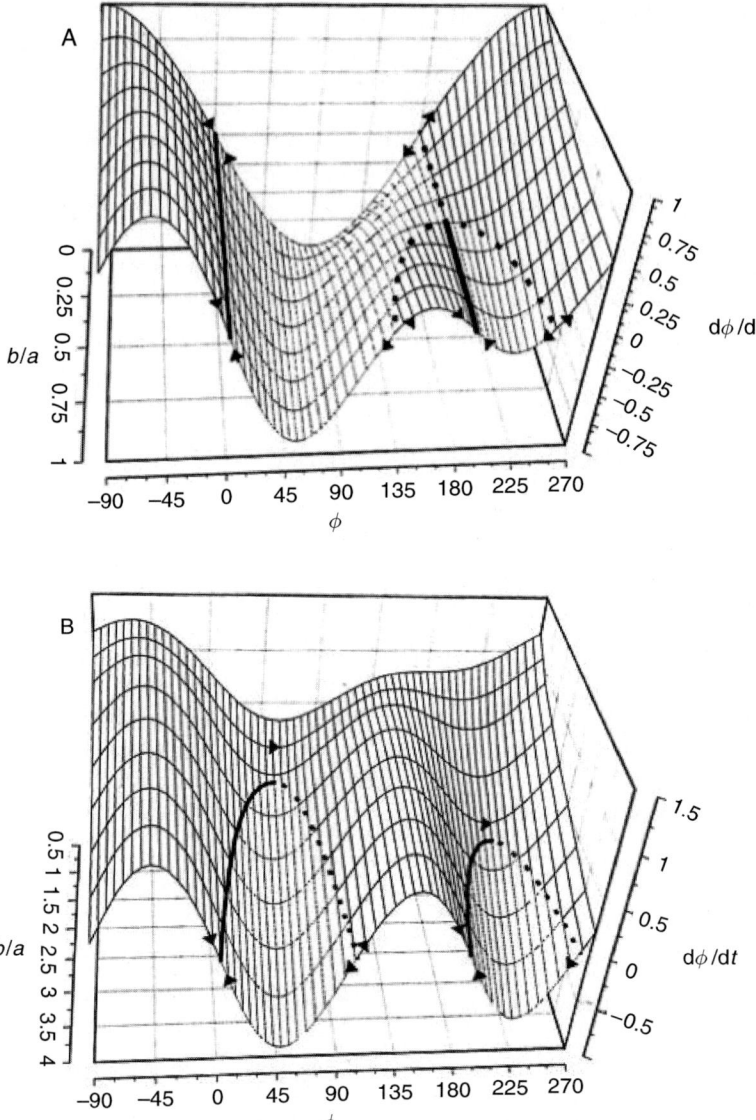

Fig. 2. A. The elementary coordination law, (3), expressed as a vector field. *Converging arrows* and *solid lines* index attracting fixed points; *diverging arrows* and *dotted lines* index repelling fixed points. The defining bifurcation of the symmetric equation (3), the subcritical or inverse pitchfork, results from a decrease in b/a. B. The imperfect elementary coordination law (4) expressed as a vector field. The saddle node bifurcation characteristic of the asymmetry induced by $\delta \neq 0$ is shown for both in-phase and antiphase coordinations (Adapted with permission from Kelso [18])

4 Predictions and Methods

The elementary coordination law has proven to be more than a compact and convenient way to describe interlimb synchrony. It provides novel predictions that can be evaluated experimentally, especially when fluctuations in coordination and differences between the two limbs are taken into account, respectively, by adding a Gaussian white noise ξ_t of strength $\sqrt{Q}$ and including an imperfection parameter δ (see summaries in Amazeen et al. [1], Kelso [19], Turvey and Carello [39]). Equation (3) can be expressed as

$$\dot{\phi} = -a \sin \phi - 2b \sin 2\phi + \delta + \sqrt{Q}\,\xi_t. \qquad (4)$$

Prominent among the major predictions of (4) are the following. First, the in-phase and antiphase point attractors ($\dot{\phi} = 0$) are shifted from their canonical (alias, perfect) values by $\delta \neq 0$ (where the canonical values are those for $\delta = 0$) with the induced deviation from $\phi = \pm\pi$ greater than the induced deviation from $\phi = 0$ and scaled to the magnitude of $\delta \neq 0$. Second, the influence of a given imperfection $\delta \neq 0$ on the in-phase and antiphase point attractors is magnified by reductions in b/a. (Both of the preceding predictions are visible in Fig. 2B.) Third, accompanying the predicted shift in attractors due to δ is a predicted increase in the variability of relative phase as measured by the standard deviation, SDϕ. The magnitude of SDϕ is predicted to be larger for $\delta \neq 0$ than for $\delta = 0$. Fourth, the rate of increase of SDϕ with deviation from $\delta = 0$ is predicted to be greater for the attractor with canonical value $\phi = \pi$. The preceding increases and magnitude differences for the two attractors arise from the degrees of flattening of the potential wells surrounding the attractors. As a potential well becomes shallower, random kicks of fixed strength will displace the coordination further from the well's lowest point and prolong the relaxation time (that is, the strength of the attractor is reduced as the walls of the surrounding well become less steep). The variability in relative phase accompanying the dynamics of (4) is given by

$$\mathrm{SD}\phi = \sqrt{\frac{Q}{2\,|\lambda|}}, \qquad (5)$$

where λ is the attractor strength evaluated as the slope at $\dot{\phi} = 0$ [34].

Testing the four predictions has been done most intensively under experimental conditions in which the participant, on a given trial, is simply required to maintain an in-phase or antiphase coordination for a fixed duration at a fixed movement frequency (both controlled by the experimenter). It is a paradigm that focuses on steady states rather than transitions among states.

In the most common variant, the coordination is between handheld pendulums [22]. By manipulations of the length of the shaft and the mass and position of rings attached to the shaft, the experimenter can systematically vary the equivalent simple pendulum length and thereby the handheld pendulum's natural (undamped, undriven) frequency [8]. For purposes of giving a specific definition to δ and a means of controlling δ, the experimenter can fashion pendulums that are identical in their natural frequencies ($\delta = 0$) or differ in their natural frequencies ($\delta \neq 0$). To round out the method, if handheld pendulums permit systematic manipulations of the parameter δ in (4), a metronome can be used to permit systematic manipulations in the a and b parameters of (4) [15, 33].

The described method has confirmed all four of the above predictions (see [1] for a summary).

5 Breaking Reflectional Symmetry

The predictions identified in Sect. 4 and partially depicted in Fig. 2 provide a point of entry into the symmetry of interlimb coordination patterns. As will become apparent, the pivotal issues will be the nature and status of δ.

5.1 Frequency Differences

Figure 3 is a schematic of the patterning of $\phi = \theta_{\text{left}} - \theta_{\text{right}}$ (with θ the phase angle) and SDϕ for in-phase coordination under the experimental manipulation of δ, interpreted as $\Delta\omega = \omega_{\text{left}} - \omega_{\text{right}}$, and orientation (both up or both down) of the oscillating components [2]. The top row of Fig. 3 depicts the breaking of (3)'s reflectional symmetry. For simplicity of illustration, the figure does so only in terms of positional changes in the nadir of V due to

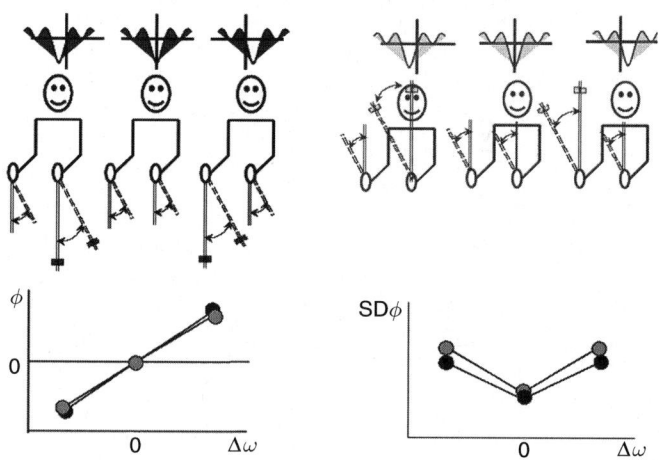

Fig. 3. Effect of $\delta = \Delta\omega$ on the coordination equilibria for coupled ordinary (*black*) and vertical (*gray*) handheld pendulums (see text for details)

the $\Delta\omega$ conditions shown in the middle row. The $\Delta\omega$ conditions consist of a zero instance and two nonzero instances that differ solely in the assignment of pendulums to hands. With respect to $|\Delta\omega|$ they are equal. As the bottom row indicates, the relative effects of temporal symmetry (indexed by $\Delta\omega$) on the steady state coordination (indexed by ϕ) and its variability (indexed by SDϕ) are indifferent to orientation.

The absence of differences in the values of ϕ due to orientation is of relevance to the basic experimental method. The neuromuscular organization for oscillating an inverted pendulum cannot be identical to that for oscillating a hanging pendulum. The difference follows from the fact that gravity tends to remove the inverted pendulum from the vertical, in contrast to its effect in the ordinary hanging situation in which gravity tends to return the pendulum to the vertical. That the attractors (equilibria) of the coupled vertical pendulums and the coupled hanging pendulums were the same for the two nonzero instances of $\Delta\omega$ affirms the identification of the undamped, undriven (unforced) frequencies as the more appropriate components of $\Delta\omega$ than the damped, driven (muscularly forced) frequencies.

5.2 Frequency and Orientation Differences

Figure 4 is a schematic of the patterning of ϕ and SDϕ for in-phase coordination under the joint manipulation of frequency and orientation differences [2]. For the participants in the left panel, the vertical pendulum is in the right hand (R-up); for the participants in the right panel, the vertical pendulum is in the left hand (L-up).

The changes in V induced by the experimental conditions depicted in Fig. 4 – and, thereby, the changes in ϕ and SDϕ – are not predictable by adding $\delta = \Delta\omega$ and $\sqrt{Q}\,\xi_t$ to (3). It should be apparent from comparison with Fig. 3 that δ interpreted as $\Delta\omega = \omega_{\text{left}} - \omega_{\text{right}}$ is incomplete. In the absence of

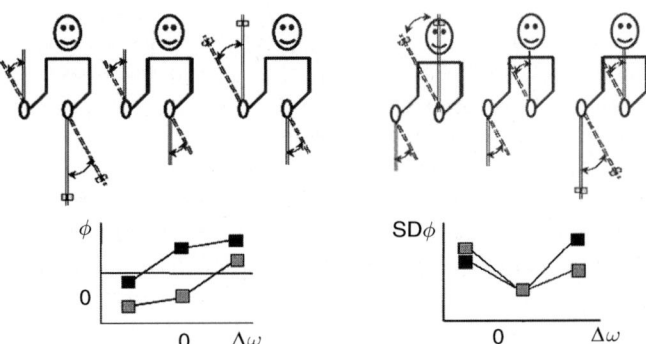

Fig. 4. Effect of differences in eigenfrequencies and orientations on the coordination equilibria of coupled ordinary and vertical handheld pendulums. *Black squares* are R-up conditions and *gray squares* are L-up conditions (see text for details)

a frequency difference ($\Delta\omega = 0$), the orientation difference shifts the attractor of the in-phase coordination pattern from the canonical value of $\phi = 0$ rad. In both the R-up and the L-up case, the relative phase defined conventionally as $\phi = \theta_{\text{left}} - \theta_{\text{right}}$ expresses a phase lead of the down pendulum relative to the up pendulum.

Nonintuitively, perhaps, the strictly orientation-induced phase shift leaves unchanged the stability benefits of zero $\Delta\omega$ relative to nonzero $\Delta\omega$. As can be seen in the plot of SDϕ, variability was least at $\Delta\omega = 0$ for both L-up and R-up. A rough understanding of the continued benefit of zero $\Delta\omega$ would follow from the two suppositions that the imperfections (a) are expressed in nondimensional form and (b) sum (Sect. 3). For example, an orientation imperfection of absolute value 0.5 added to absolute temporal imperfections of values 0, 1, 2, etc. would result in a net imperfection that was least for the 0 value of temporal imperfection.

The simple logic of adding dimensionless imperfections might also capture the order (but not the relative magnitudes) of the fixed points in the $\phi \times \Delta\omega$ plot. Let R-up be $\frac{1}{2}$ and R-down be $-\frac{1}{2}$, and let L-up be $-\frac{1}{2}$ and L-down be $\frac{1}{2}$. The convention equates both-up and both-down with 0, the R-up, L-down combination with 1 and the R-down, L-up combination with -1. Let the dimensionless values of $\Delta\omega$ be $-2, 0$, and $+2$. Then, summing the imperfections yields net values of $-1, 1$, and 3 for R-up conditions and $-3, -1$, and 1 for the L-up conditions.

It is worth noting in Fig. 4 that SDϕ is least for $\Delta\omega = 0$ even though at $\Delta\omega = 0$ the displacement of ϕ can exceed that at instances of $\Delta\omega \neq 0$. This latter fact – lower variability despite larger displacement from $\phi = 0$ – eliminates the interpretation of the down pendulum lead at $\Delta\omega = 0$ as due to a disguised frequency difference between the two oscillators.

5.3 Persistence of Reflectional Symmetry

Figure 5 shows the reflections of the R-up results depicted in Fig. 4. The reflections are tantamount to exchanging the left and right pendulums first in respect of $\Delta\omega$, $-1(\Delta\omega) = \omega_{\text{right}} - \omega_{\text{left}}$ and second in respect of ϕ, $-1(\phi) = \theta_{\text{right}} - \theta_{\text{left}}$. Both steps are needed for relative phase; only the first is needed for the standard deviation of relative phase.

The participants in the experimental conditions of Fig. 4 are discovering the attractors, the fixed points, of the monofrequency coordination dynamics governed by the elementary coordination law (3) when the reflectional symmetry of those dynamics is broken by frequency and orientation differences. Figure 5 indicates that the discovered fixed points are related by reflections, consistent with the Extended Curie Principle [37]: symmetries are not so much broken as shared around. Each condition on display in Fig. 4 is like a small experiment. At the condition level, broken symmetry seems the right description. At the condition level, however, one can only observe a single member

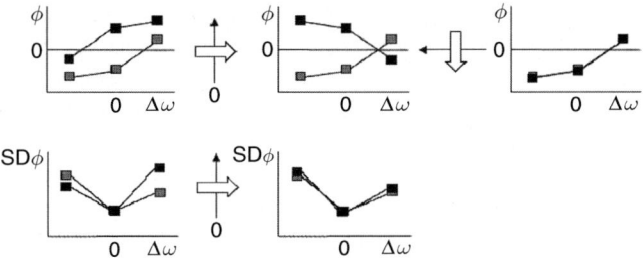

Fig. 5. Reflectional symmetry is shared. In the upper panels, the R-up conditions in Fig. 4 are reflected left to right about the axis through $\Delta\omega = 0$, and top to bottom about the axis through $\phi = 0$. In the lower panels, the R-up conditions are reflected left to right about the axis through $\Delta\omega = 0$ (see text for details)

of the symmetrically related bunch of solutions that the elementary coordination law guarantees. Looking over all of the conditions, all of the small experiments, as in Fig. 5, one can observe that the various single-condition solutions are related. To reiterate, the notion of symmetry sharing [37] seems to describe the situation better than symmetry breaking.

5.4 Frequency and Axis-of-Rotation Differences

The results made evident in the schematics of Figs. 4 and 5 should replicate under other sets of differences. In Kelso's words, '*any* situation that introduces or amplifies intrinsic differences between interacting elements may break the symmetry of the dynamics' ([18], p. 311). Figure 6 shows an adaptation of a symmetry-breaking arrangement due to Carson and colleagues [6]. The left and right hands are strictly aligned in parasagittal planes (Fig. 6A). The manipulanda that they grasp are oscillated in a paracoronal plane (Fig. 6A). The symmetry breaking of the elementary coordination dynamics occurs with respect to the axles about which the left and right manipulanda oscillate.

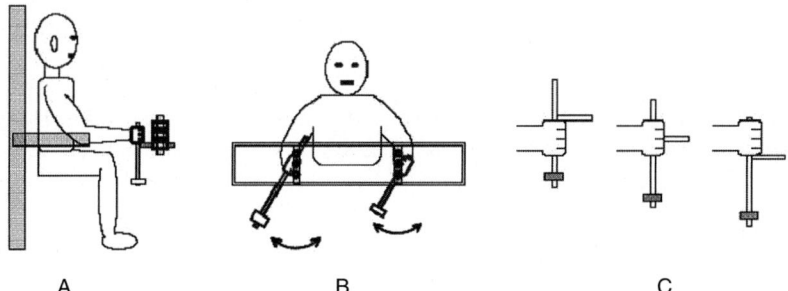

A   B   C

Fig. 6. A method for combining symmetry breaking through (**A**) a difference in rotation axis and (**B**) a difference in frequency (see text for details)

Figure 6C depicts three axles, 'top', 'middle' and 'bottom.' Reflectional symmetry is broken when the left and right axles are different. The adaptation [24] is in the addition of pendular components to the Carson manipulanda. The addition allows for the introduction of symmetry breaking through a difference in frequency.

An experiment equivalent to that of Amazeen et al. [2] can be conducted with L-top (axle) and R-top (axle) as the counterparts to L-up and R-up, respectively (Experiment 1 in [24]). Figure 7 (*left*) provides a schematic of the design. (The value of $\Delta\omega = 0$ was achieved by equating the simple pendulum lengths, the ratios of second to first moment, for the two different rotation axes and validated by freely elected, comfortable oscillations performed by a separate group of participants.)

The results shown in Fig. 7 (*right*) are of like kind with those presented in Fig. 4. In the absence of a frequency difference ($\Delta\omega = 0$), the rotation-axis difference shifts the attractor of the in-phase coordination pattern from the canonical value of $\phi = 0$ rad. In both the R-top and the L-top cases, $\phi = \omega_{\text{left}} - \omega_{\text{right}}$ expresses a phase lead of the bottom-axis pendulum relative to the top-axis pendulum (paralleling the phase lead of the down pendulum relative to the up pendulum in Fig. 4). The transformations of $-1(\Delta\omega) = \omega_{\text{right}} - \omega_{\text{left}}$ and $-1(\phi) = \theta_{\text{right}} - \theta_{\text{left}}$ confirm that the values (the attractors) for L-top and R-top are related by reflection [24]. The $-1(\Delta\omega)$ transform applied to the values of SDϕ confirms that they are similarly in agreement with the Extended Curie Principle. It should also be noted with respect to SDϕ that the axis-induced phase shift leaves unchanged the stability benefits of zero $\Delta\omega$ relative to nonzero $\Delta\omega$. As in Fig. 4, the latter suggests an adding of imperfections.

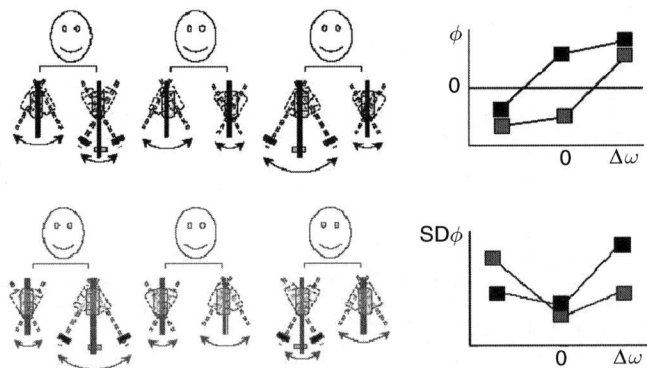

Fig. 7. Effect of differences in eigenfrequencies and rotation axes on coordination equilibria with *black squares* the R-top (axle) conditions and *gray squares* the L-top (axle) conditions (see text for details)

6 Temporal Imperfection Parameter Is Not
$\Delta\omega = \omega_{\text{left}} - \omega_{\text{right}}$

As noted above, the term appended to (3) that detunes the system from its required coordination is expressed most generally as an imperfection parameter [38]. In investigations of excitable media, as exemplified by the Belousov-Zhabotinsky reaction, a speck of dust in the petri dish introduces an imperfection with symmetry-breaking consequences for the patterns engendered by the reaction. In investigations of biological oscillations, as exemplified by the rhythmic intersegmental patterns of the lamprey eel [21, 28] and the rhythmic interlimb patterns of the human [9], the imperfection is most typically identified with an arithmetic difference between the natural (preferred) frequencies of the involved parts. That is, the imperfection parameter is most typically interpreted as nonzero values of $\Delta\omega = \omega_i - \omega_j$. This particular imperfection, of course, has multiple incarnations. For example, the arithmetic difference $\Delta\omega = \omega_i - \omega_j = 1$ rad/s is satisfied by $(2-1)$, $(3-2)$, etc., corresponding to different magnitudes of the ratio $\Omega = \omega_j/\omega_i$, namely 0.5, 0.66, etc. An obvious test of the proposal that the arithmetic difference $\Delta\omega$ is the

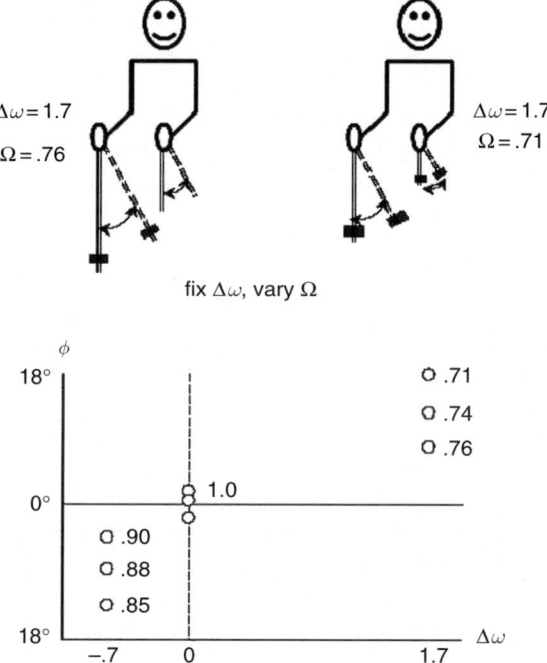

Fig. 8. A comparison of the effects on coordination equilibria of different ratios Ω of eigenfrequencies for the same arithmetic difference $\Delta\omega$ of eigenfrequencies (see text for details)

time-oriented imperfection parameter of biological rhythmic systems is to compare different Ω magnitudes for the same $\Delta\omega$ [7, 36]. The different magnitudes should affect the HKB dynamics in precisely the same way.

Figure 8 (top) shows the experimental design for achieving variations of Ω within a given $\Delta\omega$ and Fig. 8 (bottom) shows the results for three arithmetic differences of -0.7, 0, and 1.7 rad/s [7]. The values of Ω are expressed as elements of $[0, 1]$. For fixed $\Delta\omega$, as Ω approaches 1, the shift from the required relative phase value (here π) decreased systematically, replicating the finding in a reciprocal experiment that realized a fixed Ω through different magnitudes of $\Delta\omega$ [36]. Apparently, whatever is the *temporal imperfection parameter* for biological movement systems, it incorporates both absolute and relative differences in the natural frequencies of the involved segments.

7 When the Imperfection Parameter Is Strictly Spatial

Returning to the apparatus and manipulations schematized in Figs. 6 and 7 (*left*), the degree of separation between axles or axes of rotation can be varied without introducing differences in the eigenfrequencies of the left and right manipulanda [24]. The results shown in Fig. 7 (*right*) for $\Delta\omega = 0$, $\Omega = 1$, clearly indicate that the aforementioned separation of axles breaks the reflectional symmetry of (3). To a first approximation (given that other factors, as yet identified, are likely to be involved), the separation of axles is a matter of *spatial imperfection*. It can be identified as $\Delta\sigma = \sigma_{\text{left}} - \sigma_{\text{right}}$ to parallel $\Delta\omega$ where σ symbolizes a spatial quantity and bottom axis is the counterpart of higher ω. That being the case, one can ask whether the adding of nonzero $\Delta\sigma$ to (3) affects coordination dynamics in similar fashion to the adding of nonzero $\Delta\omega$.

Two specific expectations are of special significance. Both were detailed – in terms of δ – in Sect. 4. First, if $\Delta\sigma$ is nonzero (the left and right manipulanda are about nonparallel axes as in Fig. 9A), then the shift in relative phase from a required value of $\phi = 0$ or $\phi = \pi$ induced by $\Delta\sigma$ should magnify systematically with increase in movement frequency (as dictated by the frequency of the pacing metronome). As shown in Fig. 2B, solving (4) for a fixed $\Delta\omega \neq 0$ and decreasing values of b/a yields fixed points (the stable values of ϕ) that deviate increasingly from the required canonical value, 0 or π. Figure 9B shows the pattern of predicted changes in ϕ (and observed changes in ϕ, e.g., Pellecchia et al. [26]) when $\Delta\omega > 0$ and $\Delta\omega < 0$ for the same $|\Delta\omega|$. Figure 9C shows the observed changes in ϕ when $\Delta\sigma > 0$ (arbitrarily, bottom axle left hand, top axle right hand) and $\Delta\sigma < 0$ for the same $|\Delta\sigma|$ ([24] Experiment 2). The comparison of Fig. 9C with Fig. 9B reveals that the combined effects of $\Delta\sigma$ and movement frequency are qualitatively identical to the combined effects of $\Delta\omega$ and movement frequency.

The second, specific, and significant prediction of (4), with δ interpreted as $\Delta\omega$, is that for a fixed movement frequency larger values of $\Delta\omega \neq 0$ induce

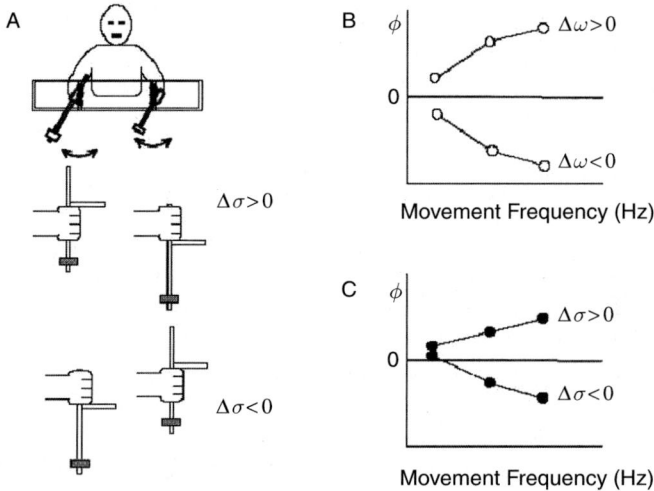

Fig. 9. The combined effects of $\Delta\sigma$ and movement frequency are qualitatively identical to the combined effects of $\Delta\omega$ and movement frequency (see text for details)

larger deviations from the canonical values of ϕ ($\phi = 0$ and $\phi = \pi$) and larger magnitudes of SDϕ. Figure 10B shows the predicted (and observed, e.g., Sternad et al., [35]) patterns for ϕ and SDϕ. The method for producing magnitudes of $\Delta\sigma$ that parallel increasing magnitudes of $\Delta\omega$ is schematized in Fig. 10A. As depicted in Fig. 10C, the method reveals that magnifying $\Delta\sigma$ has the same qualitative consequences for coordination dynamics as magnifying $\Delta\omega$ (Park [24] Experiment 3). Spatial imperfections and temporal imperfections have equivalent consequences. An important theoretical consequence of this latter conclusion is that it brings into question efforts to incorporate $\Delta\sigma$ into (1) as a formative influence on the fundamental potential function [10].

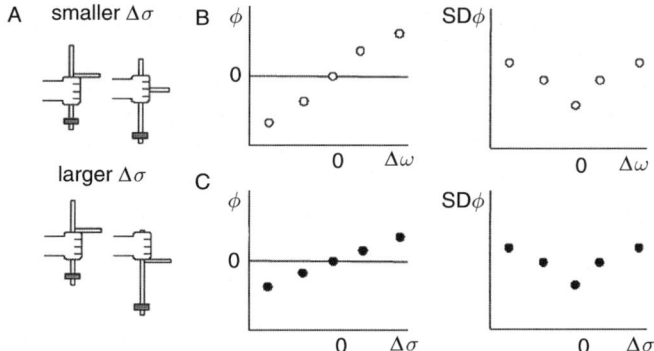

Fig. 10. Magnifying $\Delta\sigma$ has the same qualitative consequences for coordination dynamics as magnifying $\Delta\omega$ (see text for details)

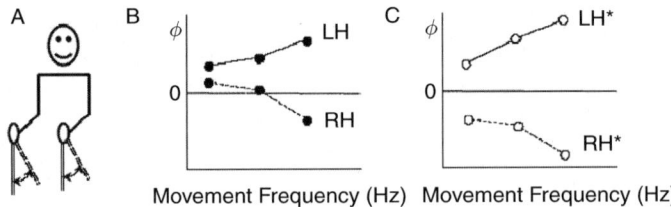

Fig. 11. The shift in equilibria due to a functional imperfection, handedness, is magnified by movement frequency (RH = right handers, LH = left handers; see text for details)

8 When the Imperfection Parameter Is Functional

The left hand and right hand are functionally distinct [13]. A functional hand preference (handedness) introduces an imperfection in the elementary coordination law. The method to assess the effect of this imperfection involves comparing left-handers and right-handers in an experimental context in which obvious temporal and spatial imperfections are absent (Fig. 11A). Under the convention of $\phi = \theta_{\text{left}} - \theta_{\text{right}}$, the imperfection shows up as a small left-hand lead ($\phi > 0$, $\phi > \pi$) in left-handers (LH) and a small right-hand lead ($\phi < 0$, $\phi < \pi$) in right-handers (RH) [40]. The question to be asked is whether the addition of handedness affects the coordination dynamics of (3) in similar fashion to the explicit addition of nonzero $\Delta\omega$ and nonzero $\Delta\sigma$. Specifically, with fixed 'positive' and fixed 'negative' handedness, one can ask whether the attractors (the values of ϕ elected by the participants) change with increasing movement frequency in the manner depicted in Figs. 9B and C. That is, relative to a required phase relation of 0 or π rad, will the attractors become increasingly more positive for LH and increasingly more negative for RH? The question was posed and answered by Treffner and Turvey [41]. The affirming results are depicted in Fig. 11B for LH and RH participants identified through conventional handedness tests and in Fig. 11C for subsets (LH*, RH*) of the same participants selected according to a criterion of a consistent preferred-hand lead in the rhythmic task of Fig. 11A.

One way to address the symmetry breaking induced by handedness is through the addition to (3) of the first two cosine terms in the Fourier series [40, 41]. The other option is to treat handedness as strictly just another instantiation of δ in (4). Closely similar predictions follow [29].

9 Not Any Difference Is Necessarily an Imperfection Parameter

Section 6 identifies the temporal imperfection as a difference between left and right eigenfrequencies expressed jointly by their absolute and relative differences. In their determination of this important and quite unexpected fact,

Sternad et al. [36] also noted differences that did not materialize as imperfection parameters, namely differences in mass and differences in moment of inertia. Confirmation by Peper et al. [27] that left–right asymmetry in moment of inertia is not an asymmetry that affects the fixed points and their stabilities was made in the context of assessing whether left–right asymmetry in muscular torques was the proper imperfection parameter. Peper et al.'s finding was expressed through the effector-oriented notion of a difference in left–right low-frequency control gain.

The absence of an obvious muscle-based mediator of the temporal imperfection coupled to the imperfection's quantitative form (a function of a sum and a quotient) forewarns of the potential abstractness of 'imperfection'. It also suggests that the criteria for 'imperfection' may be sufficiently severe to exclude many intuitive candidates for symmetry breaking.

10 Comparisons of Perfect and Imperfect Symmetry

10.1 Invariance of Fixed Points

A distinction between coordination abiding perfect symmetry and that abiding imperfect symmetry is shown in Fig. 2. The bold trajectories of fixed points in the $\phi \times \mathrm{d}\phi/\mathrm{d}t \times b/a$ space of Fig. 2A and B are replicated in experiments using the steady state paradigm. Figure 12 shows standard results [26]. The invariance of the fixed points prior to the subcritical pitchfork bifurcation is consonant with the cognate notions of pattern-energy factorization (e.g., Bullock and Grossberg [4]) and timing-power factorization [32]: the antiphase and in-phase patterns are unchanged by movement speed. These factorizations, consonant with the perfect symmetry of (3), are not general. They do not hold for imperfect symmetry as revealed by the noninvariance of stable fixed points on the trajectories toward the annihilation of stable and unstable fixed points. In the latter case, as expected from (4) and shown in Fig. 12, the specific phase relation (the pattern, the timing) of required in-phase or required antiphase coordination changes systematically with increase in movement frequency (the energy, the power).

10.2 Invariance Under Noncoincidence of Spatial and Muscular Reference Frames

Figure 12 provides a stepping-stone to another contrast between perfect and imperfect symmetries, namely their sensitivity to the relation between spatial (allocentric) and muscular (egocentric) definitions of relative phase. Spatially, in-phase (antiphase) coordination refers to left and right motions in the same (opposite) directions at the same points in time. Muscularly, in-phase (antiphase) coordination refers to simultaneous (alternating) contractions of

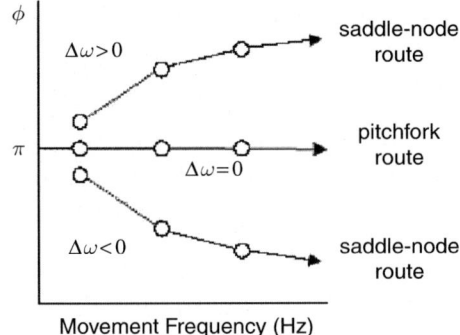

Fig. 12. Invariant equilibria up to bifurcation in the perfect case $\Delta\omega = 0$ suggests timing-power invariance; changing equilibria up to bifurcation in the imperfect case $\Delta\omega \neq 0$ suggests otherwise (see text for details)

left and right homologous muscles (e.g., ulnar muscles of left and right forearms). The experimental arrangements shown in Fig. 6 and Fig. 13 identify how the two definitions can be rendered noncoincident, anti-symmetric (see [31], for a third arrangement). In Fig. 13, upper panels, the transformation of monofrequency coordination from a plane parallel to the body's sagittal plane to a plane parallel to the body's coronal plane is a transformation from coincidence to noncoincidence of spatial and muscular definitions of relative

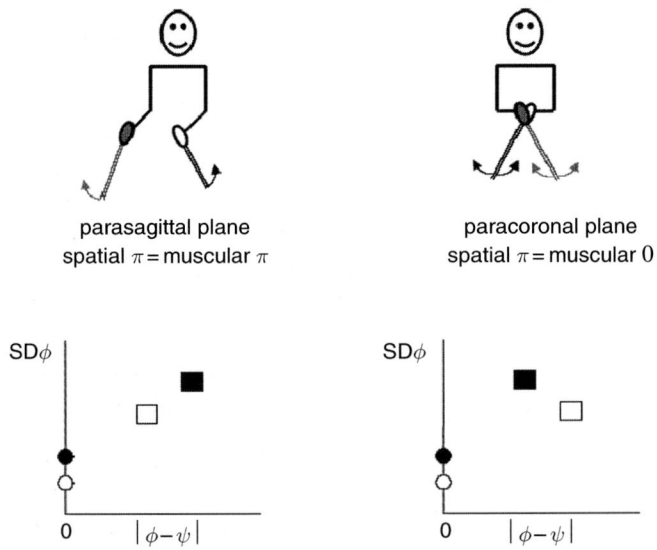

Fig. 13. Noncoincidence of spatial and muscular definitions of relative phase leaves the stability order of in-phase (*white*) and antiphase (*black*) invariant for perfect (*circles*) but not for imperfect (*squares*) symmetry (see text for details)

phase. For the coordination depicted in the top left figure, spatial antiphase is muscular antiphase. For the coordination depicted in the top right figure, spatial antiphase is muscular in-phase.

The lower panels in Fig. 13 summarize the consequences [23, 24, 25]. In both panels, relative phase is defined muscularly. For coordination under conditions of perfect symmetry, the transformation leaves invariant the relations between the in-phase and antiphase fixed points and their stabilities predicted by (3) and (5). In contrast, under conditions of imperfect symmetry, the transformation alters the relations between the in-phase and antiphase fixed points and their stabilities predicted by (4) and (5). Those predicted relations are satisfied in full when the spatial and muscular definitions of ϕ are coincident. When they are not coincident, the antiphase fixed point is displaced (ψ is required phase) less by the imperfection parameter than the in-phase fixed point (contrary to prediction) without becoming more stable in the sense of smaller SDϕ (contrary to prediction). The anomalous nature of the preceding outcomes is underscored by defining relative phase spatially. Then, the antiphase pattern is displaced more by δ than the in-phase pattern (in agreement with prediction) but possesses greater stability than the in-phase pattern (contrary to prediction).

The results of dissociating of spatial and muscular definitions of relative phase invite the following hypothesis [23, 24, 25]. When the symmetry is imperfect, coordination dynamics are factored into (a) dynamics that determine the location of the attractor (the mean value of relative phase on which a required in-phase or required antiphase coordination settles for a given set of conditions) and (b) dynamics that determine the strength of the attractor. Factor (a) is the dynamics associated with the spatial (allocentric) definition of relative phase. Factor (b) is the dynamics associated with the muscular (egocentric) definition of relative phase.

10.3 Invariance of b/a

In the context of (3)'s perfect symmetry, the ratio of the coefficients of the sine functions constitutes the control parameter. The ratio is a unitary quantity. Regardless of how it is composed, regardless of the particular values of b and a, a given ratio of b to a exerts an invariant effect (same fixed point, same stability). This invariance is not preserved in the context of (4)'s imperfect symmetry. In (4), the same ratio composed in different ways – through different b and a pairs – leads to different fixed points, different stabilities. The experimental implications of this formal feature of (4) for the dynamics of coordination with asymmetries are currently unknown. Furthermore, the formal implication for defining the principal control parameter or parameters based on the a and b coefficients has yet to be explored.

10.4 Invariance of the Pitchfork

As noted, the elementary coordination law is intimately connected to the sub-critical pitchfork bifurcation. The pitchfork (in either its supercritical or its subcritical form) is a natural accompaniment of reflectional or Z_2 symmetry. When the latter symmetry is broken by an imperfection, the resulting bifur-cation is most usefully referred to as an imperfect pitchfork bifurcation. It has multiple incarnations (including the saddle node) dependent on the extant region of the *control parameter* × *imperfection parameter* space occupied by the coordination.

Park et al. [25] analyzed phase transitions between coordination patterns in a paracoronal plane. On a given trial in Park et al.'s Experiment 2, $\Delta\omega$ was fixed and movement frequency was increased in increments of 0.24 Hz from 0.73 to 1.92 Hz in six plateaus of 10-s duration, with the first plateau consid-ered as preparatory. Under conditions of imperfect symmetry, $\Delta\omega \neq 0$, abrupt (pitchfork) transitions were prominent in 25% of the trials for muscular an-tiphase coordination and 0% of the trials for muscular in-phase coordination, rates that matched those for the condition of (experimentally) perfect sym-metry, $\Delta\omega = 0$. In contrast, prominent gradual (saddle node) transitions were far fewer and limited to imperfect symmetry. Only 5% of the imperfect sym-metry trials exhibited a continuous rather than discontinuous passage from π to 0 rad (for an example, see Mitra et al. [23], Fig. 11). Phase slippage and phase wandering, as alternative forms of transitions and adaptations, were strongly evident, however. For imperfect symmetry, they dominated 57% of the antiphase trials and 23% of the in-phase trials; for perfect symmetry the corresponding numbers were 5 and 0%, respectively. In concert with the bi-furcations and slippage/wandering was the change in relative attractiveness of $\phi = 0$ and $\phi = \pi$ with increases in movement frequency. An evaluation of the probability masses about 0 and π revealed that, for both perfect and imperfect symmetries, antiphase became systematically less attractive than in-phase.

The preceding pattern of transitions within the handheld pendulum paradigm replicates many of those identified for monofrequency coordina-tion of homologous (e.g., left arm–right arm) and nonhomologous (e.g., left arm–left leg) limbs [20]. It is a pattern that points to the enrichment of op-portunities for adapting the elementary coordination law that arises when the symmetry of the law is broken.

11 The Challenge of Imperfection

There are four primary questions in regard to the imperfect symmetry of coordination dynamics:

1. Is there an upper bound on the number of imperfection parameters?
2. Are imperfection parameters, whatever their number, added to the ele-mentary coordination law?

3. How can the forms of the imperfection parameters be determined?
4. Are imperfection parameters derivable from coupled oscillators?

11.1 Imperfection Parameters: Number and Additivity

In its most general form, the elementary coordination law L has the structure

$$\dot{\phi} = L(\phi, c, \delta_1, \ldots, \delta_k). \tag{6}$$

In (6), c is the control (or primary bifurcation) parameter and δ_k are the imperfection (or auxiliary, or unfolding) parameters. δ_k are nonredundant and in their absence L is in its perfect form.

The control parameter is a variable that when scaled upward or downward induces changes in the number and/or type of fixed points. Although commonly identified in HKB dynamics with velocity (movement frequency), c can also be identified with amplitude (movement excursion) [30]. The nature of c can be expected to be task-dependent. (Its amplitude version is needed, apparently, for the coordination dynamics of bimanual circle drawing.) Like control parameters, imperfection parameters can also be expected to be task-dependent, but the concern, voiced in Sect. 3, was that for any single coordination task k could be indefinitely large. The lesson of Sect. 9, however, is that 'imperfection' need not equal 'imperfection parameter'.

One line of argument identifies the *minimal* size of δ_k. Equation (6) is readily cast as the equation for the imperfect pitchfork bifurcation by setting it equal to zero. The casting brings into focus the fixed points and the phase diagram that captures all of the bifurcations within a space defined by c and one or more imperfection parameters. Locally, the pitchfork is equivalent to the normal form [38]

$$p(x, c) = \pm x^3 \pm cx \tag{7}$$

The prototypical unfolding of the pitchfork is given as

$$P(x, c, \alpha, \beta) = x^3 - cx + \alpha + \beta x^2 \tag{8}$$

where α and β are imperfection parameters [3, 12]. Equation (8) informs that a minimum of two imperfection parameters along with the primary bifurcation parameter are needed to encompass the bifurcation possibilities of a given system with Z_2 symmetry – that is, $x = -x$, and manifest in the elementary coordination law as $V(\phi) = V(-\phi)$.

Equation (8) is informative in two further senses. First, it gives emphasis to the implication of (4) that imperfection parameters are basically additions to the original, perfectly symmetric system. Figures 9 and 11 suggest that whether the imperfection parameter is time related, space related, or functional, its influence on coordination equilibria combines with that of the control (primary bifurcation) parameter. Additionally, Figs. 4 and 7 suggest that temporal asymmetry and spatial asymmetry add to produce the pattern of fixed points related through reflection. Second, (8) indicates that the effect of one imperfection parameter (see β) is conditional on the order parameter or collective variable (x in (8), ϕ in (3) and (4)). The implication is that the addition of the effects of imperfection parameters will not be perfectly linear.

11.2 Imperfection Parameters: Composition and Derivation

The conventional interpretation of δ as $\Delta\omega$ is an appropriate platform for reviewing the challenges of composition and derivation. As detailed in Sect. 6, *temporal imperfection* $= f(\Delta\omega, \Omega)$. An interpretation solely in terms of an arithmetic difference between the left and right eigenfrequencies will not suffice.

The standard route for deriving the form of temporal imperfection was alluded to in Sect. 6. The derivation is strictly in terms of *phase dynamics*. The oscillators are presumed to be limit cycles, with the coupling between them dependent solely on their phase angles, θ_1 and θ_2. The coupling strength k is not so strong as to compromise limit cycle behavior but sufficient to displace each oscillator from its preferred frequency ω. Accordingly, motion equations for the two oscillators can be written in the form

$$\dot{\theta}_1 = \omega_1 + k_{12}\sin(\theta_1 - \theta_2),$$
$$\dot{\theta}_2 = \omega_2 + k_{21}\sin(\theta_2 - \theta_1) \tag{9}$$

with 2π periodic functions ensuring consistent coupling effects from one cycle to the next. A single motion equation in relative phase follows from letting, $\dot{\theta}_1 - \dot{\theta}_2 = \dot{\phi}$, $\omega_1 - \omega_2 = \Delta\omega$, and $k_{12} + k_{21} = k$, namely

$$\dot{\phi} = \Delta\omega - k\sin\phi. \tag{10}$$

Despite the simplicity of the steps, and the monostability of the product, the foregoing is nonetheless suitably representative of a general orientation toward deriving systems of coupled oscillators [16, 21, 28]. Equation (10) is reached identically by Haken [14] (pp. 239–241) via a more intricate sequence grounded in the Hopf bifurcation; and a concerted effort to derive (4) explicitly from the component oscillators by Kelso and colleagues [9] converges similarly on $\delta = \Delta\omega$.

At issue, simply, is whether the theory of coupled oscillators can deliver the precise form of the temporal imperfection parameter (revealed empirically, Sect. 6). A reasonable conclusion at this stage is that the theory cast in terms of phase dynamics cannot do so. It is confined by its basic premises to expressions of the kind shown in (9).

The challenge for the phase coupling orientation is magnified by the equivalence for coordination dynamics of time-dependent and time-independent definitions of δ, as highlighted in Fig. 9. The question posed by the equivalence is one of articulating the kinds of assumptions, paralleling those prefacing (9), which would yield a nontemporal $\Delta\sigma$ in place of the temporal $\Delta\omega$. At a minimum, the issue of the composition of imperfection parameters emphasizes the need for the theory of coupled oscillators for biological phenomena to consider amplitude dynamics in addition to, or instead of, phase dynamics.

12 Postscript

The challenges of imperfect symmetry for the theory of coupled oscillators (and, by implication, for neural explanations) raise the possibility that the more prudent strategy is to persist with Lesson 2 (see Introduction). Specifically, extend the phenomenological and aggregated description of coordinated movement's highest level. An additional good motivation for doing so is that many naturally arising imperfection parameters will seldom be knowable in precise form and may often be random quantities.

A first step would be a formal examination of a parameterized family of imperfect coordination dynamics characterized by a minimum of two dimensionless imperfection parameters. Essentially, it would be an investigation of (8) for the particular case of the elementary coordination law (3). For example, for fixed values of the nondimensional δs, the equilibrium paths would be computed for changing values of the primary bifurcation parameter. The outcome would be a classification of the singularities in the set of stationary states – a bifurcation diagram over the parameter space. The second step would be a systematic experimental investigation of the space with an appropriate model system that permits reasonably precise manipulations of the primary and imperfection parameters.

An expected benefit of furthering investigation in the spirit of Lesson 2 is the clarification of results that have been, or can be, interpreted as either falling outside the scope of the elementary coordination law or not readily addressed by it. The results in question may well be anomalous-looking by virtue of the nonobvious parameter space from which they were derived.

In sum, there is considerable warrant for expecting further advances in the qualitative understanding of coordination dynamics from a modeling strategy in concert with Lesson 2. Key to this strategy is a minimality principle: include in the modeling as few processes, variables, stipulations, and explanatory hypotheses as possible. It is a principle consonant with Lesson 1: be prepared to discover that the laws governing complex systems may assume surprisingly simple forms.

References

1. Amazeen PG, Amazeen EL, Turvey MT (1998a) Dynamics of human intersegmental coordination: theory and research. In: Rosenbaum DA, Collyer CE (eds) Timing of behavior: Neural, computational and psychological perspectives. MIT Press, Cambridge MA, pp. 237–259
2. Amazeen PG, Amazeen EL, Turvey MT (1998b) Breaking the reflectional symmetry of interlimb coordination dynamics. J Mot Behav 30:199–216
3. Ball R (2001) Understanding critical behavior through visualization: a walk around the pitchfork. Comput Phys Commun 142:71–75
4. Bullock D, Grossberg S (1988) Neural dynamics of planned arm movements: Emergent invariants and speed-accuracy properties during trajectory formation. Psychol Rev 95:49–90

5. Bunge M (1977) Treatise on basic philosophy, Vol. 3. Ontology I. Reidel, Boston
6. Carson RG, Riek S, Smethurst CJ, Lison, JF, Byblow, WD (2000) Neuromuscular-skeletal constraints upon the dynamics of unimanual and bimanual coordination. Exp Brain Res 131:196–214
7. Collins D, Sternad D, Turvey MT (1996) An experimental note on defining frequency competition in intersegmental coordination dynamics. J Mot Behav 28:299–304
8. Den Hartog JP (1961) Mechanics. Dover, New York
9. Fuchs A, Jirsa VK, Haken H, Kelso JAS (1996) Extending the HKB model of coordinated movements to oscillators with different eigenfrequencies. Bio Cybern 74:21–30
10. Fuchs A, Jirsa VK (2000) The HKB model revisited: How varying the degree of symmetry controls dynamics. Hum Mov Sci 19:425–449
11. Goldenfeld N, Kadanoff LP (1999) Simple lessons from complexity. Science 284:87–89
12. Golubitsky M, Schaeffer DG (1985) Singularities and groups in bifurcation theory, Vol. 1. Springer, New York
13. Guiard Y (1987) Asymmetric division of labor in human skilled action: The kinematic chain as a model. J Mot Behav 19:486–517
14. Haken H (1987) Advanced Synergetics. Springer-Verlag, Heidelberg
15. Haken H, Kelso JAS, Bunz H (1985) A theoretical model of phase transitions in human movements. Bio Cybern 51:347–356
16. Holmes P, Full RJ, Koditscheck D, Guckenheimer, J (2006) The dynamics of legged locomotion: Models, analyses, and challenges. Soc Ind Appl Math 48:207–304.
17. Kelso JAS (1984) Phase transitions and critical behavior in human bimanual coordination. Amer J Physiol Regul Integr Comp Physiol 15:R1000–1004
18. Kelso JAS (1994) Elementary coordination dynamics. In: Swinnen S, Heuer H, Massion J, Casaer P (eds) Interlimb coordination: Neural, dynamical, and cognitive constraints. Academic Press, San Diego, CA, pp. 301–320
19. Kelso JAS (1995) Dynamic patterns: the self-organization of brain and behavior. MIT Press, Cambridge, MA
20. Kelso JAS, Jeka J (1992) Symmetry breaking dynamics of human multilimb coordination. J Exp Psychol Hum Percep Perf 18:645–668
21. Kopell N (1988) Toward a theory of modeling pattern generators. In: Cohen AH, Rossignol S, Grillner S (eds) Neural control of rhythmic movements in vertebrates. Wiley, New York pp. 369–413
22. Kugler P, Turvey MT (1987) Information, natural law, and the self-assembly of rhythmic movement. Erlbaum, Hillsdale, NJ
23. Mitra S, Amazeen PG, Turvey MT (1997) Dynamics of bimanual rhythmic coor-dination in the coronal plane. Motor Control 1:44–71
24. Park H (2003) Induced spatial and temporal symmetry breaking in coordination dynamics. Ph.D. thesis, University of Connecticut
25. Park H, Collins D, Turvey MT (2001) Dissociation of muscular and spatial constraints on patterns of interlimb coordination. J Exp Psychol Hum Percep Perf 27:32–47
26. Pellecchia G, Shockley K, Turvey MT (2005) Concurrent cognitive task modulates coordination dynamics. Cognit Sci 29:531–557
27. Peper CE, Nooij SAE, van Soest J (2004) Mass perturbation of a body segment: 2 effects on interlimb coordination. J Mot Behav 36:442–449

28. Rand RH, Cohen AH, Holmes PJ (1988) Systems of coupled oscillators as models of central pattern generators. In: Cohen AH, Rossignol S, Grillner S (eds) Neural control of rhythmic movements in vertebrates. Wiley, New York pp. 333–367

29. Riley MA, Amazeen EL, Amazeen PG, Treffner PJ, Turvey MT (1997) Effects of temporal scaling and attention on the asymmetrical dynamics of bimanual coordination. Motor Control 1:263–283

30. Ryu YU, Buchanan JJ (2004) Amplitude scaling in a bimanual circle-drawing task: Pattern switching and end-effector variability. J Mot Behav 36:265–280

31. Salesse R, Ouiller O, Temprado JJ (2005) Plane of motion mediates the coalition of constraints in rhythmic biological coordination. J Mot Behav 37:454–464

32. Schmidt RA (1988) Motor control and learning. Human Kinetics, Champaign IL

33. Schmidt RC, Shaw BK, Turvey MT (1993) Coupling dynamics in interlimb coordination. J Exp Psychol Hum Percep Perf 19:397–415

34. Schöner G, Haken H, Kelso JAS (1986) A stochastic theory of phase transitions in human hand movement. Bio Cybern 53:442–452

35. Sternad D, Amazeen E, Turvey MT (1996) Diffusive, synaptic and synergetic coupling: an evaluation through in-phase and antiphase coordination. J Mot Behav 28:255–269

36. Sternad D, Collins D, Turvey MT (1995) The detuning term in interlimb coordination dynamics. Bio Cybern 73:27–35

37. Stewart I, Golubitsky M (1992) Fearful symmetry. Blackwell, Oxford

38. Strogatz SH (1994) Nonlinear dynamics and chaos. Reading MA, Addison Wesley,

39. Turvey MT, Carello C (1996) Dynamics of Bernstein's level of synergies. In: Latash M, Turvey MT (eds) Dexterity and its development. Hillsdale NJ, Erlbaum

40. Treffner PJ, Turvey MT (1995) Handedness and the asymmetric dynamics of bi-maual rhythmic coordination. J Exp Psychol: Hum Percep Perf 21:318–333

41. Treffner PJ, Turvey MT (1996) Symmetry, broken symmetry, and handedness in bimanual coordination dynamics, Exp Brain Res 107:463–478

Landscapes Beyond the HKB Model

Karl M. Newell[1], Yeou-Teh Liu[2], and Gottfried Mayer-Kress[1]

[1] Department of Kinesiology, The Pennsylvania State University, University Park, PA 16802, USA
[2] Graduate Institute of Exercise and Sport Science, National Taiwan Normal University, Taipei, Taiwan

1 Introduction

The change in behavior over time that reflects human motor learning and development has long been thought of as a consequence of the evolving cycle of system stability and instability [16, 17, 37]. Over the last 25 years or so, the dynamics of this pathway of change in behavior has been modeled using the physical construct of potential wells that has been used so regularly in physics and the study of systems for which its' potential energy can be represented as a multi-dimensional landscape. In the study of both biology and behavior many approaches have followed the initial schematic of Waddington [39] in developing a landscape for change in behavior over time.

In human movement the most studied landscape, and one that has gone beyond landscapes merely as metaphor, has been that arising from the Haken, Kelso and Bunz [9] HKB model of bimanual limb control. This model builds from the theory of synergetics [6, 7] and the earlier experimental findings of transitions in bimanual finger movements [10, 11] to provide a landscape of the intrinsic dynamics in bimanual coordination. This theoretical model has been developed over the years into a prototype of how to consider experimentally the dynamics of movement coordination and is the foundation of what has become known as coordination dynamics [12, 13]. Indeed, the HKB model has been the dominant theoretical and experimental paradigm in behavioral motor control over the last 20 years, and that it retains this position today is a special 60th birthday present for Scott Kelso.

In this paper we consider issues in developing the role of potential well landscapes in motor learning and development beyond that provided by the HKB model. A particular emphasis is the challenge of considering this approach to the modeling of the change in movement behavior over time – a hallmark feature to be considered by any theory of human learning and development. Our paper seeks to build on the early efforts in this regard to extend the HKB model that were outlined by Scott Kelso, Gregor Schöner and Pier Zanone [33, 34, 35, 36].

2 Historical Background of the HKB Landscape

The HKB model of bimanual coordination was developed within the context of synergetics as the self-organization framework for complex systems [6]. This theoretical framework and basic model for the study of dynamic stability and instability had been investigated previously in the pattern formation and change of a variety of non-living phenomena. The emphasis in synergetics is on how the spatial and temporal patterns at the macroscopic level of a system arise from the many degrees of freedom at more microlevels of the system. The HKB model of bimanual coordination was the initial application of the synergetics framework to human movement [9], since when a range of applications has been studied [8, 12, 13, 32]. The theoretical background and experimental formulations of the HKB model are well documented and, therefore, we only provide enough detail here to motivate our subsequent development of a more general view on landscapes in motor learning and development.

The theory of synergetics provides the theoretical framework and experimental tools to model the ebb and flow of the change in the qualitative and quantitative properties of movement coordination modes. A key insight is that dynamic instabilities provide a way to capture the principles of the formation and change of macrolevel coordinative states. Furthermore, the critical properties of these coordination states are to be found in properties of the system behavior and not through the standard application of Newtonian mechanics to human movement, as in the traditions of biomechanics.

In the view of synergetics, the information in the many degrees of freedom of the multiple levels of a system can be compressed at the macroscopic level into order parameter(s) that capture the collective behavior (macroscopic state(s)) of the system. It is assumed that the collective behavior of the system is low dimensional having a single or possibly only a small set of order parameters and that the time scale of change of the order parameter is relatively slow within the performance of the given action or movement sequence. The system can in many situations be moved through a range of stable order parameter states by a relevant control parameter. It is the critical fluctuations that arise from the transition between states that provide the fundamental information about the organization of the system and the relation between the order and control parameters.

A challenge in motor control is finding an action or task protocol that is rich enough to capture the properties of the fundamental dynamics of a system but not so complicated that it precludes modeling. The HKB model is based on the bimanual oscillatory motions of the index fingers of the two hands [10, 11], a protocol that has since been shown to capture many of the dynamic properties of complex systems. The order parameter in this task is the relative phase between the two fingers while the control parameter is the task frequency, usually specified to the subject by a metronome.

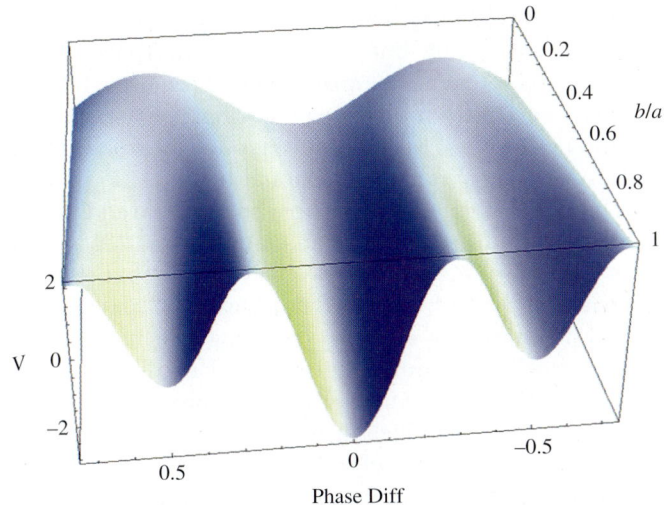

Fig. 1. Landscape of the intrinsic dynamics of the HKB model

Figure 1 shows a potential landscape of the relative stability and instability of bimanual coordination in the HKB model as a function of the relative phase and b/a (a parameter that determines the frequency of oscillation, large values of b/a correspond to low frequencies and vice versa, see (1) below). This landscape reveals many properties of the dynamics of bimanual coordination including the following: (1) the basic stable states are in-phase ($0°$ relative phase) and anti-phase ($180°$ relative phase), (2) increasing the frequency of oscillation eventually leads the anti-phase condition to a transition into in-phase motion due to the loss of stability, and (3) there is a general symmetry across relative phases away from zero, though this can be modified by several influences. More particularly, the experimental protocol has shown a rich array of dynamic properties including, multistability, critical fluctuations, bifurcations, hysteresis and symmetry breaking [12].

The equation for the potential (V) landscape formed from the model is:

$$V(\phi) = -a\cos\phi - b\cos 2\phi \qquad (1)$$

where ϕ is relative phase between the fingers and the b/a ratio represents a control parameter (frequency) value. This basic version of the HKB model [9] has been extended to encompass both an asymmetry term for the two oscillators (fingers) and a stochastic term [12]. In the HKB landscape of Fig. 1 we see the symmetry assumption in place for positive and negative relative phases.

It is important to note that the landscape in Fig. 1 arises from a particular instruction given by the experimenter to the subject as to how to change finger movement behavior in the face of the scaling over time of movement frequency. The subject was told that "as the metronome beat goes faster if they felt the finger pattern begin to change, they should not consciously try to prevent it from happening but rather adopt the pattern that was most comfortable under the current conditions. Furthermore, if the pattern does change, don't try to go back to the original pattern but stay in the one that is most comfortable. Above all, try to keep the one-to-one relationship between your rhythmical motions and the metronome beat" ([12], p. 47). Specifically and in more technical language, this experimental paradigm seeks to determine the intrinsic dynamics of the bimanual coordination system – namely, those patterns that arise due to non-specific changes in a control parameter.

One significant aspect of the HKB landscape of the intrinsic dynamics shown in Fig. 1 is that it is seen as reflecting the background tendencies of the system that will determine the potential of the system to learn new coordination modes. Expressed another way, it shows the layout of the existing capacities of the system upon which future states of the system will be learned. In short, it captures the dynamics of what is already learned – a critical feature for a dynamical approach to learning, retention and transfer. This property of the HKB landscape is one of its most central contributions to motor learning and control and a feature we discuss in more depth later in the paper.

3 Landscapes in Motor Learning and Development

There have been several proposals to consider the role of potential well landscapes in elucidating the issue of change in motor learning and development. The origins of these interpretations can be found in the epigenetic landscape of Waddington [39]. The foundational landscape metaphor of Waddington in Fig. 2 shows the pathways of developmental change over time.

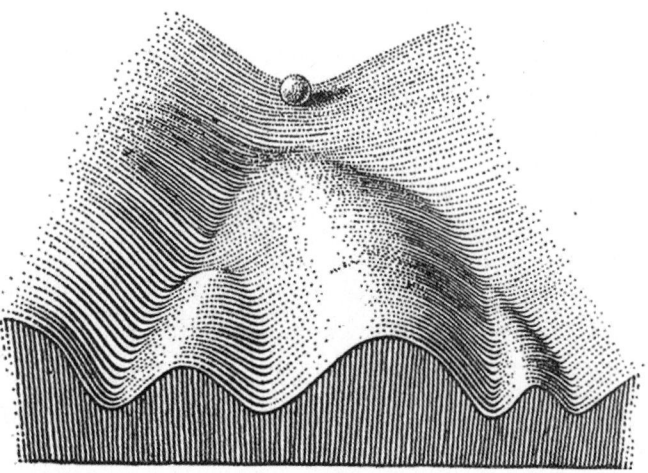

Fig. 2. Waddington's [39] schematic of the epigenetic landscape

In Fig. 2 the front to back dimension is time or developmental age, the horizontal axis instantiates the particular activities that hold dynamic equilibria and the slope of the surface captures the rate of development change. The stability of the system is inferred from the depth of the landscape wells. The change in the landscape over time reveals the emergence and dissolution of the stability of activities. The landscape is a low-dimensional description of the basic processes of development that are derived from the high-dimensional biological system.

The key concept is that the motion of the ball reflects the developing phenotype and the dynamic process of development. Waddington's [39] theoretical framework emphasized the epigenetic control of the evolving landscape but this metaphor has subsequently been broadened to include the confluence of organismic, environmental and task constraints [3, 15]. Indeed, the dynamic metaphor of the Waddington [39] landscape has been used in a number of areas of learning and development including: growth processes [21], connectionist models of learning [4], motor learning [26, 28, 30] and motor development [3, 15, 27].

A recent use of the landscape metaphor is that of Muchisky et al. [23] who have elaborated the presentation of the basic Waddington landscape to more directly address the processes of change in motor development. Figure 3 captures for Muchisky et al. [23] the evolving landscape for the development of the fundamental movement skills in infancy, though it could also be construed as a landscape for the learning of motor skills in general through the lifespan. The figure provides an intuitive metaphor for the dynamic change of motor development revealing both the emergence and dissolution of the dynamic stability of activities (as defined by the horizontal dimension of the collective variable for those activities) over time.

A common feature of the landscapes mentioned above is that they subscribe to the view that the wells and peaks of the landscape represent stable and unstable dynamics, respectively. The landscapes reflect how different activities or the scaling of the same activity have different levels of stability. This leads to the core position of stability and instability as constructs in the preservation and change of movement coordination modes (whether expressed as synergies, coordinative structures, motor programs and so on).

There is, however, a major difference between the HKB [9] landscape and the landscape of Waddington [39], including those who followed him with epigenetic landscapes. In the Waddington landscape time is an explicit dimension. This provides the basis to characterize the dynamics of the change in behavior over time and to examine the constructs of learning and development. Time is only implicit in the HKB model (and hence landscape) in the sense that with task frequency of finger oscillation changing typically only intermittently over time as a control parameter there is not a direct link to real time. The HKB model in its core form [9] is, therefore, a motor control model that does not capture the dynamics of change over time that reflect learning and development.

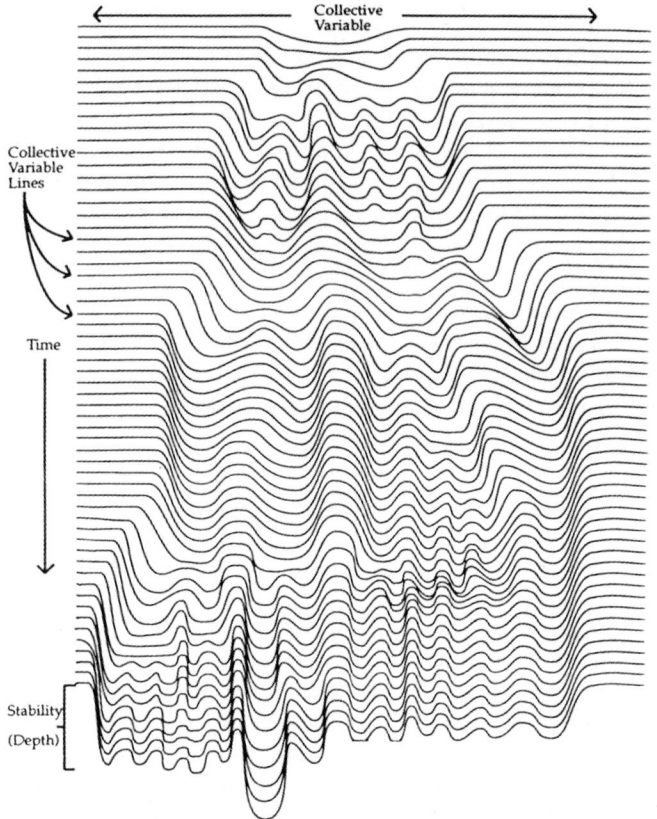

Fig. 3. The multiple time scales of motor development (from [23])

The interpretation of the stability and instability of the developmental epigenetic landscapes, such as that of Muchisky et al. [23], is also not as straightforward as expressed above [27, 29]. A particular limitation with reference to the emerging theme of this paper is that the stability regimes expressed by the collective variables at a given developmental time (the wells on the same horizontal line) do not represent the actual stability of each activity at the same moment in real time. The landscape, for example, of Muchisky et al. [23] only shows the potential scope of stable states at a point in time of development. Also, without considering the stable states of activity in a multidimensional framework, it is difficult to assess the relations *between* the stable states, because the order in which the states or activities are represented on the horizontal dimension is at best confined to intuition in the metaphorical model. In short, the back to front time line of development is not a true line of real time and the landscape does not capture the relative stability of emerging activities at a given moment in time.

A question that emerges from a consideration of these approaches to dynamical landscapes in motor learning and development is if and how the formal HKB model of coordination dynamics can be developed to accommodate the formation of new stable states over time. This was the focus of the initial papers on learning in coordination dynamics from Kelso, Schöner and Zanone.

4 Learning in Coordination Dynamics

The acquisition of a new coordination mode, that is a qualitative coordinative state that one has not produced before, is a core issue in the study of motor learning, particularly the early phases of learning the qualitative aspects of an action. Classic examples include the fundamental movement pattern sequence that emerges in infancy and some particular task demands in music and sport. Nevertheless, this problem has received through the years little theoretical or experimental attention because most tasks used in experimentation of motor learning have merely required the subject to learn to scale an already acquired coordination mode to the environmental and task demands [24].

The introduction of the coordinative structure framework to motor learning and control [16, 17] provided the preliminaries to a self-organizational theoretical perspective to the acquisition of new coordination modes. Schöner and Kelso [34, 35] developed the background perspective into particulars of a learning coordination dynamics approach to this problem. The cornerstone of this approach was the role of intention and behavioral information in creating new stable states to the landscape of intrinsic dynamics shown in the HKB model (Fig. 1). Intention and behavioral information act as additional forces in the landscape and through the repetition of practice over trials there is the emergence of new stable states where none had previously existed. For example, in bimanual finger coordination a prototypical new task is the learning of the 90° relative phase relation [40].

The contributing forces to the development of new stable states in the landscape are modeled in this extension of the HKB framework by assuming that these contributing forces are additive. For example, the influence of the relative roles of the intrinsic dynamics and the behavioral information in learning a new phase relation ψ can be modeled by:

$$V(\phi) = -a\cos\phi - b\cos 2\phi - c\cos\frac{\phi - \psi}{2} \qquad (2)$$

where ϕ is relative phase, the b/a ratio represents a (frequency) control parameter value, c is the force of behavioral information and ψ is the order parameter value specified by this information [34]. This equation has been extended to also include the contribution of intention and memory in learning dynamics and, in principle, it holds the basis for adding other contributing factors in learning and development, modeled as forces in the evolution of the landscape.

One important consequence of this model equation is that because of the large scales of the added function to the intrinsic dynamics the learning of a new relative phase changes the properties of the whole landscape and not just the formation of a new stable well that maps to the task demand. Thus, for example, learning to stabilize a new coordination mode at 90° relative phase has a negative influence on the stability of 180° anti-phase motion [40, 41]. The influence of these inter-task influences in the HKB model are, however, not fully established and other experimental findings in this framework have begun to challenge the assumption that the learning of a new relative phase changes the layout of the whole landscape [5, 18]. Clearly, these experimental findings on the relation between the intrinsic dynamics and the learning of new stable states relate to the traditional theory and experiment on the learning, retention and transfer of motor skill [1, 25, 31].

5 Generalizing from the HKB Landscape

The HKB landscape is one-dimensional in that phase difference is the only variable that influences stability of movement patterns. In generalizing from the HKB landscape of intrinsic dynamics we might ask what other dimensions can influence the outcome of the movement pattern and are sufficiently important to add to the landscape. In our view there are basically two strategies to represent these additional influences. These are adding the dimension of time and the dimension of initial bias to the landscape.

5.1 Time Dimension

The first option in representing extra dimensions in a landscape model is to display a parameter that is associated with practice or developmental time, in close analogy to the original Waddington landscape. In order to implement more realistic assumptions regarding time in the model we modified in several respects the Schöner and Kelso [34] landscape model of (2) for learning a new relative phase (Fig. 4). In this new representation we show increasing practice time/skill level along the y-axis. In addition, and in line with the original HKB model, our first assumption – suggested by empirical evidence – is that learning the 90° task also facilitates learning the −90° task [40, 41]. That means we create a second potential well at −90° almost as deep as the one at 90°, corresponding to the task where the alternate finger leads the oscillation.

Note, that this cannot be done with the Ansatz of a cosine function that was chosen in the original HKB model. The original choice of a periodic function like the cosine function for the HKB model was motivated by the periodic nature of the oscillatory finger movement. The first term in the HKB potential $V_{\mathrm{HKB}}(\phi) = -a\cos\phi - b\cos 2\phi$ has a period $T = 2\pi$ and generates a stable minimum at the in-phase oscillation $\phi = 0°$. The second term has a period that is half as long as $T = \pi$. It creates new potential wells at $\phi = \pm 180°$. In order to create new minima at $\phi = \pm 90°$ the natural choice would be to add a term with period $T = \pi/2$. But this periodic function would also deepen the potential at the in-phase ($\phi = 0°$) and anti-phase ($\phi = 180°$) minima so that

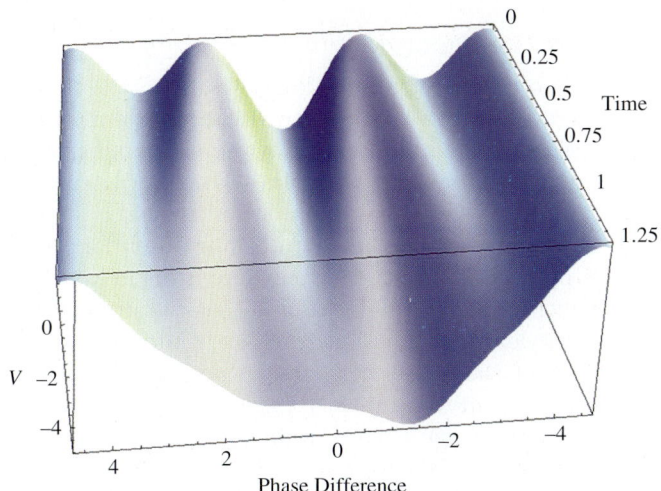

Fig. 4. The landscape from (3) and the learning of 90° relative phase. This shows the emergence of the 90° over time and the partial symmetry at −90°. Note this landscape is for performance at a single frequency that is low to modest and would not induce a transition from anti-phase to in-phase

learning of the 90° task would require initial conditions close to 90°. In order to make the 90° fixed point globally attractive, the attractors at 0 and 180° need to be destabilized. One possible choice is the shifted cosine function as in the last term of (2). It has twice the original period of $T = 4\pi$ and, therefore, has no longer the symmetry of the original HKB model, e.g., the values at +180° and −180° no longer match.

In Schöner [33] the periodicity of the contribution to the potential representing environmental information (equations (8) and (9) in [33]) is the same as in the first learning extension of the original HKB model, namely $T = 2\pi$, and its phase is shifted so that the potential minimum is located at the phase difference ψ to be learned. Because of the periodicity of the cosine function this shift will introduce a maximum in the potential at a phase difference of $\psi = 180°$. This leads to a testable experimental prediction in the case of $\psi = 180°$. We would expect that learning the $\psi = 90°$ task would interfere with the $\psi = 90° + 180° = -90°$ task.

In Newell et al. [26] we introduced a general, localized potential well model based on an inverted Gaussian function. The advantage in this approach is that we have independent control about location, depth and width of the potential well. The symmetry of the Gaussian function guarantees that its influence on the original model is not biased in one direction or the other.

The analytical expression for our Gaussian model potential $V_G(\phi, t)$ is given by:

$$V_{\mathrm{G}}(\phi, t) = V_{\mathrm{HKB}}(\phi) - \beta t e^{-\frac{(\phi-\psi)^2}{(\sigma+\zeta t)^2}} - \gamma t e^{-\frac{(\phi+\psi)^2}{(\sigma+\xi t)^2}} \tag{3}$$

where $V_{\mathrm{HKB}}(\phi) = -a\,\cos\phi - b\cos 2\phi$ is the original HKB [9] potential of 1 with $a = 1$, $b = 1.5$; and β and γ are parameters that determine the depth of the wells at $\pm\psi$, σ indicates the initial width of the wells and ζ, ξ determine the rate at which their width increases with time. For completeness sake we want to mention that our local model has a discontinuity at $\phi = 180°$ which can be removed by adding a smooth perturbation to our potential function $V_{\mathrm{G}}(\phi, t)$. Note, that for simplicity we have assumed here a linear time dependence of the depth and width of the wells. Of course that feature can be modified to reflect more realistic learning functions such as (multiple) exponentials.

The above form of the model contains an additional assumption namely that as we practice the movement at $\psi = +90°$ we also create a potential well at $-90°$ and at the same time destabilize the intrinsic dynamics of the attractors at 0 and $180°$. This is in contrast to the Schöner and Kelso [34] and Schöner [33] model, the latter of which would predict negative interference with learning the $\psi = -90°$ task.

5.2 Initial Bias Dimension

The second option in the generalization of the HKB landscape is to add one extra independent dimension to the graphical representation of the landscape. From experiments of Liu et al. [19, 20] with a roller ball learning task we know that the movement outcome will be dependent on whether the hand is kept at rest at task onset or if the participant is allowed to perform preparatory movements. For the HKB task we also expect that there are factors other than initial phase difference that will determine the movement outcome. Thus, the second strategy is to add a dimension to the landscape that represents the initial bias to subsequent control and this dimension could be influenced by a range of learning variables. Schöner [33] conducted simulations where the initial conditions (namely, relative phase) were manipulated but initial bias is used here in a more general way.

In Liu et al. [19] we have tested the influence of the initial bias of preparatory movements on the performance of a roller ball task. The task involves rhythmical hand movements to keep a ball – contained in a shell – spinning. Similar to the HKB situation we have a control parameter that corresponds to the initial frequency of the ball rotation ("ball speed"). We have two attractors in that system: "failure" indicates the attractor where the participant cannot sustain a minimal target ball speed. The "success" attractor indicates a movement pattern, where the participant can keep the ball spinning over the required test period of the experiment. The difficulty of the task increases as the ball speed decreases. Therefore, we can prepare a participant in a success state if we choose the initial ball speed high enough. In preparation of the task at a lower ball speed we have two basic initial conditions for the participants: (1) a resting hand condition where the participant does not move until an auditory start signal indicates that the target ball speed has been reached; (2) the participant continues a hand movement close to the movement that led to a "success" state at a slightly higher ball speed but without applying torque to accelerate the ball. In both cases the ball will slow down because of internal friction.

Starting from a state of success, participants are asked to wait until the ball has reached the (lower) target speed while following instructions (1) or (2). As soon as they hear the start signal, participants will actively move the ball in order to sustain its speed and prevent being attracted to the "failure" state. For a given ball speed (task difficulty) and skill level of the participant it was observed that initial bias (2) significantly increased the probability to be attracted by the "success" mode.

For a 90° learning task in the HKB context we would expect a similar dependence of success on initial condition/bias. Here the situation is much less complex than in the roller ball task, where the exact pattern of successful wrist movement is not known in detail. In the HKB model it is assumed that the phase difference is the only order parameter and, therefore, the movement can be described by a single, ordinary differential equation. We know, however, that ignoring inertial forces in the finger oscillation (an implication of the above assumption) is only an approximation that leads to an error that increases with oscillation frequency. For a given frequency of finger oscillation a successful 90° movement will be characterized not only by the correct phase difference but also, for example, by a peak velocity that is associated with the movement frequency. Our prediction would be that an initial bias where we oscillate the fingers close to the target frequency but with the correct phase difference will have a greater chance of success at the (higher) target frequency than an initial state of correct phase difference but fingers at rest.

In Fig. 5 we illustrate this process for the example of learning a 90° task, i.e., the fingers are supposed to oscillate with a 90° phase difference. Note, that the symmetry of this movement pattern is quite different from the original two patterns of 0° and 180° phase difference. For both of those movement tasks the fingers are treated symmetrically, i.e., there is no difference between +0° and −0° or +180° or −180° phase difference. For the 90° pattern, there is, however, a clear distinction to the −90° movement. One finger will "lead" the movement, whereas the second finger will "follow". It seems that the resulting movement patterns and learning rates can also differ depending on the role of the dominant hand vs. non-dominant hand [38].

The sequence in Fig. 5 shows progressive phases over time of learning the 90° task. The y-axis represents a dimension of initial bias in the sense that positive values of y represent a supportive bias whereas negative values correspond to negative interference with the task. Thus, we have an interpretation of the horizontal dimensions ("Phase Difference" and "Initial Bias") as initial conditions for our dynamical model and the vertical dimension ("V") is interpreted as a form of generalized potential in the sense that at any point in the landscape the movement will follow the steepest slope at that point in the

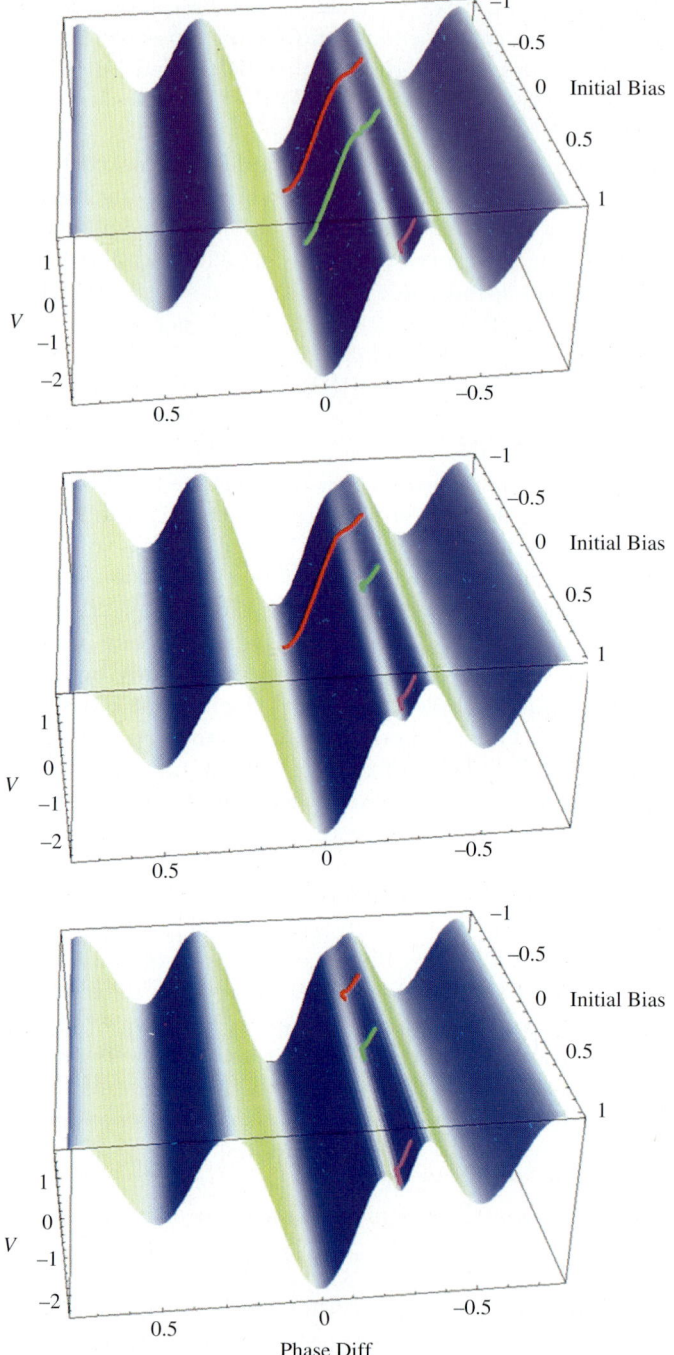

Fig. 5. The figures each show the additional dimension of initial bias instead of time for three stages during practice, early, intermediate, late. The *colored lines* show the gradient descent of single trials with different initial bias conditions

landscape. This type of dynamics is known as "gradient descent". Of course in a full simulation we would also add a stochastic component to this deterministic dynamics. The curves shown in the landscapes of Fig. 5 correspond to trajectories that are computed as solutions to the differential equations describing the gradient descent dynamics. The explicit form of the potential landscapes in Fig. 5 is given by:

$$V_i(\phi, y) = V_{\text{HKB}}(\phi, y) - c_i(y + 2)e^{-\frac{(\phi - \phi_0)^2}{\sigma^2}} \qquad (4)$$

where i is a, b, or c and $c_a = 0.380821$, $c_b = 0.381$, $c_c = 0.5047$. The learning target is a phase difference of $\phi_0 = 90°$ and the width of the potential well is given by $\sigma = 0.05$.

In Fig. 5 we see that, unless we have a strong supportive bias for the given initial conditions shown in the figure, the 90° phase difference cannot be stabilized and sustained. Instead we will observe the transition to the in-phase oscillation. In Fig. 5 we can also observe that the role of initial bias becomes less effective with increase in practice time and that eventually even strong interference (initial condition with negative y value) cannot destabilize the 90° movement. In this situation the task can now be performed securely from a range of initial conditions. In summary, the series of figures show the independent roles of the landscape and the initial bias and how these can change over different time scales.

We want to emphasize here that the landscapes presented represent hypothetical models that illustrate the implications of theoretical assumptions that can lead to experimental tests. In Fig. 5, for instance, the implicit assumptions are that the intrinsic dynamics of the original HKB model has a persistent influence on the observed dynamics and that the intention of learning a 90° task leads to a very localized change of behavior in terms of initial phase difference. If the participant starts the movement close to an in-phase or anti-phase pattern, the intrinsic dynamics would be strong enough to stabilize that movement pattern. Existing experimental results, however, have not given an extensive examination of these assumptions that arise from our model.

There is a long history of the study of preparatory processes in motor control but they have not been expressed directly in dynamical accounts of motor learning and control [14]. This seems to have arisen due to the emphasis on autonomous equations and the study of ongoing oscillatory movements [2] with the result in practice that the processes involved in the initiation of a movement have been largely omitted from the HKB model and dynamical approaches in general. This perspective has also led to little effort to study discrete movements from a dynamical perspective. The HKB model has a formal basis for the exhibition of hysteresis in the transition but there are many factors in learning and performance that influence or bias the dynamics of the initial conditions that are not included in the standard interpretation of hysteresis.

6 Issues on Intrinsic Dynamics and Learning

The contribution of intrinsic dynamics in the learning of new coordination modes is central in the coordination dynamics framework. Yet, the original equation (2) for learning dynamics [33, 34] includes factors beyond the intrinsic dynamics such as the influence of the effects of environmental constraints and learning strategies, including for example, informational support, in motor learning. Thus, by definition in the HKB framework it is the *additive* effects of these factors and their forces at play that determine the evolving landscape. There are a number of issues open to investigation from this perspective in motor learning from both within and without the assumptions of the coordination dynamics framework that need to be considered in generalizing the HKB framework.

6.1 The Task Dependency of Intrinsic Dynamic Effects

Even within the principles of the HKB model framework the relative influence of the intrinsic dynamics will vary in the learning of certain novel coordination modes. The task differences and individual differences within task will be considerable in determining the influence of the intrinsic dynamics. Thus, the effects of intrinsic dynamics may be manifest in the finger coordination task but in, for example, slow positioning or even slow oscillatory movements the influence may be very small to non-existent. It is also of note that the boundaries to the frequency dimension of the HKB model do not appear to have been established either in an absolute or body relative fashion.

A pattern to the task dependency of the role of intrinsic tendencies would be consistent with the findings from the many years of studying transfer in motor skills [1, 31], where positive, neutral and negative transfers have all been demonstrated. One strength of the HKB model is that it provides a vehicle to assess the relation between existing tendencies for particular task conditions and the learning of a new task. However, the role of the intrinsic dynamics in the transfer to the learning of a new task needs to be considered in a broad range of contexts and in terms of their interactive effects with informational support and other learning strategies present for learning.

6.2 Intrinsic Dynamics and Information

The intrinsic dynamics represented in the HKB model of Fig. 1 and (1) are emergent from a particular set of environmental and instructional conditions used in the basic finger experiments [10, 11]. If the same basic experiment was run with different performance conditions, particularly those that provided stronger informational support for performance (though still non-specific behavioral information in the HKB sense), it seems likely that a different landscape of intrinsic dynamics would emerge. This may not be simply an overall reduction in the elevation of the landscape with the same parameter conditions but an entirely different landscape reflecting the interaction between behavioral information and intrinsic dynamics.

Experiments by Mechsner et al. [22] have shown that the type of informational support can influence the learning of a new relative phase and change the impact of the intrinsic dynamics of the HKB model. This pattern of findings and others [5] lend support to the interactive nature of the information and the intrinsic tendencies for learning. The additive or interactive contributions of learning to the formation of the landscape need to be investigated.

The above points raised for future experimental consideration reflect challenges within the principles of the HKB framework, but there are additional points that challenge some of the assumptions of this framework.

6.3 The Scaling Control Parameter Protocol

The HKB landscape is formed from the scaling of oscillation frequency in an intermittent fashion. The typical experiment scales the control parameter of frequency about every 15–20 s over the range of frequencies studied. Presumably, however, the HKB model predicts the same pattern of intrinsic dynamics if the frequency and relative phase conditions are run as a set of blocked trials at a series of single learning conditions but this has not been tested. The single frequency and relative phase condition approach eliminates the presence of inter-task interference that arises from the scaling protocol and the effects of which may be in the landscape of intrinsic dynamics. Some studies using a more standard single condition learning approach to a new relative phase pattern have not shown the same degree of transfer from the intrinsic dynamics [5, 22].

6.4 Are Intrinsic Dynamics Neutral to the Intention Under Which They Were Measured?

It was noted at the outset that the intrinsic dynamics of the HKB model were derived from a particular experimental instruction to the subject in the running of the experiment. The assumption is that the resulting landscape represents the intrinsic dynamics for finger coordination dynamics. There are, however, other non-specific parameter manipulations than frequency (amplitude, for example) that could be made and one could ask whether the resultant intrinsic dynamics will be the same as those of the HKB model under the frequency manipulation?

The core question, however, is whether the intrinsic dynamics are peculiar or general to the intention? If the intrinsic dynamics are not neutral to intention it suggests that these effects are interactive with the different conditions of running the coordination experiments. Such a relation would not negate the influence of intrinsic tendencies to learning but would imply that the particular effects do not generalize in the same additive way to every novel to-be-learned phase relation in the landscape.

7 Concluding Comments

In this paper we have outlined an approach that builds from the principles of intrinsic dynamics and the HKB model. In particular, we sought to develop a more general model for the change over time in learning and development for the two degree of freedom finger task. This exercise reveals that the consideration of time and the change in behavior over time provides challenges to some assumptions of the HKB model and its extensions.

It should be noted in closing, however, that both our Gaussian model and the original HKB model work on the assumption of the additivity of the intrinsic dynamics to the other terms in the equations. This is a useful working assumption but as we outlined the principles that support this assumption may not hold. One of the strengths of the HKB model is that it provides the basis to take us beyond the metaphor of landscapes in motor learning and control. In turn, it also provides the basis for the consideration of landscapes beyond the HKB horizon.

A fundamental challenge in the context of motor learning remains open for future empirical investigations: namely, that of precise and testable interpretation of the landscape and its connection with experimental observations. Both the Kelso–Schöner (K-S) and our approach assume gradient descent dynamics, i.e., the system changes its state according to the local slope in the landscape. But the interpretation diverges as it comes to modifications of the original HKB landscape. Whereas K-S emphasize influences from environmental information and memory we keep the origin of the influence open and focus on properties of potentiality and manifestation of movement patterns. In the original Waddington landscape as well as in the standard HKB model, the minima of the landscape indicate movement patterns that are a potentiality for the participant under the given conditions. If the frequency is low in the HKB case we have the skill and option to choose in-phase or anti-phase oscillations. The potentiality of anti-phase disappears as the frequency becomes too high.

On the other hand, once we decided that our intention is to perform anti-phase oscillations at low frequency, the probability that an experimenter will observe a transition to in-phase oscillation becomes zero. This process is very similar to the situation in quantum systems where an unobserved system is in a superposition of possible states but once an observation has been made the wave function collapses into a single state, the one that will be observed classically.

Acknowledgments

This work was supported in part by NSF #0114568.

References

1. Adams JA (1987) Historical review and appraisal of research on learning, retention and transfer of human motor skills. Psychol Bull 101:41–74
2. Beek PJ, Turvey MT, Schmidt RC (1992) Autonomous and nonautonomous dynamics of coordinated rhythmic movement. Ecol Psychol 4:65–96
3. Connolly KJ (1986) A perspective on motor control. In: Wade MG, Whiting HTA (eds) Motor development in children: Aspects of coordination and control. Martinus Nijhoff, Dordrecht, pp. 3–22
4. Elman JL, Bates EA, Johnson MH, Karmiloff-Smith A, Parisi D, Plunkett K (1996) Rethinking innateness: A connectionist perspective on development. The MIT Press, Cambridge, MA
5. Fontaine RJ, Lee TD, Swinnen SP (1997) Learning a new bimanual coordination pattern: Reciprocal influence on intrinsic and to-be-learned patterns. Can J Exp Psychol 51:1–9
6. Haken H (1983) Synergetics: An introduction, 3rd edn. Springer-Verlag, Berlin
7. Haken H (1984) Advanced synergetics, 2nd edn. Springer-Verlag, Berlin
8. Haken H (ed) (1988) Neural and synergetic computers. Springer-Verlag, Berlin
9. Haken H, Kelso JAS, Bunz H (1985) A theoretical model of phase transitions in human hand movements. Biol Cybern 51:347–356
10. Kelso JAS (1981) On the oscillatory basis of movement. Bull Psychon Soc 18:63
11. Kelso JAS (1984) Phase transitions and critical behavior in human bimanual coordination. Am J Physiol Regul, Integr Comp Physiol 15:R1000–R1004
12. Kelso JAS (1995) Dynamic patterns: The self-organization of brain and behavior. MIT Press, Cambridge, MA
13. Kelso JAS, Engstrøm, DA (2006) The complementary nature. MIT Press, Cambridge, MA
14. Kornblum S, Requin J (eds) (1984) Preparatory states and processes. Lawrence Erlbaum Associates, Hillsdale, NJ
15. Kugler PN (1986) A morphological perspective on the origin and evolution of movement patterns. In: Wade MG, Whiting HTA (eds) Motor development in children: Aspects of coordination and control. Martinus Nijhoff, Dordrecht, pp. 459–525
16. Kugler PN, Kelso JAS, Turvey MT (1980) On the concept of coordinative structures as dissipative structures: I. Theoretical lines of convergence. In: Stelmach GE, Reqiun J (eds) Tutorials in motor behavior. North-Holland, New York, pp. 1–49
17. Kugler PN, Kelso JAS, Turvey MT (1982) On the control and coordination of naturally developing systems. In: Kelso JAS, Clark JE (eds) The development of movement control and co-ordination. Wiley, New York, pp. 5–78
18. Lee TD, Swinnen SP, Verschueren S (1995) Relative phase alteration during bimanual skill acquisition. J Mot Behav 27:263–274
19. Liu Y-T, Mayer-Kress GJ, Newell KM (2005) Evidence for phase transitions and bi-stability in a rhythmic motor learning task. Paper presented at NASPSPA, St. Pete Beach, Florida, USA
20. Liu Y-T, Mayer-Kress GJ, Newell KM (2006) Qualitative and quantitative change in the dynamics of motor learning. J Exp Psychol Hum Percept Perform 32:380–393
21. Meinhardt H (1982) Models of biological pattern formation. Academic Press, New York
22. Mechsner F, Kerzel D, Knoblich G, Prinz W (2001) Perceptual basis of bimanual coordination. Nature 414(6859):69–73

23. Muchisky M, Gershkoff-Stowe L, Cole E, Thelen E (1996) The epigenetic landscape revisited: A dynamic interpretation. In: Rovee-Collier C (ed) Advances in infancy research, Vol. 10. Ablex, Norwood, NJ, pp. 121–159
24. Newell KM (1985) Coordination, control and skill. In: Goodman D, Franks I, Wilberg R (eds) Differing perspectives in motor learning, memory and control. North-Holland, Amsterdam, pp. 295–317
25. Newell KM (1996) Change in movement and skill: Learning, retention, and transfer. In: Latash M, Turvey M (eds) Dexterity and its development. Erlbaum, Hillsdale, NJ
26. Newell KM, Liu Y-T, Mayer-Kress G (2001) Time scales in motor learning and development. Psychol Rev 108: 57–82
27. Newell KM, Liu Y-T, Mayer-Kress G (2003) A dynamical systems interpretation of epigenetic landscapes for infant motor development. Infant Dev Behav 26:449–472
28. Newell KM, Liu Y-T, Mayer-Kress G (2005) Learning in the brain-computer interface: Insights about degrees of freedom and degeneracy in a landscape model of motor learning. Cognit Process 6:37–47
29. Newell KM, Liu Y-T, Mayer-Kress G (in press) Time scales in connectionist and dynamical systems approaches to learning and development. In: Spencer JP, Thomas MS, McClelland JL (eds) Toward a new grand theory of development? Connectionism and dynamic systems theory reconsidered. Oxford University Press, New York
30. Newell KM, Kugler PN, van Emmerik REA, McDonald PV (1989) Search strategies and the acquisition of coordination. In: Wallace SA (ed) Perspectives on coordination. North Holland, Amsterdam, pp. 86–122
31. Osgood CE (1949) The similarity paradox in human learning: A resolution. Psychol Rev 56:132–143
32. Rensing L, an der Heiden U, Mackey MC (eds) (1987) Temporal disorder in human oscillatory systems. Springer-Verlag, Berlin
33. Schöner G (1989) Learning and recall in a dynamic theory of coordination patterns. Biological Cybernetics 62:39–54
34. Schöner G, Kelso JAS (1988) A synergetic theory of environmentally-specified and learned patterns of movement coordination. I. Relative phase dynamics. Biological Cybernetics 58:71–80
35. Schöner G, Kelso JAS (1988) A synergetic theory of environmentally-specified and learned patterns of movement coordination. II. Component oscillator dynamics. Biol Cybern 58:81–89
36. Schöner G, Zanone PG, Kelso JAS (1992) Learning as a change of coordination dynamics. J Mot Behav 24:29–48
37. Thelen E, Smith LB (1994) A dynamic systems approach to the development of cognition and action. MIT Press, Cambridge, MA
38. Treffner PJ, Turvey MT (1996) Symmetry, broken symmetry, and handedness in bimanual coordination dynamics. Exp Brain Res 107:463–478
39. Waddington CH (1957) The strategy of the genes. George Allen & Unwin, London
40. Zanone P, Kelso JAS (1992) Evolution of behavioral attractors with learning: Non equilibrium phase transitions. J Exp Psychol Hum Percept and Perfor 18:403–21
41. Zanone P, Kelso JAS (1997) Coordination dynamics of learning and transfer: Collective and component levels. J Exp Psychol Hum Percept and Perfor 23:1454–1480

Behavioral Dynamics of Visually Guided Locomotion

William H. Warren[1] and Brett R. Fajen[2]

[1] Department of Cognitive and Linguistic Sciences, Brown University, Providence, RI 02912, USA
[2] Department of Cognitive Science, Rensselaer Polytechnic Institute, Troy, NY 12180-3590, USA

1 Introduction

In their seminal research on coordination dynamics, Scott Kelso and his colleagues have developed an account of biological coordination in which the organization of action is viewed as a species of self-organizing pattern formation. Motor patterns are understood as a reflection of stability properties in a complex dynamical system, such as phase attraction and repulsion, bifurcation, and meta-stability. Moreover, the observation that similar coordination phenomena occur between visually coupled and neurally coupled oscillatory systems, from a swarm of Malaysian fireflies to the limbs of two people, led Scott to the view that coordination dynamics is fundamentally informational rather than physical in nature [29, 30].

The motivation behind the present work is to extend such dynamical principles to everyday behavior in which perception and action must be coordinated with a complex changing environment [49]. Such varied interactions implicate a role for informational variables that specify not only coordination conditions but also the state of the agent–environment system, in the spirit of Gibson [21] and Lee [32]. The dynamics of such behavior is often nonstationary, evolving as the interaction between agent and environment unfolds. We use the term *behavioral dynamics* to mark this emphasis on adaptive behavior by an agent in an environment, which are coupled by perceptual information.

Visually guided locomotion is paradigmatic of this kind of interactive adaptive behavior, for it is exhibited by mobile animals ranging from invertebrates to primates and thus holds promise for uncovering general principles of perception and action. Our present focus is on the phenomenon of *path formation*, the emergence of a locomotor path as an agent travels through a structured environment. This problem has been approached from a local vantage point, such as analyzing the information a particular agent uses to control locomotion [20, 48, 52], as well as from a global vantage point, such as modeling the evolution of trail networks [23, 24]. We are interested in the bridge between

these two levels: how global paths arise from local agent–environment interactions on the basis of simple laws for steering and obstacle avoidance. Our
analysis aims for a general characterization of locomotor behavior that is
not tied to particular informational variables and assumes only that certain
agent–environment relations are perceptually available.

The problem is nicely illustrated by the game of American football, in
which a player must steer toward goals, avoid stationary and moving obstacles, intercept moving targets, and evade pursuit. Similar challenges are faced
on a daily basis by animals in the wild, people walking through public spaces,
and autonomous mobile robots. It is commonly assumed that selecting a route
through a cluttered scene requires advance path planning on the basis of a
detailed internal model of the world. On the other hand, it has been shown
that global locomotor behavior such as flocking and schooling can arise from
local interaction rules without explicit planning [43]. However, with notable
exceptions [28], the agent-based simulation approach tends toward computer
demonstrations rather than empirical models of agent behavior. We seek to
derive minimal "rules" for human locomotion experimentally and determine
whether they can account for path formation, route selection, pedestrian interactions, and ultimately herding and crowd behavior, as self-organized emergent behaviors, without explicit path planning.

2 Modeling Approach

We begin by decomposing locomotion into a set of elementary behaviors that
can be modeled individually. As a first approximation, these include (a) steering to a stationary goal, (b) avoiding a stationary obstacle, (c) intercepting
a moving target, and (d) avoiding a moving obstacle. Additional components
may be required to account for behavior such as evading a pursuer or herding,
for example, a component for following neighbors, but the aim is to empirically determine the minimum rule set. Our strategy [14] is to model each
elementary behavior as a nonlinear dynamical system and then attempt to
predict human behavior in more complex environments by linearly combining
these components.

The modeling approach is inspired by the work of Schöner et al. [22, 46],
who developed a dynamical control system for mobile robots. Goal-directed
behavior can be described by a few behavioral variables that define a *state
space* for the system. For locomotion, the behavioral variables are defined as
the current *heading* (ϕ) or direction of travel with respect to an allocentric
reference axis, and the current *turning rate* ($\dot{\phi}$), assuming for the moment a
constant speed of travel v (see Fig. 1). From the agent's current (x,z) position,
a goal lies in the direction ψ_g at a distance d_g. The simplest description
of steering toward a goal is for the agent to bring the *target-heading angle*
to zero ($\beta = \phi - \psi_g = 0$) as it moves forward, which defines an attractor
in state space at $[\phi, \dot{\phi}] = [\psi_g, 0]$. Conversely, for an obstacle that lies in a

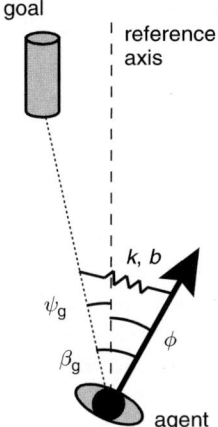

Fig. 1. Definition of variables: heading direction (ϕ) and goal direction (ψ_g) in an allocentric reference frame, target-heading angle (β), metaphorical spring stiffness (k) and damping (b) coefficients

bearing direction (ψ_o) with respect to the reference axis, at a distance d_0, the simplest description of obstacle avoidance is to magnify the *obstacle-heading angle* ($\phi - \psi_o > 0$), which defines a repeller at $[\phi, \dot{\phi}] = [\psi_o, 0]$. The challenge is to formalize a system of differential equations that generates a trajectory through the state space $[\phi, \dot{\phi}]$ – that is, a sequence of headings and turning rates – that captures observed locomotor behavior. Schöner et al.'s [46] control algorithm was predicated on a first-order dynamical system, which brings the turning rate to zero linearly as the target-heading angle goes to zero. The advantage of this system is that local asymptotic stability is assured as the environment becomes more complex and model components are combined. In studies of biological systems, however, one is directly measuring exhibited behavior, from which hypothetical control laws must be inferred (see [14] for details). Thus, the first task is to formalize the behavioral dynamics, a description of the time evolution of exhibited behavior. Because the agent is physically embodied, changing the direction of travel of an inertial body involves angular acceleration, implicating at least a second-order dynamical system to describe steering behavior.

2.1 Intuitive Model

To get an intuition for the model, imagine that the agent's current heading direction is attached to the goal direction by a damped spring (Fig. 1). Angular acceleration toward the goal direction would thus depend on the stiffness of the spring (k_g) and be resisted by the damping (b). Damping may be physically interpreted as the resistance of a body in motion to deflection of its path, which must be overcome by centripetal force. For obstacle avoidance, imagine that the heading direction is attached to the obstacle direction by another spring, but one with a stiffness of opposite sign, such that the heading is repelled from the obstacle direction.

One advantage of the model is that it scales linearly with the complexity of the scene, simply adding one spring term for each object. The direction of travel at any moment is determined by the resultant of all spring forces acting on the agent at that point. A route through a complex scene thus unfolds as the agent tracks the local heading attractor, whose current direction is determined by the combined influence of the goal and obstacle.

Our research strategy is to collect basic data on each elementary behavior, formalize a model component that closely describes the mean data, and fix parameter values by fitting the model to the mean data. We then test the model by using it to predict human paths in novel configurations of objects with fixed parameters. Of particular interest is how well the model will generalize to more complex environments by linearly combining the components.

2.2 Method

The experiments described here were carried out in the Virtual Environment Navigation Lab (VENLab) at Brown, a $12\,\mathrm{m} \times 12\,\mathrm{m}$ area in which a participant can walk freely while wearing a tethered head-mounted display (HMD). The display is stereoscopic ($60°$ horizontal $\times$ $40°$ vertical), with a frame rate $\geq 60\,\mathrm{Hz}$ and a total latency of 50–70 ms. Head position is recorded at $60\,\mathrm{Hz}$ using a sonic/inertial tracking system. To establish the initial conditions for each trial, the participant walks 1 m toward an orientation marker on a textured ground plane; then colored goal and obstacle posts appear ($2.5\,\mathrm{m}$ tall $\times$ $0.1\,\mathrm{m}$ radius). The participant's task is simply to walk to the goal while detouring around the obstacle(s) at a roughly constant speed.

3 Component 1: Stationary Goal

Let us begin with the simple task of steering to a stationary goal. In [14], we proposed a model that describes the angular acceleration of heading as a function of the current target-heading angle $(\phi - \psi_\mathrm{g})$ and goal distance (d_g), while the mass term is suppressed:

$$\ddot{\phi} = -b\dot{\phi} - k_\mathrm{g}(\phi - \psi_\mathrm{g})(\mathrm{e}^{-c_1 d_\mathrm{g}} + c_2). \tag{1}$$

Essentially, this system *nulls the target-heading angle* by defining an attractor in heading whose strength decreases exponentially with goal distance.

The "damping" term $b\dot{\phi}$ acts as a frictional force that is proportional to the turning rate, which tends to keep the path straight and prevents the heading from oscillating about the goal; parameter b expresses the ratio of damping to the body's moment of inertia in units of s^{-1}. The "stiffness" term

$k_g(\phi - \psi_g)$ reflects the empirical finding that attraction toward the goal direction increases linearly with the target-heading angle; parameter k_g expresses the ratio of stiffness to moment of inertia, in units of s^{-2}. Interestingly, this linear relationship has been observed (over some range) in both humans [14] and houseflies [42], suggestive of the generality of steering dynamics. To capture the observation that people turn faster toward nearer goals (so as not to miss them), the stiffness is modulated by goal distance so that acceleration increases exponentially with nearer goals ($e^{-c_1 d_g} + c_2$). The constant c_1 determines the decay rate with distance in units of m^{-1}, and c_2 determines a minimum value so acceleration does not go to zero at large distances, and is dimensionless. These parameter dynamics make the equation nonlinear.

The perceptual variables assumed by the model are (i) the target-heading angle, which may be specified by various combinations of visual, proprioceptive, and podokinetic information, and (ii) the goal distance, which may be specified by a variable as simple as declination angle [37]. Alternatively, because distance is equivalent to time-to-contact at a constant speed of travel, obstacle avoidance could depend on perceived time-to-contact. However, preliminary evidence suggests that obstacle avoidance appears to be controlled by perceived distance rather than perceived time-to-contact [3]. To simulate a locomotor path, as the agent moves to a new (x,z) position, the perceptual variables of target-heading angle and goal distance are updated, and the next heading and turning rate are computed according to (1).

To specify this function, we collected descriptive data on humans walking to a goal as initial conditions were varied [14]. In these experiments, we manipulated the initial target-heading angle ($0 - 25°$) and initial goal distance (2–8 m). Participants turn onto a straight path to the goal (Fig. 2) and do so more rapidly when the goal is at a larger target-heading angle or a shorter distance. The time series of target-heading angle (Fig. 2b) also converges to zero from all initial conditions, illustrating that the goal direction behaves like an attractor of heading. Moreover, the angular acceleration increases with initial target-heading angle and decreases with goal distance. The model closely fits the mean time series of target-heading angle in each condition, with a mean $r^2 = 0.98$ for parameter values of $b = 3.25$, $k_g = 7.50$, $c_1 = 0.40$, and $c_2 = 0.40$. As shown in Fig. 3, model simulations generate locomotor paths and time series that are highly similar to the human data. The model thus accurately describes the behavioral dynamics of steering to a stationary goal, such that the goal direction serves as an attractor of heading.

3.1 Alternative Steering Strategies

Our model steers to a stationary goal by nulling the target-heading angle, but other strategies have also been proposed. Llewellyn [35] originally proposed that one could steer to a target by canceling the drift of the target in the field of view, on the assumption that one is not rotating. In the general case, this is equivalent to nulling *change* in the target-heading angle ($\dot{\beta}$) rather than the

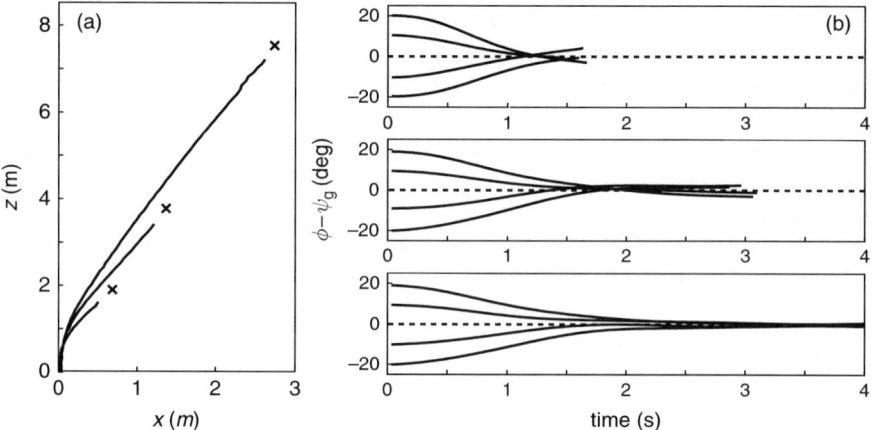

Fig. 2. Human data for walking to a goal at 2, 4, and 8 m. (a) Mean paths for an initial target-heading angle of 20°. (b) Mean time series of target-heading angle, with initial angles of 10 and 20° at each target distance. Reprinted from [14], copyright 2003 by the American Psychological Association

target-heading angle itself (β). Versions of this *constant target-heading angle strategy* have been developed by Rushton, Wen, and Allison [45] and Wilkie and Wann [52, 53]. We recently modeled this strategy using a dynamical system similar to (1) that nulled $\dot{\beta}$, without a distance term [16]. Simulations revealed that this strategy successfully gets the agent to the goal, but generates equiangular spiral paths under many conditions, and fails to reproduce

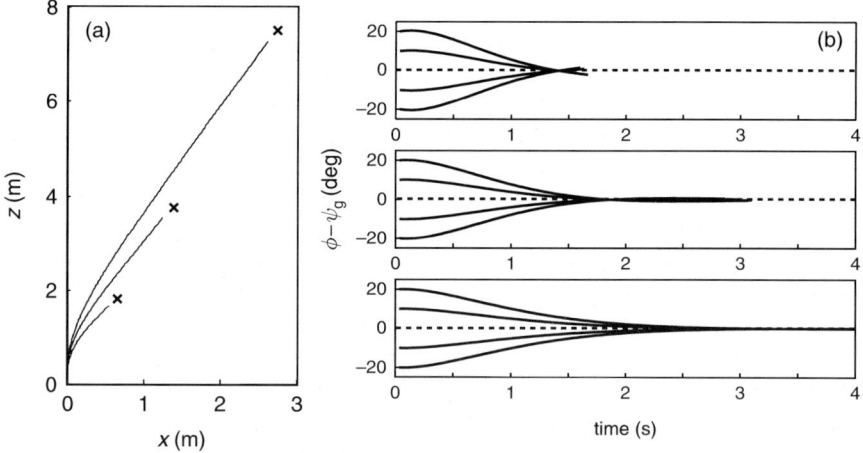

Fig. 3. Model simulations of walking to a goal, same conditions as Fig. 2. (a) Paths for an initial target-heading angle of 20°. (b) Time series of target-heading angle. Reprinted from [14], copyright 2003 by the American Psychological Association

Fajen and Warren's [14] human data. The time series of target-heading angle can get stuck at a constant nonzero value, rather than going to zero as in human steering.

We should point out that (1) is a dynamical system that corresponds to a standard proportional-derivative controller that nulls an error variable, in this case the heading error. However, the addition of a nonlinear stiffness term that depends on distance is nonstandard. Furthermore, the obstacle components to be described below do not correspond to PD controllers. A dynamical systems framework appears to offer a more flexible modeling framework than the typical control theoretic one and leads the researcher to consider more generic topological features of behavior.

4 Component 2: Stationary Obstacle

Next consider the task of obstacle avoidance. To model the repulsion of heading from the direction of an obstacle, we proposed an equation of similar form but flipped the sign on the stiffness term [14]:

$$\ddot{\phi} = -b\dot{\phi} + k_0(\phi - \psi_o)e^{-c_3|\phi-\psi_o|}e^{-c_4 d_0}. \tag{2}$$

In addition, rather than a linear repulsion function, the repulsion decays exponentially to the left and right of the obstacle. Thus, when the agent is heading slightly to the right of the obstacle, it has a high angular acceleration to the right (in a positive direction) that falls off as the obstacle-heading angle increases; conversely, when heading slightly to the left of the obstacle, it has a high acceleration to the left (in a negative direction) that falls off with angle. The amplitude of this repulsion function is influenced by the parameter k_o, its decay rate and amplitude by constant c_3 (in units of rad^{-1}), and it asymptotes to zero. As before, the stiffness decreases exponentially to zero with obstacle distance, where parameter c_4 is the decay rate (in units of m^{-1}). In simulation, the influence of this component is set to zero once the agent passes the obstacle.

To specify this equation, we recorded participants detouring around a single obstacle en route to a goal, under various conditions. The initial angle between the obstacle direction and the path to the goal (1–$8°$), as well as the initial obstacle distance (3–$5\,$m), were manipulated. Human paths deviated around the obstacle (Fig. 4a), with an angular acceleration that decreased exponentially with both obstacle-heading angle and distance. The time series of the obstacle-heading angle (Fig. 4b) shows that that the curves diverge from zero in all conditions such that the obstacle direction behaves like a repeller. To fit the model to the mean time series in each condition, we kept the previous parameter values for the goal (including damping) and added the obstacle stiffness component to it, yielding parameters of $k_o = 198.0$, $c_3 = 6.5$, and $c_4 = 0.8$, with a mean $r^2 = 0.975$. Simulations reproduced the mean human paths (Fig. 5a) and time series (Fig. 5b) quite well. The model thus captures

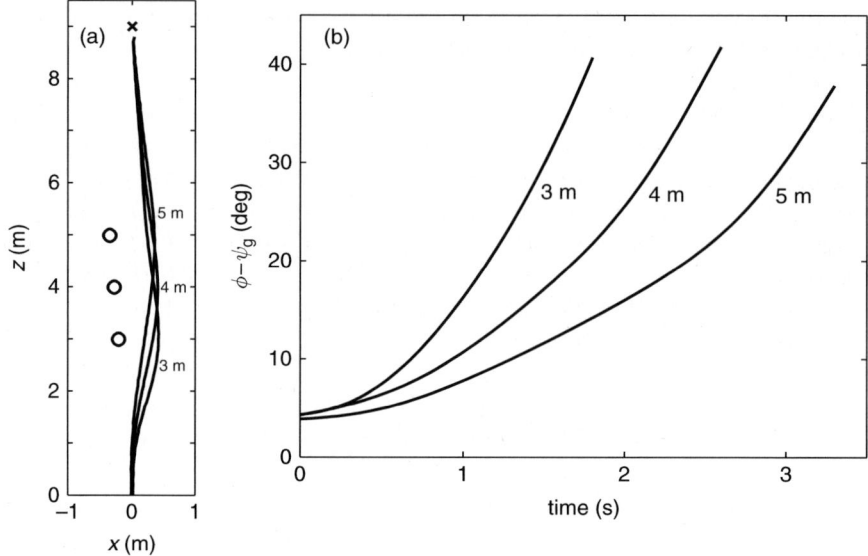

Fig. 4. Human data for obstacle avoidance, with an initial obstacle distance of 3, 4, or 5 m and an initial obstacle-target angle of −4°. (**a**) Mean human paths. (**b**) Mean time series of obstacle-heading angle. Reprinted from [14], copyright 2003 by the American Psychological Association

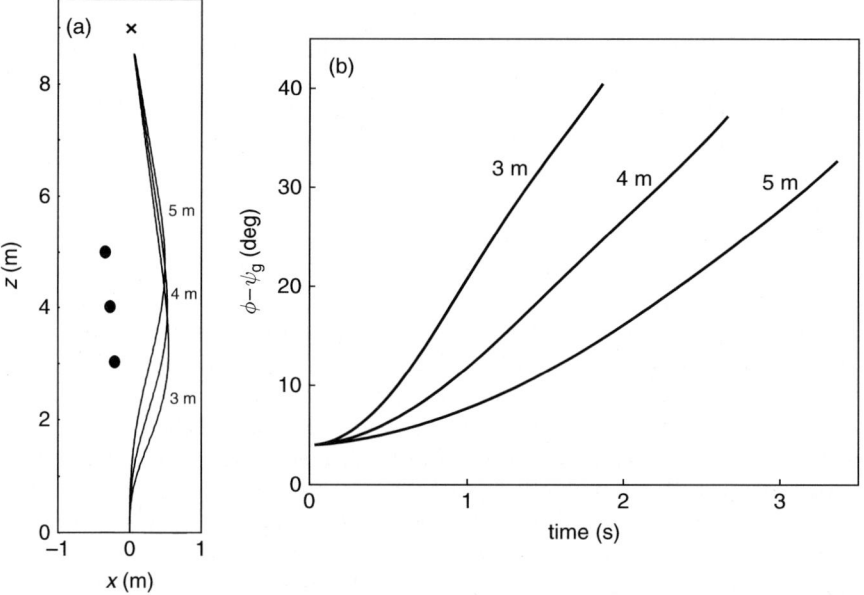

Fig. 5. Model simulations of obstacle avoidance, same conditions as Fig. 4. (**a**) Model paths. (**b**) Time series of obstacle-heading angle. Reprinted from [14], copyright 2003 by the American Psychological Association

the behavioral dynamics of obstacle avoidance such that the obstacle direction behaves like a repeller of heading.

4.1 Online Control

It is important to emphasize that the model does not rely on a full internal world model but only on a limited perceptual sample of the environment. The influence of obstacles decreases to a negligible value at a distance of 3–4 m and an angle of $\pm 60°$ about the heading direction. Moreover, there is no advance path planning, rather the path unfolds as the agent moves through the environment and the local attractor shifts in phase space. Thus, the model demonstrates the sufficiency of online control as a model of human locomotor behavior: steering can be based on avoiding the upcoming obstacles in near space, simply by tracking the local attractor as the dynamics evolve.

4.2 Robustness and Variability

Because the distance functions are gradually decreasing exponentials, the model is not sensitive to error in perceived distance. Adding a 10% Gaussian constant error on each trial into the perceptual variables (goal and obstacle distance and direction) yields a standard deviation of only 2 cm in lateral position midway through a trial for the goal configuration, and 6.6 cm in lateral position around an obstacle, demonstrating the model's robustness.

Moreover, the distribution of inside and outside paths around the obstacle can be closely reproduced by adding 10% Gaussian error into perceptual variables and parameters on each trial and matching the standard deviation of initial position and heading direction observed in the human data. This suggests that inter-trial variation can be accounted for in terms of biological noise.

4.3 Validity of VR

The fact that these experiments were carried out in a virtual environment raises a question about the validity of the method, and hence the model. Fink et al. [18] evaluated the VR methodology by having participants avoid an obstacle post en route to a goal post in matched virtual and real environments. Four new object configurations were used to test the ability of the model to generalize to new conditions. The qualitative shapes of the paths in the two environments were very similar. Using fixed parameters, the model fit the mean paths in both environments with $r^2 = 0.95$ across all configurations.

However, the paths did deviate slightly farther around the virtual obstacle than the physical obstacle, by 16 cm or about one head width. This may reflect uncertainty about the location of the obstacle relative to the observer's position in the virtual world, perhaps due to a reduced field of view or the fact that the observer's own body is not visible in the display. The slight difference is reflected in the fact that separate model fits to the physical and virtual paths yielded different parameter values and closer paths in both conditions, with $r^2 = 0.98$ in each.

The steering dynamics model thus generalizes quite well to novel object configurations and from virtual to physical environments, with the caveat that parameter values may need to be adjusted to account for quantitative details in a physical environment.

4.4 Reduced Model

Fink et al. [18] also performed analyses of individual participants, fitting the model to each participant's mean data. The results revealed some redundancy in the model parameters such that different combinations of parameter values produced highly similar paths. First, the obstacle stiffness parameter k_o is redundant with parameter c_3, because they both affect the amplitude of the repulsion function. As long as k_o is fixed at a high value, c_3 is sufficient to determine the deviation around an obstacle. Second, the goal parameter c_2 is a constant that insures a minimum goal attraction at long distances, and can be fixed at a small value. Third, assuming that the damping parameter b reflects the physical resistance to path deviation at a walking speed, it can also be fixed (or made a function of participant mass and speed).

By holding these parameters constant at Fajen and Warren's [14] values ($b = 3.25, c_2 = 0.4, k_o = 198$), we can thus reduce the number of free parameters in the model from seven to four: two for the goal component and two for the obstacle component. When this reduced model was fit to the mean paths in the physical and virtual environments, it was able to account for the range of paths in Fink et al.'s [18] data set with $r^2 > 0.98$, as well as the mean paths of each individual participant with $r^2 > 0.95$. Thus, the reduced model is sufficient to account for steering and obstacle avoidance in a variety of physical and virtual configurations with only four free parameters.

4.5 Alternative Obstacle Avoidance Strategies

Most other obstacle avoidance strategies have been developed as part of pedestrian models or robotic control systems, rather than as empirical models of human locomotor behavior. Helbing's *social force model* [25, 26], for example, treats goals, waypoints, obstacles, and other pedestrians as forces that act on an agent's path. Collisions are prevented by specifying minimum distances, and herding is produced by varying the likelihood of following a neighbor. The motivation for such pedestrian models is to capture the statistical properties of traffic flow and crowd behavior using multi-agent simulations, and the locomotor rules for individual agents are seldom evaluated experimentally. Our approach is to proceed in the opposite direction, beginning with an empirical model of individual behavior to provide a valid basis for the simulation of crowds.

In robotics, a standard control technique is the *potential field* method [31]. Goals and obstacles behave like attractors and repellers in 2D space, rather than in 1D heading. A goal is defined by a potential well whose attraction increases with the square of distance, whereas obstacles are defined by potential hills whose repulsion decreases sharply with distance according to an inverse square law. The gradient resulting from the linear sum of these fields determines a force vector that is used to control the robot's direction and speed of motion. In a direct comparison of the two approaches, Fajen et al. [17] determined that the steering dynamics model produces smoother and shorter paths than the potential field model and tends to avoid many of the local minima that are created in a spatial potential field, although spurious attractors can still occur (see below).

Another common approach to robotic control is the Vector Field Histogram method [2, 47]. In VMH+, a thresholded polar histogram of obstacle density in local space is used to steer in an obstacle-free direction and eliminates directions that violate the robot's inertial or mechanical constraints. This concept has also been applied to human locomotion through a field of obstacles [41], although the model is not well specified and makes assumptions about a number of free parameters. Huang et al. [27] implemented versions of the VMH+ method and the steering dynamics model in a mobile robot platform. They similarly concluded that the steering dynamics model generated much smoother paths. Most importantly, the paths produced by the steering dynamics model are also closer to the human data than paths generated by either of these robotic methods.

5 Component 3: Moving Target

We now attempt to extend the model to a dynamic environment, beginning with the task of intercepting a moving target. According to a rule of thumb familiar to sailors and pilots, if another craft remains at a constant compass bearing you are on a collision course. This observation suggests an effective interception strategy for achieving such a collision: steer so that the bearing direction of the moving target (ψ_t) remains constant [5]. For constant speed motion, this *constant bearing strategy* will yield the shortest straight interception path.

We [16] modeled the constant bearing strategy with an equation of similar form to (1), but which nulls change in the target's bearing direction $(\dot{\psi}_t)$ rather than nulling the target-heading angle:

$$\ddot{\phi} = -b\dot{\phi} - k_t(-\dot{\psi}_t)(d_t + 1).$$ (3)

Note that the distance term (d_t+1) acts to increase the influence of the moving target as its distance increases. This offsets the fact that, for a given target speed, $\dot{\psi}_t$ decreases to zero with distance due to perspective. The constant 1.0 insures that the stiffness will not drop below k_t as the target is approached, and hence it is not a free parameter.

However, the direction of the target from the observer may be affected not only by target motion but also by observer rotation. The effect of rotation can be eliminated by determining the target's relative motion with respect to a visible external reference frame, such as a fixed background or distant landmark. Dragonflies use just this strategy, adopting the horizon as a fixed reference frame. Specifically, they intercept overhead prey by maintaining a constant visual angle between the target and the horizon [36]. On the other hand, Fajen and Warren [15] found that human interception paths were similar whether or not a visible background was present (textured floor, walls, ceiling), as well as when motion was artificially added to the background. This suggests that a visible reference frame is neither necessary nor used when it is available.

But there are other means of compensating for body rotation. An equivalent version of the model nulls change in the target-heading angle ($\dot{\beta}$) while factoring out the influence of turning rate ($\dot{\phi}$). Because by definition $\beta = \phi - \psi_t$ (see Fig. 1), we can substitute $-\dot{\psi}_t = \dot{\beta} - \dot{\phi}$ into (3), yielding

$$\ddot{\phi} = -b\dot{\phi} - k_t(\dot{\beta} - \dot{\phi})(d_t + 1). \tag{4}$$

Information for turning rate $\dot{\phi}$ is available, and the evidence suggests that during locomotion, humans rely on proprioceptive information rather than visual or vestibular information for body rotation [1, 12, 15, 53].

The existing human data are consistent with some form of interception, rather than a *pursuit strategy* in which the agent heads straight at the target. Participants riding down a track [33, 34] or walking on a treadmill [6, 7, 8] adjust their speed to keep the bearing angle and target-heading angle (which are equivalent in this case) approximately constant. However, in those experiments, participants could only vary their speed, not their direction of travel.

We collected data for the general case of interception in the open field by presenting moving targets to participants walking in the virtual environment [15]. The target appeared at a distance of 3 m in depth, either directly ahead of a walking subject or 25° to one side, and moved laterally on one of three trajectories at a constant speed (0.6 m/s). In all conditions, participants successfully intercepted the target by turning onto a straight trajectory that led the target, while walking at a fairly constant speed (Fig. 6a). The time series of target-heading angle tended to plateau at a large positive angle, ahead of the target. These data are consistent with an interception strategy and inconsistent with simple pursuit.

To determine whether the constant bearing model can account for human interception, we fit (4) to the time series of target-heading angle [16]. Because the human paths suggested a latency to detect and respond to the motion of the target, we phased in the perceived target motion as a sigmoidal function of

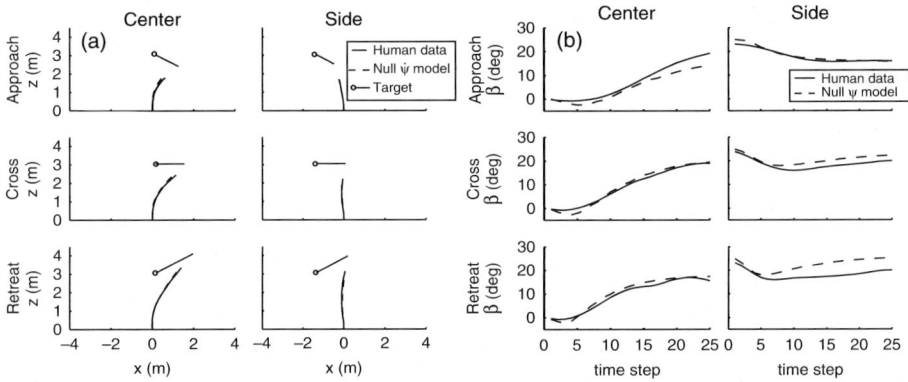

Fig. 6. Human data and model simulations of moving target interception, for the constant bearing strategy (null $\dot{\psi}$). In the center condition, the target appeared at an initial target-heading angle of $0°$, and in the side condition at an angle of $\pm 25°$ (data are collapsed). The target trajectory either approached at an angle of $120°$ from the initial heading direction, crossed at $90°$ to the initial heading, or retreated at $60°$, with a constant speed of $0.6\,\text{m/s}$. (a) Mean paths. (b) Mean time series of target-heading angle. Reprinted from [16], copyright by Pion Ltd

time, with a latency of $0.5\,\text{s}$. The fits yielded parameter values of $b = 7.75\,\text{s}^{-1}$, $k_t = 6.00\,\text{m}^{-1}\text{s}^{-1}$, and $c_1 = 1.0\,\text{m}$, with a mean $rmse = 2.15°$ and $r^2 = 0.87$ across all conditions. Simulated paths matched the human data very closely (Fig. 6a), and the time series show a similar tendency to plateau at a positive value, leading the target (Fig. 6b). The constant bearing model thus closely approximates human target interception.

5.1 Alternative Interception Strategies

Fajen and Warren [16] also tested several competing interception strategies. First was the simple pursuit strategy, in which the agent heads directly toward the moving target, bringing the target-heading angle to zero [44]. Since the agent must turn to track the target while traveling forward, this yields a continuously curved path that chases after the target, rather than an interception path. We modeled the pursuit strategy by applying our stationary goal model (1), which nulls the target-heading angle β. The model generates paths that are more curved than the participants' interception paths (Fig. 7), as well as time series of target-heading angle that are highly divergent. The pursuit strategy clearly fails to capture human interception behavior.

The main alternative is the *constant target-heading angle strategy*, which nulls change in the target-heading angle, $\dot{\beta}$ [6, 7, 8, 13, 45]. We modeled this strategy using a dynamical system similar to (4) that nulls $\dot{\beta}$ without factoring out the turning rate $\dot{\phi}$. Simulations revealed that there are two distinct solutions: a "lead" solution that heads a constant angle in front of

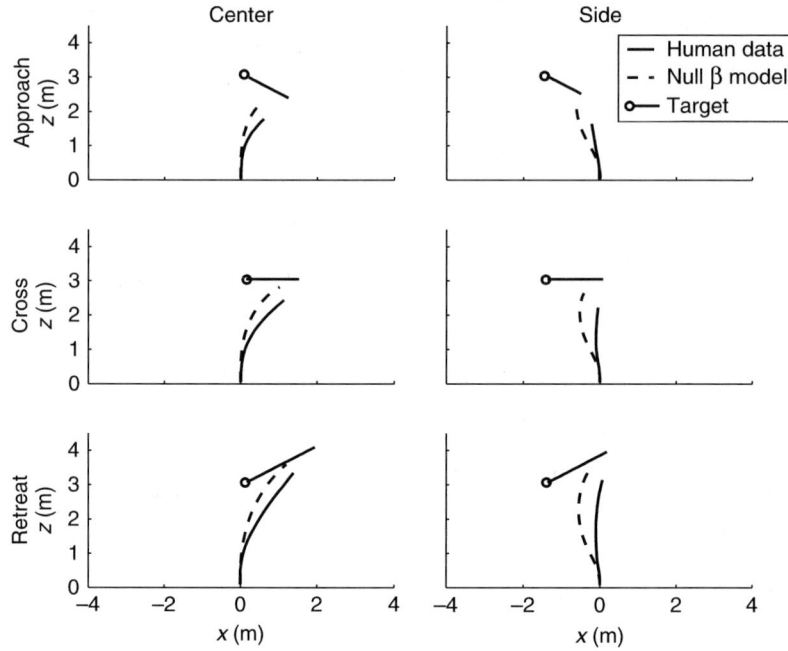

Fig. 7. Mean human paths and model simulations of target interception, for the pursuit strategy (null β). Same conditions and human data as in Fig. 6. Reprinted from [16], copyright by Pion Ltd

the moving target and a "lag" solution that heads a constant angle behind the target, while simultaneously turning. The lead solution yields a turn onto a straight interception path similar to the constant bearing strategy, but the lag solution produces a continuously curved path that chases the target.

Which solution is manifested depends on initial conditions. For example, with a rightward-moving target, the lead solution occurs when the target appears to the left of the agent's heading (Fig. 8, side condition), whereas the lag solution occurs when the target appears directly in front of the agent (Fig. 8, center condition). But participants always adopt a lead solution in these conditions. Thus, nulling change in the target-heading angle is an inadequate model of human interception behavior.

A third possible alternative is to compute the *required interception angle* based on perceived distal target velocity and perceived walking speed. Our simulations show that this model reproduces the human data quite well, but it depends on accurate perception of velocity and is not robust [16]. A 10% error in perceived target velocity and walking speed can produce as much as a 50% error in the computed interception angle. Computing the interception angle thus does not appear to be a viable strategy.

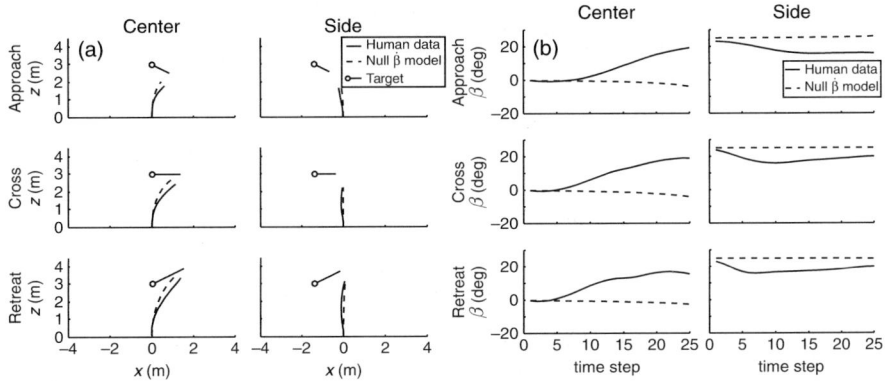

Fig. 8. Human data and simulations of target interception, for the constant target-heading angle strategy (null $\hat{\beta}$). Same conditions and human data as in Fig. 6. (a) Mean paths. (b) Mean time series of target-heading angle. Reprinted from [16], copyright by Pion Ltd

5.2 Can People Anticipate Target Motion?

The constant bearing strategy yields what appears to be anticipatory behavior simply by nulling change in the current target bearing, without actually predicting the target's trajectory in 3D space. This yields some counterintuitive predictions. First, if the target accelerates, the model predicts that people will follow a continuously curved path, rather than a straight interception path that anticipates the target's motion. Second, if the target travels on a circular trajectory, the model predicts that people will chase after the target on a continuously curved path, rather than anticipating its trajectory and taking a shortcut across the circle.

Owens and Warren [39] tested the latter prediction by studying human interception of targets moving on circular trajectories, with random variation in left/right directions and radius. Sure enough, participants chased these targets on smoothly curved paths similar to those predicted by the constant bearing strategy (Fig. 9). With repeated exposure to the same target trajectory, just over half of the participants do learn the heuristic of cutting across the circle after about a half-dozen trials, although they pick up the target on the far side using the constant bearing strategy again [40]. But with repeated exposure to only two trajectories having different radii or left/right directions, participants fail to learn this heuristic – even after 20 trials [38]. Thus, although the constant bearing strategy can be cognitively overridden in a single familiar case, the results suggest it is a basic control law that resists cognitive intervention. We are currently investigating a similar question for moving obstacles, specifically, whether participants can learn the characteristic trajectories of identifiable obstacles, as drivers seem to do when anticipating young children or animals darting into the street.

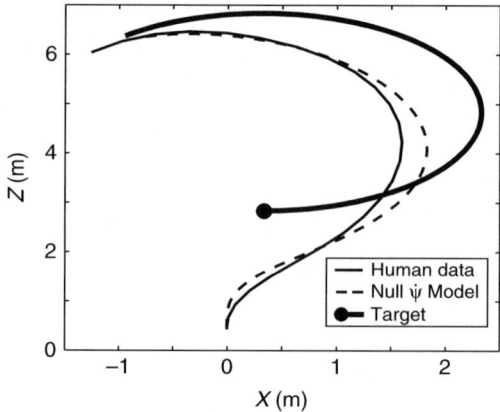

Fig. 9. Mean human and model paths for intercepting a target on a circular trajectory, for the constant bearing strategy (null $\dot{\psi}$)

The interception model may even allow us to test what trajectories are adaptive for evading a pursuer. A rabbit's irregular zig-zag path, for example, may be the best countermeasure to a fox that relies on the constant bearing strategy. Conversely, constant bearing may be the best strategy available to a pursuer whose targets travel on unpredictable trajectories.

5.3 Speed Control

In our work up to this point, we had found that participants adopt a fairly constant walking speed with stationary goals and obstacles, as well as with a moving target. However, previous results showed that when steering is restricted on a track or treadmill, speed can be adjusted to intercept a moving target [6, 33]. A complete interception model must thus account for speed control as well as heading control. One possibility is that heading and speed are jointly regulated to achieve a constant target bearing. Alternatively, they could be independently regulated by different visual information, but effectively coupled through the environment, for the heading that achieves a constant bearing will depend on the current walking speed.

To examine the conditions under which heading direction and walking speed are varied during interception in the open field, Chardenon and Warren [9] manipulated the speed and trajectory of a linearly moving target. They found that participants vary their heading under all conditions but only accelerate when needed to catch the target. In trying to simulate this data, the most successful model independently controlled angular acceleration (heading) with the constant bearing strategy and controlled linear acceleration (speed) using the time-to-contact with the target (TTC). Specifically, they developed another second-order system in which linear acceleration depends on an empirical TTC function. If the agent is more than two-thirds of a second from

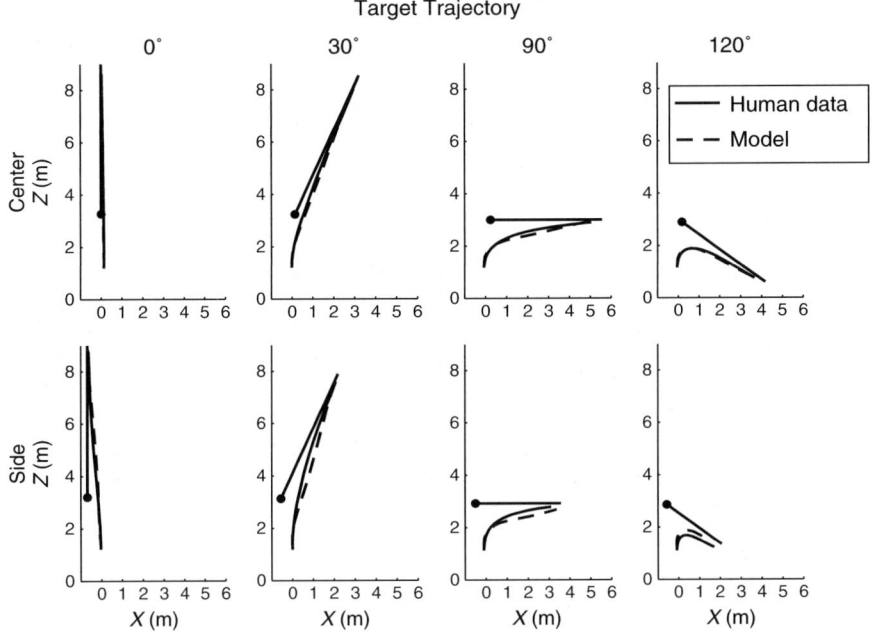

Fig. 10. Mean human and model paths for speed control during fast target interception. In the center condition, the target appeared at an initial target-heading angle of 0°, and in the side condition at an angle of ±20° (data are collapsed), with an initial distance of 2 m and a constant velocity of 1.5 m/s. Target trajectory was 0, 30, 90, or 120° from the initial heading direction

the target (TTC < -0.66 s), acceleration increases; and if the target is getting away so its image is contracting (TTC > 0 s), acceleration also increases. In between, as the agent is closing in on the target ($-0.66 <$ TTC < 0), the agent decelerates. Simulations of this model closely reproduced human paths to fast (Fig. 10) as well as slow targets and approximated the time series of agent speed. We thus consider the model to be a first approximation of speed control during target interception.

6 Component 4: Moving Obstacle

Finally, consider avoiding a moving obstacle. Analogous to the stationary case, Cohen and Warren [11] modeled moving obstacle avoidance using the inverse of the stiffness term for a moving target:

$$\ddot{\phi} = -b\dot{\phi} + k_m(-\dot{\psi})e^{-c_5|\dot{\psi}|}e^{-c_6 d_m}. \tag{5}$$

In effect, this creates a repeller in the direction of the interception path so that the agent avoids a constant bearing. As with stationary obstacles, the amplitude of the repulsion function is influenced by the parameter k_m and its lateral decay rate by the constant c_5 (in units of rad^{-1}). The distance term again causes stiffness to decrease exponentially to zero with obstacle distance, where parameter c_6 is the decay rate (in units of m^{-1}). The influence of this component is set to zero once the obstacle is passed.

To specify this equation, Cohen, Bruggeman, and Warren [10] recorded participants detouring around a single moving obstacle on their way to a stationary goal. In one experiment, they varied the obstacle speed (0.4–0.8 m/s) and trajectory (70–110°, where a 0° trajectory lies on a line from the agent to the goal). Participants switched from a path ahead of the moving obstacle on most trials to a path behind it when obstacle speed reached 0.6 m/s (see Fig. 11). In a second experiment, as the obstacle's approach angle increased

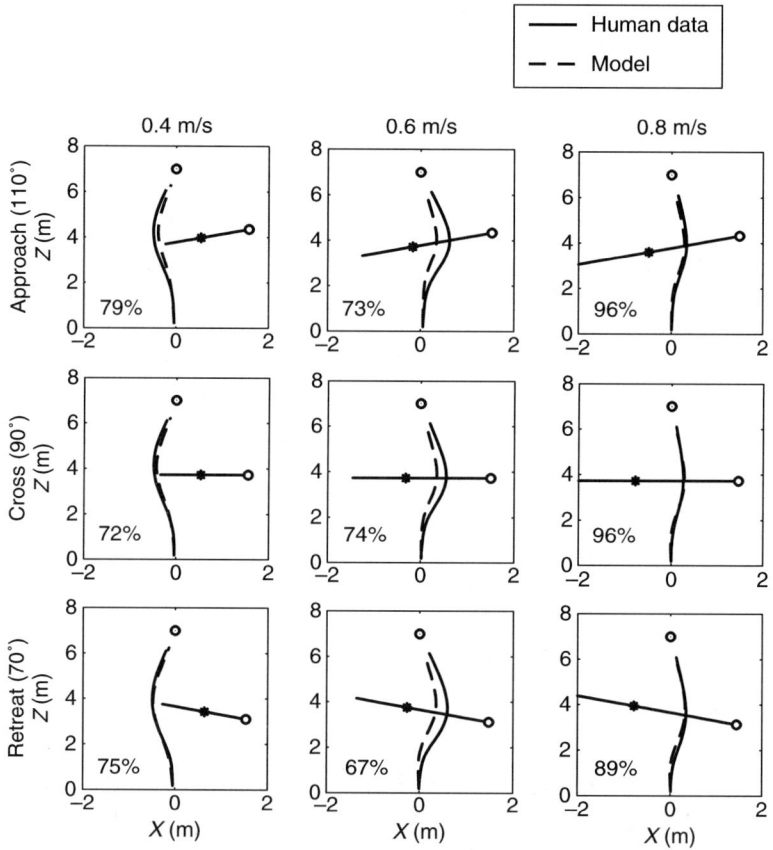

Fig. 11. Mean dominant human paths and model simulations for moving obstacle avoidance. Obstacle trajectory and speed were varied. Asterisks indicate the location of the obstacle when the participant passed its Z position. The percentage of trials in each condition on which participants took the dominant path is indicated

(from 100 to 140°), participants switched from a behind path to an ahead path at 140°. Avoidance behavior thus appears to depend on the obstacle's speed and trajectory in a systematic fashion.

We fit the mean time series of heading for the dominant (ahead or behind) mean path in each condition by adding the moving obstacle component to (1) for the goal, keeping the goal parameters fixed (including damping). In some conditions, participants tended to slow down or speed up as they approached the obstacle. Thus, in lieu of a model of speed control for moving obstacles, we input the mean time series of subject speed in each condition to the model. The resulting parameter values over both experiments were $k_m = 41.0$, $c_5 = 4.75$, and $c_6 = 0.5$. Simulations of the first experiment not only accurately reproduced the dominant human paths in each condition but also switched from ahead to behind paths at the same obstacle speed (Fig. 11), with a mean $rmse = 2.68°$ and a mean $r^2 = 0.94$. In the second experiment, the model and human paths were quite similar, but the switching point was slightly shifted for the model (at 130° rather than 140°); excluding this anomalous 130° condition, the mean $rmse = 4.35°$ and the mean $r^2 = 0.89$. The model thus captures the behavioral dynamics of moving obstacle avoidance reasonably well. As with stationary obstacles, there was a distribution of ahead and behind trials in each condition. When we performed trial-by-trial simulations with the participant's initial position and initial heading on each trial, we were able to reproduce the empirical pattern of ahead and behind paths quite well. Thus, the distribution of paths seems to fall out of the variation in initial conditions.

7 Routes as Emergent Behavior

Now that elementary goal and obstacle components have been formulated, can they be used to predict more complex behavior? In the model, routes emerge from the agent's interaction with the environment, rather than being explicitly planned in advance. The question at issue is whether this is sufficient to account for more complex human behavior.

7.1 Switching Behavior with One Obstacle

We begin with the simplest case of route selection, testing predictions about switching from a longer "outside" path to a shorter "inside" path around a stationary obstacle [14]. For an obstacle lying just to the left of a direct line to the goal, an agent initially heading to the left of the obstacle is repelled leftward on an outside path around it. But when the goal is closer, it becomes more attractive and at some point overcomes the repulsion so that the agent crosses in front of the obstacle and takes an inside path to the goal. The model predicts outside paths for small goal-obstacle angles and inside paths for large angles; in an intermediate range, it predicts switching from an outside to an inside path as the goal gets closer.

The human data bear out these predictions. The proportion of "inside" trials rose significantly as the goal-obstacle angle increased and as goal distance decreased. However, to precisely reproduce the intermediate range of goal-obstacle angles at which switching occurred, we had to increase the decay rate of the repulsion function with distance from $c_4 = 0.8$ to 1.6. This could be due to the fact that we only tested outside routes in our initial experiments, and thus did not sample a sufficiently wide range of conditions to identify general parameter values. We thus fixed $c_4 = 1.6$ for all subsequent simulations. Parameter c_4 might be thought of as a "risk" parameter, for higher values cause the repulsion function to decay more rapidly with distance, allowing a closer approach to obstacles. The agent's body size is thus implicitly represented in the model by the repulsion parameters.

Route switching results from competition between the attraction of the goal and the repulsion of the obstacle, which evolves as the agent moves through the environment. The choice of whether to steer right or left appears as a *tangent bifurcation* in the system's dynamics (Fig. 12). Specifically, as the agent moves with respect to the object configuration (Fig. 12a), there may be a stable fixed point corresponding to an outside heading, a stable fixed point corresponding to an inside heading (Fig. 12b,d), or bistability – stable outside and inside headings that straddle a repeller (Fig. 12c). In the bistable region, the realized route depends on the agent's initial heading. The "choice" of a particular route can thus be understood as a consequence of bifurcations in the model's dynamics and the agent's initial conditions.

7.2 Switching Behavior with Two Obstacles

The next step in complexity is a configuration of two obstacles en route to a goal, creating three possible routes [17]. In this experiment, the goal and the near obstacle were in fixed positions, while we manipulated the lateral position of the slightly farther obstacle. As the obstacle-target angle for the far obstacle increases from 0 to 10°, the model predicts a particular sequence of route switching: from a path outside the far obstacle, to a path outside the near obstacle, to one that shoots the gap between them (Fig. 13). The human data demonstrated exactly this sequence of switching behavior, a particularly striking example of the predictive power of the model. Moreover, the distribution over the three routes can be closely reproduced by adding 10% Gaussian noise to perceptual variables and parameters. Once again, the "choice" of route can be understood as a consequence of bifurcations in the underlying dynamics, together with the agent's initial conditions.

7.3 Field of 12 Obstacles

As mentioned above, an advantage of the model is that it scales linearly with the complexity of the scene by linearly combining nonlinear components for

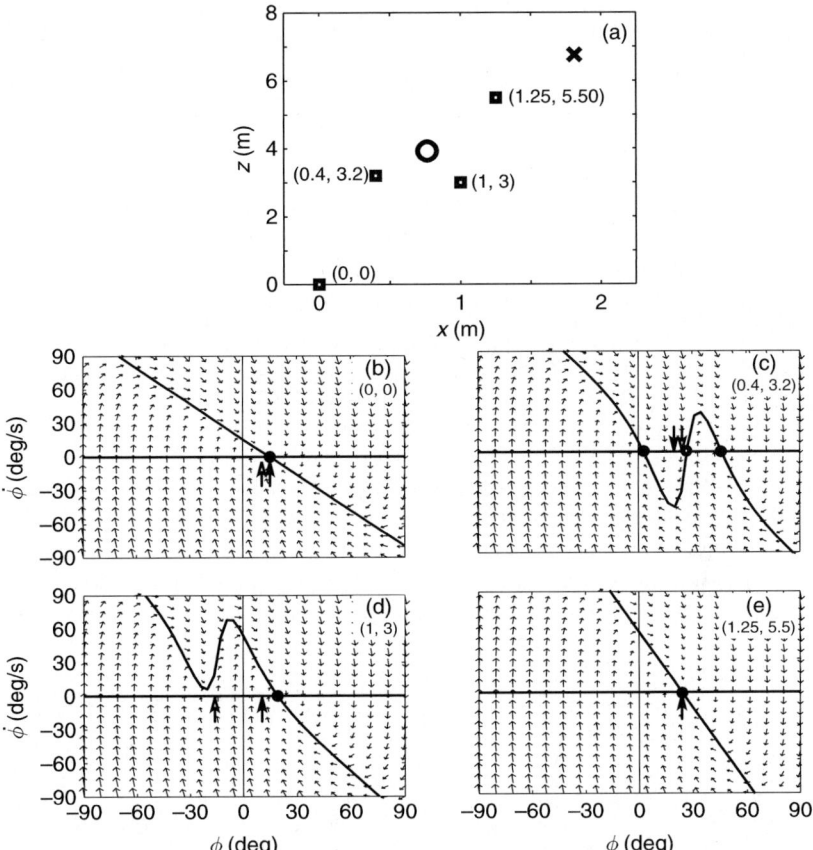

Fig. 12. Vector fields (ϕ vs. $\dot{\phi}$) for steering around an obstacle (O) en route to a goal (X). (a) Four sample positions on typical routes. (b–e) Vector fields corresponding to the four agent positions. The vertical component of each vector represents the angular acceleration and the horizontal component represents the angular velocity at a point in phase space. *Solid curves* correspond to nullclines at which angular acceleration is zero. *Filled circles* represent point attractors, *open circles* repellers, *filled arrows* indicate the goal direction, and *open arrows* the obstacle direction. Reprinted from [14], copyright 2003 by the American Psychological Association

each object. A strong test of this principle is route selection through a complex scene, such as a large field of obstacles [50]. We compared the model with human data for walking through a random field of 12 obstacle posts to reach a visible goal post; eight different obstacle arrays were tested. The challenge with such a complex environment is that the number of bifurcation points increases with the number of obstacles, creating a combinatoric explosion of possible routes.

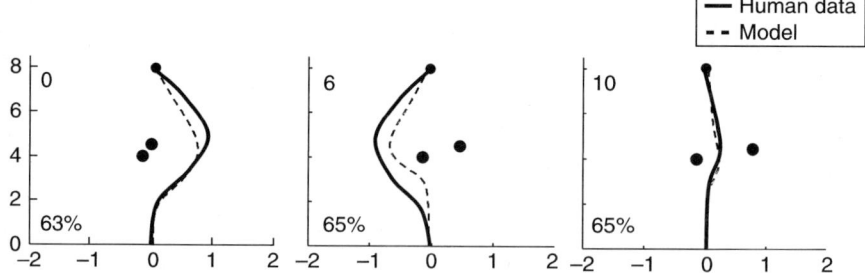

Fig. 13. Route switching with two obstacles. Mean dominant human paths and model simulations with initial far obstacle-target angles of 0, 6, and 10°. The percentage of trials in each condition on which participants took the dominant path is indicated

The model does a reasonable job of reproducing the most frequent human route through the obstacle field, with no free parameters (Fig. 14). On half of the obstacle arrays, the model was identical to the median human route, two arrays differed by only one obstacle, one array differed by two obstacles, and one by four obstacles. Considering the number of possible routes through

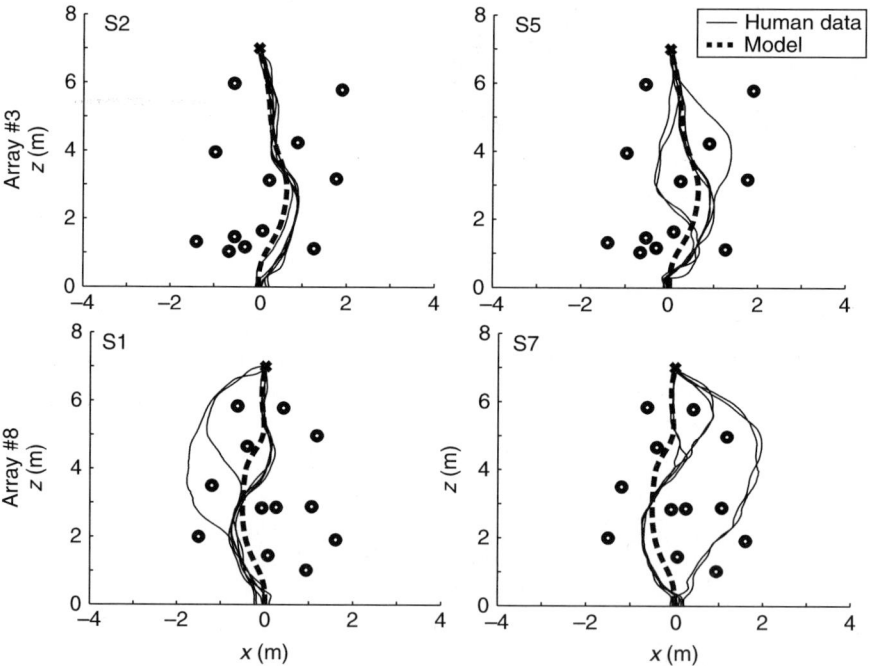

Fig. 14. Routes through a field of 12 obstacles (O) to a goal (X), for two sample obstacle arrays. *Solid curves* represent six individual trials from each participant, *dashed curves* represent model simulation, with starting point at (0,0) and initial heading toward goal

the array, it is striking that the model was this close to the median human route with no tweaking. Presumably, performance could be improved further by adapting the model to the multiple obstacle task, rather than generalizing from the single obstacle case.

Of course, there was also variation in human routes across trials. Given the cascade of bifurcation points through the obstacle configuration, some of this variation can be attributed to the sensitivity of the route to initial conditions and noise. To simulate these effects, we again added 10% Gaussian constant error into the perceptual variables and parameter values on each trial and matched the human standard deviation of initial position and heading. The resulting distribution of observed routes was similar to the distribution in the human data for many configurations, although the model did not account for all exhibited paths. We are continuing to examine the multiple obstacle case in order to determine whether the variation in human routes results from sensitivity to initial conditions or calls for adjustments to the model for this task.

The model offers an existence proof that humanlike routes through complex scenes can be generated in a purely online manner, without explicit path planning. Faced with an explosion of possible paths, each need not be evaluated and a route planned in advance. Rather, the agent's route is an emergent consequence of local interactions between the agent's heading and the next few upcoming obstacles.

8 Integrating Behaviors in a Dynamic Environment

Most recently, we have been investigating whether the principle of linear composition of nonlinear terms can account for the integration of elementary locomotor behaviors in dynamic environments. We have begun by examining the possible combinations of a stationary or moving goal with a stationary or moving obstacle, with the aim of scaling up to the football scenario – multiple moving and stationary obstacles. Having already described stationary goals with both types of obstacles, we turn to moving targets.

8.1 Moving Target with a Stationary Obstacle

To investigate how stationary obstacle avoidance is integrated with moving target interception, Bruggeman and Warren [4] collected human data for this case. The speed of a target moving in the frontal plane (0.6 and 0.9 m/s) and the initial target-obstacle angle (0–18°) were manipulated and the data compared to model predictions, with fixed parameters. Two aspects of the locomotor path can be evaluated, the shape of the dominant path and the conditions for route switching from one side of the obstacle to the other.

The model predicts a switch from an inside path to an outside path as the target-obstacle angle increases (Fig. 15). With the fast target, participants

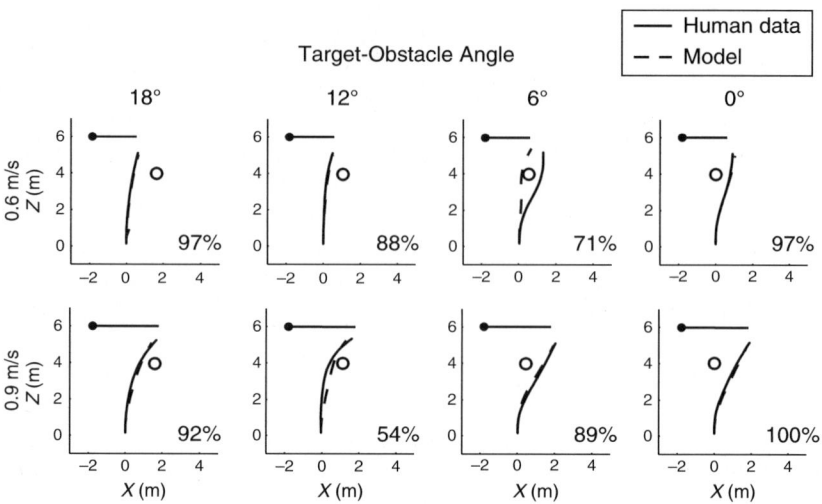

Fig. 15. Integrating target interception and obstacle avoidance: Mean dominant human paths and model simulations for a moving target with a stationary obstacle. Target speed and initial target-obstacle angle were varied, the target trajectory was 90° to the initial heading direction. The percentage of trials in each condition on which participants took the dominant path is indicated

switched at the predicted target-obstacle angle (12°) on average, while with the slow target the qualitative pattern was similar but the switching point for participants (12°) occurred at a slightly larger angle than the model (6°). Moreover, the model fit the time series of heading for the dominant human path in each condition with a mean $r_2 = 0.88$; the exception was the one divergent 6° condition, $r_2 = 0.18$.

The model thus predicts the switching behavior and shape of the dominant path quite accurately. This suggests that the integration of target interception and obstacle avoidance can be captured by the linear combination of model components, at least in a simple two-object configuration.

8.2 Moving Target with a Moving Obstacle

Now consider the integration of target interception with moving obstacle avoidance. Cohen et al. [10] varied the speed of a target (0.6 and 0.9 m/s) and the speed of an obstacle (0.3–0.9 m/s) moving in the same direction in two different depth planes.

The model predicts route switching from a path ahead of the obstacle to a path behind it as obstacle speed increases (see Fig. 16). With the slow target, participants switch at the predicted obstacle speed (0.75 m/s), while with the fast target they exhibit the same qualitative pattern but switch at a slightly lower speed (0.9 m/s) than the model. In addition, the model fit the

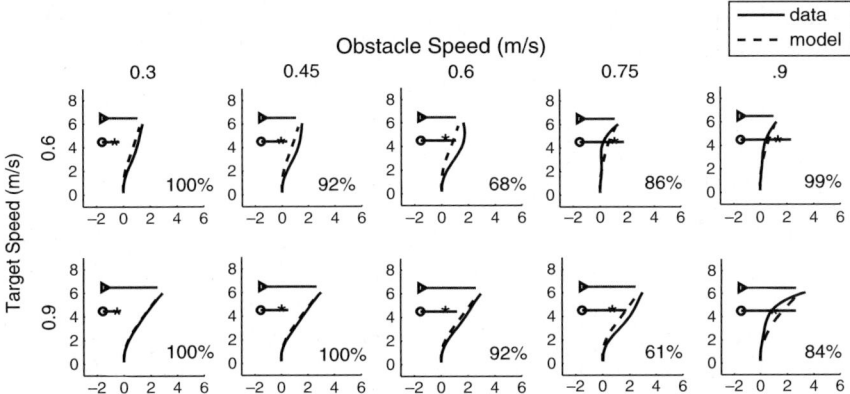

Fig. 16. Integrating target interception and moving obstacle avoidance: Mean dominant human paths and model simulations for a moving target with a moving obstacle. Target speed and obstacle speed were varied, their trajectories 90° to the initial heading direction. The percentage of trials in each condition on which participants took the dominant path is indicated

time series of heading for the dominant path in each condition with a mean $r_2 = 0.85$; the exception was the one divergent 0.9 m/s condition, $r_2 = 0.79$. The model thus predicts the pattern of route switching and the shape of the dominant path in each condition quite successfully.

Taken together, these findings indicate that the attraction and repulsion functions we derived for single objects can be linearly combined to account for the integration of steering and obstacle avoidance with pairs of objects. The model does a good job of predicting both the quantitative path shape and the qualitative switching behavior.

9 Limitations and Challenges

The steering dynamics model is still in development and, as currently constituted, faces certain limitations and challenges. We briefly review them here and suggest possible avenues.

9.1 Fat Obstacles

An important limitation of the model is that it currently treats all goals and obstacles as points and does not represent wide obstacles or extended barriers. One approach to this "fat obstacle" problem is to treat a barrier as a set of points at finite intervals such that the repulsion function is summed over a larger visual angle. This might also be achieved by convolving the repulsion function over horizontal visual direction so that the entire visual angle of the obstacle is repulsive. Interestingly, summing the repulsion function makes the empirical prediction that the end of a barrier should be more repulsive than a post at the same location, whereas averaging the repulsion function predicts no such effect.

Huang et al. [27] took another approach, replacing the obstacle distance term (which modulates the amplitude of the repulsion function) with one based on the tangent of the obstacle's visual angle. As an obstacle is approached, its horizontal visual angle increases, driving the repulsion amplitude – and hence the agent's angular acceleration – toward infinity. Distant wide obstacles can thus have the same influence as near small obstacles, yielding conservative behavior and large detours. We are planning human experiments that will enable us to evaluate the biological relevance of these various solutions to the fat obstacle problem.

9.2 Body Size, Personal Space, and Handedness

The converse of the fat obstacle problem is that the model also treats the agent as a point. It is known, however, that body width alters the affordances of the environment for locomotion. As observed by Warren and Whang [51], for example, when the width of an opening reaches 1.3 times shoulder width, walkers of all sizes initiate body rotation to pass through the aperture. Despite the fact that body width is not explicitly represented in the model, it is implicit in the parameter values that define the spread and amplitude of the repulsion function.

This has the important implication that what is commonly thought of as a zone of "personal space" [19] does not have hard boundaries but is rather an emergent property of the dynamic interplay of attractive and repulsive forces. Consistent with this principle, Fink et al. [18] found that the minimum obstacle clearance of human paths is not fixed but depends on the configuration of starting, goal, and obstacle positions, as predicted by the model. However, without a minimum safety margin, the model can produce collisions in certain special cases (see "spurious attractors" below). Whether a hard limit exists in human obstacle avoidance remains an empirical question.

The presumed "handedness" of locomotor behavior might be incorporated into the repulsion function in a similar fashion. It is often suggested that there are biases in the direction of obstacle avoidance such that right-handed (or footed) individuals may prefer a rightward path around an obstacle that is dead ahead, resolving the problem of an unstable fixed point. Similarly, a cultural bias to pass on the right (or left) adaptively avoids collisions and contributes to the self-organization of traffic patterns. These biases could be incorporated into the model in the form of an asymmetric repulsion function such that a greater amplitude on the left side would embody a rightward bias. Experiments are underway to investigate such directional biases in locomotor behavior.

9.3 Object Identification and Valence

At present, the identity of stationary and moving goals and obstacles is simply given to the model. This issue could be finessed by giving each object class a distinguishing feature (e.g., color, height), but that would be cheating. The problem of object recognition is not unique to locomotor control but is common to many visual tasks and is not the focus of our current research.

A related limitation is that all goals are currently treated as equivalent in their attractiveness, and all obstacles as equivalently repulsive. However, the valence of environmental objects could easily be expressed in the model's parameter values. Given that each object in a scene is represented in the model by a corresponding component, the attractiveness of an especially appealing goal might be captured by increasing the corresponding stiffness term, and the repulsiveness of a particularly dangerous obstacle by decreasing its c_3 parameter, spreading out its repulsion function. The empirical issue is determining a mapping from the valence of goals and obstacles to these parameter values, in order to test this approach.

9.4 Local Minima and Spurious Attractors

One of the most serious problems facing any locomotor model is that of getting stuck in local minima. As noted above, the steering dynamics model creates many fewer local minima than the potential field model because it controls angular acceleration rather than the velocity vector and defines attractors and repellers in heading rather than in 2D space. Nevertheless, spurious attractors can still occur due to what Fajen et al. [17] called the *cancelation effect*.

Consider the special case of an array of obstacles that is symmetrical about a direct path to the goal, such as a cul-de-sac of closely spaced posts. In this case, the contributions of obstacles on the left and right to the angular acceleration can cancel, creating a spurious heading attractor toward the center of the array. Because the agent travels at a constant speed, it will attempt to pass between obstacles whether or not the gap is wide enough for the agent to fit through (Fig. 17). Thus, rather than getting stuck in a local minimum, the current model can lead to a collision.

Adding noise will allow the model to escape marginally stable fixed points in simple arrays, but the risk of a collision remains a possibility. One solution is to add a speed control law that brakes at a critical distance or time-to-contact, in which case the agent would avoid a collision but get stuck instead. The usual approach to avoiding such cul-de-sacs is to add an advance planning layer, but this is contrary to the spirit of our minimalist approach. Another possibility is to add an escape strategy that is invoked only after the agent gets stuck. How humans deal with cul-de-sacs is, once again, an empirical question that is preliminary to an appropriate model solution.

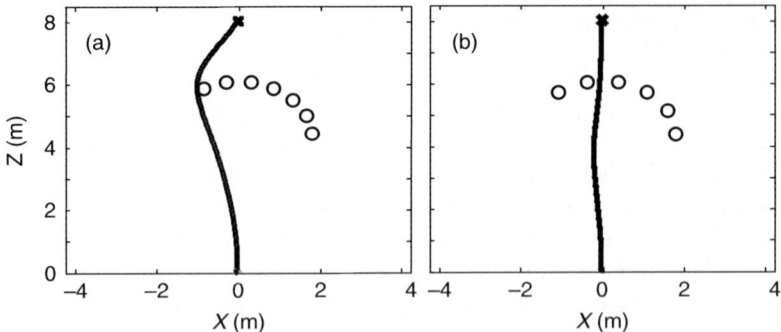

Fig. 17. Model simulations of a spurious attractor in a cul-de-sac. (a) The repulsion functions of closely spaced obstacles (O) sum to deflect the agent around the obstacle array to the goal (X). (b) The cancelation effect: with more widely spaced obstacles, the repulsion functions to the right and left of the direct path cancel and the goal attracts the agent through a narrow gap. The agent is constrained to move at a constant speed

9.5 Competing Goals

A related problem is route selection with two goals. The model will prefer a nearer goal to a farther one, but in symmetric cases a spurious attractor will appear between the two goals. The addition of nonlinear competition between goals is necessary to resolve this problem. We are preparing human experiments in which the relative distance and eccentricity of two goals are systematically varied.

9.6 Complex Dynamic Environments

We are currently experimenting with scaling up scene complexity by adding more stationary and moving obstacles to the environment. However, preliminary results suggest that, once two moving obstacles are present, human participants no longer behave systematically across trials or individuals. This inconsistency in route selection could be a consequence of an increased sensitivity to initial conditions in complex dynamic scenes, or perhaps to attentional factors. We are thus beginning to investigate the various factors that might influence route selection with multiple moving obstacles.

The next step is to examine behavioral interactions between two active agents, such as mutual avoidance in passing pedestrians or pursuit-evasion games. The goal here is to determine whether such interactive behaviors fall out of the four elementary locomotor components or whether additional components are needed, such as specific evasion or neighbor-following strategies. Thus, we seek to empirically determine the minimal set of elementary locomotor "rules" necessary to simulate the interactions of multiple agents in complex environments.

10 Conclusion

The success of the steering dynamics model thus far is that it enables us to understand the "surface structure" of exhibited locomotor paths and route switching in terms of the "deep structure" of the underlying dynamics of attractors, repellers, and bifurcations. As the agent moves through the world, the dynamics evolve, for as the directions and distances of goals and obstacles change, the attractors and repellers shift and bifurcations occur. Thus far, such an online theory of steering dynamics is empirically sufficient to account for human locomotor paths and route selection without advance planning or an internal world model.

In sum, human routes can be understood as a form of emergent behavior, which unfolds as an agent with certain steering dynamics interacts with a structured environment. The aim of this research program is to formulate the minimal set of control laws needed to characterize locomotor behavior in a complex dynamic environment. Once these basic locomotor "rules" for an individual agent are understood, they should ultimately enable us to model interactions among multiple agents, including pedestrian traffic flow and crowd behavior in specific architectural environments. Locomotion thus offers a comparatively simple model system for understanding how adaptive human behavior emerges from information and dynamics.

References

1. Bastin J, Calvin S, Montagne G (2006) Muscular proprioception contributes to the control of interceptive actions. J Exp Psychol Hum Percept Perform 32:964–972
2. Borenstein J, Koren Y (1991) The vector field histogram – fast obstacle avoidance for mobile robots. IEEE Trans Rob Autom 7:278–288
3. Bruggeman H, Rothman D, Warren WH (2006) Is obstacle avoidance controlled by perceived distance or time-to-contact? J Vis 6:136a
4. Bruggeman H, Warren WH (2005) Integrating target interception and obstacle avoidance. J Vis 5:311a
5. Chapman S (1968) Catching a baseball. Am J Phys 53:849–855
6. Chardenon A, Montagne G, Buekers MJ, Laurent M (2002) The visual control of ball interception during human locomotion. Neurosci Lett 334:13–16
7. Chardenon A, Montagne G, Laurent M, Bootsma RJ (2004) The perceptual control of goal-directed locomotion: A common control architecture for interception and navigation? Exp Brain Res 158:100–108
8. Chardenon A, Montagne G, Laurent M, Bootsma RJ (2005) A robust solution for dealing with environmental changes in intercepting moving balls. J Mot Behav 37:52–64
9. Chardenon A, Warren WH (2004) Intercepting moving targets on foot: Control of walking speed and direction. Abstracts of the Psychonomics Society 9:5
10. Cohen JA, Bruggeman H, Warren WH (2006) Combining moving targets and moving obstacles in a locomotion model. J Vis 6:135a

11. Cohen JA, Bruggeman H, Warren WH (2005) Switching behavior in moving obstacle avoidance. J Vis 5:312a
12. Crowell JA, Banks MS, Shenoy KV, Andersen RA (1998) Visual self-motion perception during head turns. Nat Neurosci 1:732–737
13. Cutting JE, Vishton PM, Braren PA (1995) How we avoid collisions with stationary and with moving obstacles. Psychol Rev 102:627–651
14. Fajen BR, Warren WH (2003) Behavioral dynamics of steering, obstacle avoidance, and route selection. J Exp Psychol Hum Percept Perform 29:343–362
15. Fajen BR, Warren WH (2004) Visual guidance of intercepting a moving target on foot. Perception 33:689–715
16. Fajen BR, Warren WH (2007) Behavioral dynamics of intercepting a moving target. Exp Brain Res 180:303–319
17. Fajen BR, Warren WH, Temizer S, Kaelbling LP (2003) A dynamical model of visually-guided steering, obstacle avoidance, and route selection. Int J Comput Vis 54:15–34
18. Fink PW, Foo P, Warren WH (2007) Obstacle avoidance during walking in real and virtual environments. ACM Trans Appl Percep 4:1–8
19. Gérin-Lajoie M, Richards CL, McFadyen BJ (2005) The negotiation of stationary and moving obstructions during walking: Anticipatory locomotor adaptations and preservation of personal space. Motor Control 9:242–269
20. Gibson JJ (1958/1998) Visually controlled locomotion and visual orientation in animals. Br J Psychol 49:182-194. Reprinted in Ecological Psychology 10: 161–176
21. Gibson JJ (1979) The ecological approach to visual perception. Houghton Mifflin, Boston
22. Goldenstein S, Large E, Metaxas D (1999) Nonlinear dynamical system approach to behavior modeling. Vis Comput 15:349–364
23. Goldstone RL, Roberts ME (2006) Self-organized trail systems in groups of humans. Complexity 11:43–50
24. Helbing D, Keltsch J, Molnár P (1997) Modelling the evolution of human trail systems. Nature 388:47–50
25. Helbing D, Molnár P (1995) Social force model of pedestrian dynamics. Phys Rev E 51:4282–4286
26. Helbing D, Molnár P, Farkas I, Bolay K (2001) Self-organizing pedestrian movement. Environ Plann B: Plann Des 28:361–383
27. Huang WH, Fajen BR, Fink JR, Warren WH (2006) Visual navigation and obstacle avoidance using a steering potential function. Robot Auton Syst 54:288–299
28. Huth A, Wissel C (1992) The simulation of the movement of fish schools. J Theor Biol 156:365–385
29. Kelso JAS (1994) The informational character of self-organized coordination dynamics. Hum Mov Sci 13:393–413
30. Kelso JAS (1995) Dynamic patterns: The self-organization of brain and behavior. MIT Press, Cambridge, MA
31. Khatib O (1986) Real-time obstacle avoidance for manipulators and mobile robots. Int J of Rob Research 5:90–98
32. Lee DN (1980) Visuo-motor coordination in space-time. In: Stelmach GE, Requin J (eds) Tutorials in motor behavior. North-Holland, Amsterdam, pp. 281–295

33. Lenoir M, Musch E, Janssens M, Thiery E, Uyttenhove J (1999) Intercepting moving objects during self-motion. J Mot Behav 31:55–67
34. Lenoir M, Musch E. Thiery E, Savelsbergh GJ (2002) Rate of change of angular bearing as the relevant property in a horizontal interception task during locomotion. J Motor Behav 34:385–401
35. Llewellyn KR (1971) Visual guidance of locomotion. J Exp Psychol 91:224–230
36. Olberg RM, Worthington AH, Venator KR (2000) Prey pursuit and interception in dragonflies. J Comp Physiol [A] 186:155–162
37. Ooi TL, Wu B, He ZJ (2001) Distance determined by the angular declination below the horizon. Nature 414:197–200
38. Owens J, Warren WH (2006) Intercepting moving targets on foot: Can people learn multiple target trajectories? J Vis 6:145a
39. Owens JM, Warren WH (2004) Intercepting moving targets on foot: Target direction change. Abstracts of the Psychonomics Society 9:54
40. Owens JM, Warren WH (2005) Intercepting moving targets on foot: Can people learn to anticipate target motion? J Vis 5:310a
41. Patla AE, Tomescu SS, Ishac MGA (2004) What visual information is used for navigation around obstacles in a cluttered environment? Can J Physiol Pharmacol 82:682–692
42. Reichardt W, Poggio T (1976) Visual control of orientation behavior in the fly: I. A quantitative analysis. Quarterly Review of Biophysics 9:311–375
43. Reynolds CW (1987) Flocks, herds, and schools: a distributed behavioral model. Comput Graphics 21:25–34
44. Rushton SK, Harris JM, Lloyd M, Wann JP (1998) Guidance of locomotion on foot uses perceived target location rather than optic flow. Curr Biol 8:1191–1194
45. Rushton SK, Wen J, Allison RS (2002) Egocentric direction and the visual guidance of robot locomotion: Background, theory, and implementation. In: Bülthoff HH, Lee S-W, Poggio TA, Wallraven C (eds) Biologically motivated computer vision, Proceedings: Lecture notes in computer science, 2525. Springer-Verlag, Berlin, pp. 576–591
46. Schöner G, Dose M, Engels C (1995) Dynamics of behavior: Theory and applications for autonomous robot architectures. Robot Auton Syst 16:213–245
47. Ulrich I, Borenstein J (1998) VFH+: Reliable obstacle avoidance for fast mobile robots. Proceedings of the IEEE International Conference on Robotics and Automation, Leuven, Belgium, pp. 1572–1577
48. Warren WH (2004) Optic flow. In: Chalupa LM, Werner JS (eds) The visual neurosciences, v II. MIT Press, Cambridge, MA, pp. 1247–1259
49. Warren WH (2006) The dynamics of perception and action. Psychol Rev 113:358–389
50. Warren WH, Fajen BR, Belcher D (2001) Behavioral dynamics of steering, obstacle avoidance, and route selection. Proceedings of the Vision Sciences Society 1:53
51. Warren WH, Whang S (1987) Visual guidance of walking through apertures: Body scaled information for affordances. J of Exp Psychol Hum Percept Perform 13:371–383
52. Wilkie RM, Wann J (2003) Controlling steering and judging heading: Retinal flow, visual direction, and extra-retinal information. Journal of Experimental Psychology: Human Perception and Performance 29:363–378
53. Wilkie RM, Wann JP (2005) The role of visual and nonvisual information in the control of locomotion. J Exp Psychol Hum Percept Perform 31:901–911

Human Trajectory Formation: Taxonomy of Movement Based on Phase Flow Topology

Raoul Huys,[1] Viktor K. Jirsa,[1] Breanna E. Studenka,[2] Nicole Rheaume,[2] and Howard N. Zelaznik[2]

[1] Theoretical Neuroscience Group, UMR 6152 Institut Mouvement et Perception, Université de la Méditerranée and CNRS, Marseille, Cedex 09 France
[2] Health and Kinesiology, Purdue University, West Lafayette, IN 47907, USA

Summary. The notion that a limited number of 'motor primitives' underwrites complex (human) movements is pertinent to various theoretical perspectives on motor control. Consequently, motor primitives have been classified according to different (and often empirically driven) criteria. Departing from the perspective that dynamical systems are unambiguously described in phase space, we propose a movement taxonomy based on phase flow topology. We denote qualitative distinct movement classes as normal forms of movement. The existence of two normal forms of movement governing discrete and rhythmic behavior has been debated repeatedly in the literature. We provide evidence testifying to the existence (and utilization by humans) of both normal forms through a computational analysis and an experimental study involving human participants. We furthermore argue that one other dynamic possibility governing movement likely exists.

1 Introduction

It is generally assumed that 'complex' movements are 'constructed' through the assembly of fundamental 'motor primitives' (the motor system's elements sometimes thought of as 'building blocks'). However, there is little consensus about the number and nature of such primitives (for a recent review, see [19]). Motor primitives have been defined in terms of 'computational elements in the sensorimotor map' [20, 65], ('hardwired') spinal fields [47, 48], trajectory segments or strokes [43], (delta-lognormal) velocity profiles [72] and dynamical features (cf. [6, 7, 55]). Clearly, these different notions can to a large extent be traced back to the different theoretical perspectives adopted by various researchers. Moreover, it seems fair to say that the search for motor primitives has often been pursued primarily along empirical lines. The fundamental question as to the classifying principle to demarcate motor primitives is often evaded. For instance, motor primitives have been defined based on the segmentation of end-effector trajectories

[43, 66, 69]. However, Sternad and Schaal [54, 59] showed that segmentation of end-effector trajectories can be due to the geometrical properties of the effector systems supporting the trajectory formation and thus does not necessarily imply segmented control. Also, it is well established that a modular organization underwrites the vertebrae spinal motor system [5]. Stimulation of certain modules (neural circuits) elicits specific, reproducible coordinated muscle activity that can be described by the corresponding force field, referred to as the spinal field. The simultaneous simulation of multiple modules has been shown to sum up linearly [47, 48]. While we do not want to downplay the importance of these findings, it should be noted that the question how the stimulation of the spinal fields is organized has not been addressed.

In the present chapter, we propose a (human) trajectory formation taxonomy embracing the perspective of coordination dynamics (cf. [31, 37]). The framework of coordination dynamics is deeply rooted in dynamical systems theory (cf. [21, 64]) and Haken's synergetics [25, 26], the theory of self-organized pattern formation. Accordingly, open systems that are far from thermal equilibrium may organize themselves by forming ordered spatial and temporal patterns due to the (non-linear) interactions among their numerous (microscopic) components. In other words, a high-dimensional system can be represented by a low-dimensional dynamics that is functionally meaningful on the behavioral level [25, 26, 30, 31, 37]. This emerging dynamics can typically be cast in terms of (a system of) differential or difference equations that are open to examination using tools from dynamic systems theory.

The classification of human trajectory formation that we propose is rooted in dynamical systems theory but phenomenological in nature: it is posited on the emergent meaningful level of organization. In that regard, while the term 'motor primitive' has been used by researchers stemming from a variety of backgrounds, it is particularly associated with approaches focusing on neural 'hardware'. Our taxonomy, in contrast, emphasizes dynamical invariance in phase flows. Therefore, we refer to the corresponding qualitatively distinct movement classes as 'normal forms of movement'. Our classification builds on and extends the recent work of Jirsa and Kelso [32]. In the present chapter we will elaborate on the concepts outlined by Jirsa and Kelso, provide evidence for the existence of (at least) two normal forms of movement, and argue in favor of one more dynamic possibility in the (human) motor repertoire. First, however, we will briefly introduce the concepts of dynamical systems theory that provide the foundation of our proposed taxonomy.

2 Invariance in Phase Flows

As is well known from dynamical systems theory, every deterministic, time-continuous and autonomous system can be unambiguously described through its flow in state space. For a movement along a single (physical) dimension, the state (or phase) space is the space spanned by the system's position x and its time derivative, velocity $\dot{x}$. (This representation is valid under the commonly adopted assumption that movement trajectories can be fully described by two state variables. We adopt this assumption in the following[1]).

The flow in phase space describes the direction of the system's evolution as a function of its current state, $(x, \dot{x})$. In two-dimensional systems, only three distinct topological structures can be realized, namely, (1) fixed points (points in phase space where $\dot{x} = 0$; if stable [unstable], all trajectories eventually converge toward it [diverge away from it]); (2) limit cycles (circular-like structures or closed orbits; if stable [unstable] all trajectories eventually converge to [diverge away from] it); and (3) separatrices (structures that divide the phase space in regions with locally distinct phase flows).[2] In the case of a stable limit cycle an unstable fixed point exists in the space surrounded by the closed orbit. Examples of these structures are depicted in Fig. 1. Although clear from the figure we remind the reader that fixed points and limit cycles describe discrete and rhythmic behavior, respectively.

Two theorems known from dynamical system theory (cf. [49, 64]) are required to fully appreciate the implications following from the fact that only a limited number of topological structures can occur in two-dimensional systems. The Poincaré–Bendixson theorem states that if a trajectory is confined to a closed bounded region that contains no fixed points, then the trajectory must eventually approach a closed orbit. In other words, chaos is ruled out,

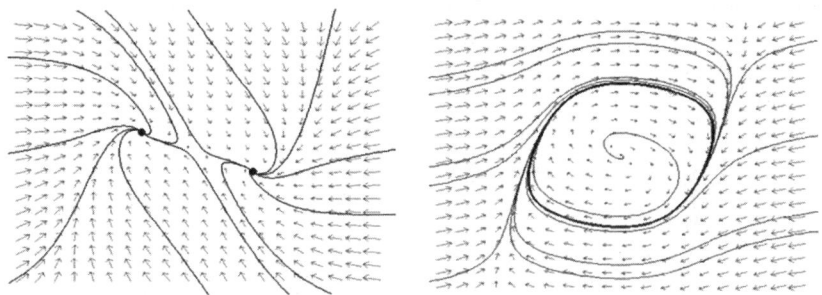

Fig. 1. Examples of phase flows (*gray arrows*) and trajectories (*gray lines*) from various initial conditions inhabited by two stable fixed points (black dots) divided by a separatrix (**left panel**) and a stable limit cycle (**black loop; right panel**). In both panels, it can be appreciated that the flow converges onto the stable structures irrespective of their initial conditions

---

[1] The assumption that human movements allow for a description with two state variables has been questioned (cf. [34, 42]). Also, Daffertshofer [12] has shown that two-dimensional (stochastic) systems cannot account for the negative lag-one correlation observed during self-paced tapping [71]. Regardless, two-dimensional systems capture by far the most phenomena, which motivates our choice in this regard.

[2] We focus on stable fixed points and limit cycles as human movement is bounded.

implying that the possible dynamics are severely limited. The Hartman–Grobman theorem states that a local phase space near a hyperbolic fixed point[3] is topologically equivalent to the phase space of the linearization, implying the existence of a continuous invertible mapping (a homeomorphism) between both local phase spaces. While the Hartman–Grobman theorem is valid for fixed points only, the notion of topological equivalence is utilized for other topological structures as well. Intuitively, dynamical systems are topological equivalent under 'distortions' if the underlying structure remains invariant. For instance, bending and stretching are allowed but disconnecting closed trajectories is not (see Fig. 2, and Steward [62], Chaps. 10 and 12, for an excellent intuitive, non-technical introduction). Dynamical systems thus belong to the same class if, and only if, they are topological equivalent.

Based on this unambiguous classification, Jirsa and Kelso [32] proposed a (dynamic) taxonomy of human trajectory formation. Next to discrete and rhythmic behavior, they recognized that movements may be initiated erroneously – too early, known as a false start (see [32]). A false start requires that a system has the ability to execute a movement that is triggered by a stimulus and occurs if such a movement is (either completely or partially) executed in the absence of the stimulus. Examples of false starts are regularly observed in the context of sports, for instance where a sprinter or ice-skater takes off before the starter's signal is given. The requirements in phase space for false starts are a stable fixed point (i.e., the start position) and a separatrix. The separatrix effectively functions as a threshold: if the system does not pass the threshold it will return to the fixed point, otherwise it will traverse in phase space according to the present flow. Further

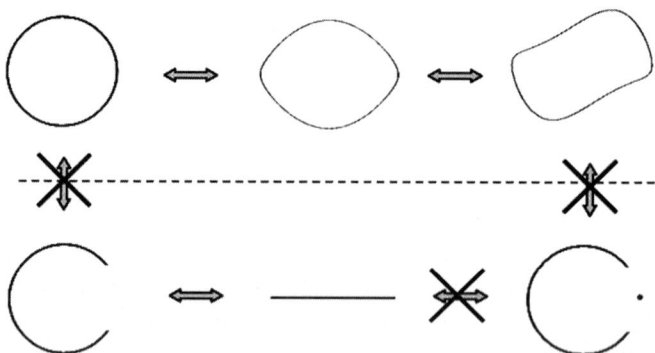

Fig. 2. The mappings in upper row and lower row are both allowed (i.e., the structures are topologically equivalent). The mapping from the upper to the lower row, however, is not (i.e., the corresponding structures are not topologically equivalent)

---

[3] A fixed point of a nth-order system is hyperbolic if all eigenvalues λ_i of the linearization have a non-vanishing real part (i.e., $\mathcal{R}(\lambda_i) \neq 0$ for $i = 1 \ldots n$), which is the case for most fixed points (and those under consideration at present).

requirements follow (among others) from the latter: the trajectories have to remain bounded. Satisfying these constraints, Jirsa and Kelso [32] arrived at the following model construct:

$$\dot{x} = [x + y - g_1(x)]\,\tau$$
$$\dot{y} = -[x - a + g_2(x, y) - I]\,/\tau$$

in which $\dot{x}$ and $\dot{y}$ represent the time derivatives of x and y, respectively, and τ and I represent a time constant and an external (instantaneous) input (external to the dynamical system in (x, y)), respectively. For an appropriate choice of $g_1(x)$ and $g_2(x, y)$, this model belongs to the class of excitable systems (cf. [18, 46, 64]). Task constraints can be implemented in this model construct via specific realizations of $g_1(x)$ and $g_2(x, y)$. The realizations chosen by Jirsa and Kelso read $g_1(x) = x^3/3$ and $g_2(x, y) = b\,y$. The parameters a and b determine whether the topological structures in phase space are a stable fixed point and a separatrix (referred to as the mono-stable condition; corresponding to an entire flexion–extension movement), two stable fixed points and a separatrix (referred to as the bi-stable condition; corresponding to a movement from one point to another) or a stable limit cycle (corresponding to cyclic movement). In the former two cases, the system will traverse through phase space only in the presence of an 'external' input I, that is, the system is non-autonomous (and strictly speaking at least three-dimensional). In the limit cycle régime the system is autonomous and no input is required.

The perspective put forward by Jirsa and Kelso [32] has two important features. First, under the realization of specific task constraints (discrete versus rhythmic movement) via the functions $g_1(x)$ and $g_2(x, y)$, the model's topology is equivalent under various instantiations of these functions (as long as the aforementioned constraints are met, see [32] for a detailed discussion). In other words, a variety of different realizations of $g_1(x)$ and $g_2(x, y)$ will affect detailed features of the phase space trajectories without affecting the system's qualitative behavior. For instance, via specific forms of $g_1(x)$ and $g_2(x, y)$ various non-linear oscillators may be realized, as for instance a van der Pol, Rayleigh, or mixed-oscillator (cf. [4, 35, 64]). While such different instantiations will impact specific (detailed) kinematic features of the system's behavior (as for instance the specific shape of the velocity profile), they are topologically equivalent (invariant): all are (stable) limit cycle oscillators (in this example). In other words, the qualitative features are – under the adoption of specific task constraints – independent of the specific realizations of $g_1(x)$ and $g_2(x, y)$. Second, as indicated, in the non-autonomous case an external input is required for motion to occur (even though in the always present case of stochastic influences, the system may exhibit false starts (cf. [32] and below). While the nature of this input is unclear, the theoretical consequence for human timing is enormous: When prepared in the autonomous régime, timing is an 'emergent' property of the system, while, when prepared in the non-autonomous regime, timing is a property of an 'external time keeper'. In that regard, the distinction between two distinct timing mechanisms (referred to as explicit and implicit [or emergent] timing) has been posited previously on the basis of behavioral experiments (cf. [29, 73]).

The phase space characteristics outlined above deviate in certain ways from those corresponding to previous models that are well known in the literature, most notably equilibrium point models [15, 16, 50, 51] and Schöner's model [56]. When cast in dynamical terms, equilibrium point models posit topological invariance throughout movement execution, although the phase space is subjected to a translation in its coordinate frame through the repositioning of the equilibrium (i.e., fixed) point [38]. This repositioning underwrites discrete as well as rhythmic movements. In Schöner's model a discrete movement is conceptualized as the destabilization of a fixed point, the onset of a limit cycle, and the subsequent stabilization of a (new) fixed point. In other words, the corresponding phase space topology is modified twice for a single discrete movement. In both conceptions, movements are executed due to changes in the phase flow on a time scale similar to that of movement execution (albeit only so for discrete movements in Schöner's model). In contrast, in the perspective outlined above (see [32]) the phase flow is invariant on this time scale.

3 Evidence for the Existence of Two Normal Forms of Movement

The framework outlined above allows (potentially) for three distinct normal forms of movement: two based on phase spaces with a single or multiple fixed points and a separatrix (the mono-stable and bi-stable condition, respectively) and one with a limit cycle. Obviously, it does not follow that these normal forms are indeed used by humans. Relative to these potential normal forms, three perspectives on the generation of discrete and rhythmic movements can be distinguished in the literature. One perspective gives primacy to discrete movements, and holds that rhythmic movements are 'merely' concatenations of discrete movements (cf. [1, 2, 15, 16, 43, 66]). The second perspective gives primacy to rhythmic movements and suggests that discrete movements are merely 'aborted' (i.e., half) cycles (cf. [23, 24, 41, 44, 56]). According to the third perspective both normal forms exist and are irreducible to each other (cf. [6, 7, 60, 61, 70]). While recent studies suggest that the 'discrete primitive' perspective is no longer tenable (cf. [55, 67]), no convincing evidence has been brought forward in favor of either of the two other perspectives.

Recently, two studies were conducted that examined the existence and utilization of the discrete and the rhythmic normal form from the perspective outlined above. In that regard, the existence of a separatrix and a fixed point is of crucial importance in examining whether human trajectory formation may be organized along the lines outlined by Jirsa and Kelso [32].[4] If there is an intention to move, false starts may be elicited through applying perturbations with a likelihood that correlates with the distance between the fixed point and the separatrix. This prediction was recently tested by Fink and colleagues [17]. In that study, the participants performed a finger flexion reaction time task in response to an auditory stimulus. A visual warning signal was given, followed by a foreperiod of either 1.5, 2.0, or 2.5 s. On some trials a mechanical perturbation was applied in the direction of the instructed movement (i.e., flexion) or in the direction away from it (i.e., extension) at 250, 200, 150, 100, or 50 ms before the auditory stimulus. The results were clear-cut: The perturbation clearly elicited false starts, more so if applied in the flexion direction than in the extension direction and with an increasing likelihood if applied longer before the auditory stimulus. In other words, evidence was found that testified to the existence of a separatrix and thus to the utilization by humans of a (discrete) normal form of movement that depends on an external stimulus (or input) for the generation of its corresponding movement.

We recently investigated the existence and utilization of discrete and rhythmic normal forms in a computational study and in an experiment involving human unimanual finger movements in comparable movement frequency régimes [27]. In the computational study, we numerically examined the 'excitator' model under a wide range of parameters (in order to scan its behavior in the mono-stable and limit cycle régime). In the experimental study, human participants executed auditory-paced flexion–extension finger movements at increasing (and decreasing) frequencies from 0.5 to 3.5 Hz under the instruction to move as fast as possible (with staccato-like movements being initiated to end/start a cycle), as smooth as possible (move so that the finger is continuously moving during the movement period interval), or without any specific instruction (see Fig. 3). We refer to these conditions as 'fast', 'smooth' and 'natural', respectively. The numerically generated data and human data were both examined in terms of kinematic measures, spectral analysis, and 2D probability distribution estimates of the system's phase space. The computational analysis indicated that at low frequencies the discrete (movement generation) system (i.e., the excitator in the mono-stable régime) produced trajectories very similar to those of the human participants under the 'natural' and 'fast' conditions at low frequencies. At higher frequencies, the timing 'imposed' by the external stimuli could no longer be achieved by the system as the re-occurring stimuli interfered with its natural coordination tendency. In contrast, in the limit cycle régime the system was able to comply with all imposed temporal demands. In other words, an externally driven movement generation system *cannot* produce movements at high frequencies while satisfying the required temporal constraints.

The trajectories produced by the human participants had a close match with the system in the limit cycle régime under the 'smooth' instruction at all

---

[4] A separatrix also exists in Schöner's model [56] under a specific parameter setting.

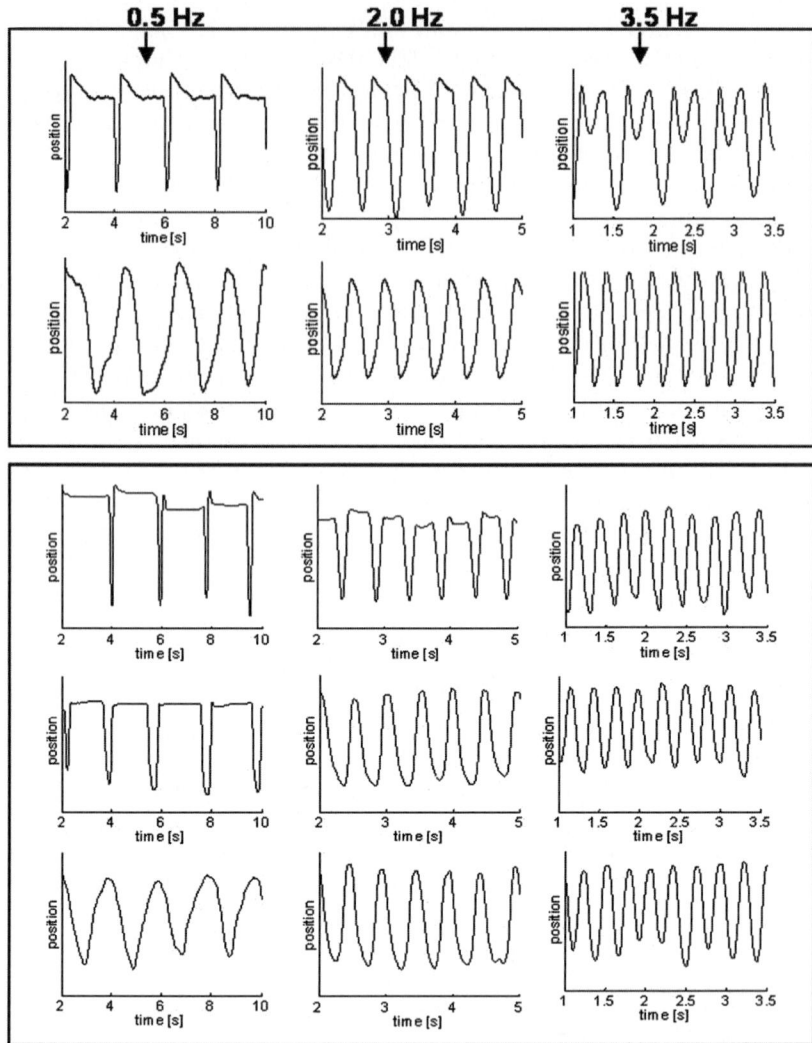

Fig. 3. Portions of representative time series of simulated data (*upper two rows*) and human tapping (*lower three rows*) at a target frequency of 0.5, 2.0 and 3.5 Hz (**left**, **middle**, and **right panel**, respectively). The panels in the first and second row represent simulations generated in the mono-stable régime and limit cycle régime, respectively. The panels in the third, fourth and fifth row represent simulations under the 'fast', 'natural' and 'smooth' conditions, respectively

frequencies. Under the 'natural' and 'fast' instructions, the trajectories produced by the participants had a close match with the excitator prepared in the discrete and limit cycle régime at low and high frequencies, respectively. Exemplary time-series of both data sets are depicted in Fig. 3. In combination, the various analyses converged onto a clear result: humans employ two different timing mechanisms as posted by Zelaznik and colleagues [73, 74] (see Figs. 4 and 5 for the corresponding spectral density estimates and phase space probability distributions of a single representative participant, respectively). The transition point from discrete to rhythmic behavior occurred at a lower frequency in the 'natural' condition, however. These results suggest that human participants naturally switch from a discrete movement generation mechanism to a rhythmic movement generation mechanism as frequency increases, even though, when specifically instructed, they are able to comply with all the temporal (task) demands moving in a rhythmic fashion.

The studies described above testify to the existence of two distinct normal forms of movement generation. In that regard, let it be noted that, although parametric changes underwrite the distinct phase flows in both cases, in the discrete case (i.e., the mono-stable condition) a movement is only generated in the face of an external trigger. The discrete and the rhythmic scenarios are clearly not reducible to each other. While we do not want to discuss the external triggering in-depth, evidence suggests that certain brain structures are involved in discrete movements that are not involved in the production of rhythmic (or continuous) movements. For instance, Spencer and colleagues [57] recently showed that patients with cerebellar lesions have deficits in producing discontinuous, but not continuous, movements. The latter finding may be taken to suggest that the 'external' stimulus is generated in the cerebellum, which has been suggested by several authors [29, 28, 53]. However, in an fMRI study Schaal and colleagues found that, notwithstanding the involvement of distinct brain areas in discrete and rhythmic movements, the main difference in cerebellar activity was that cerebellar activity was found to be bilateral during discrete movements while it was primarily ipsilateral during

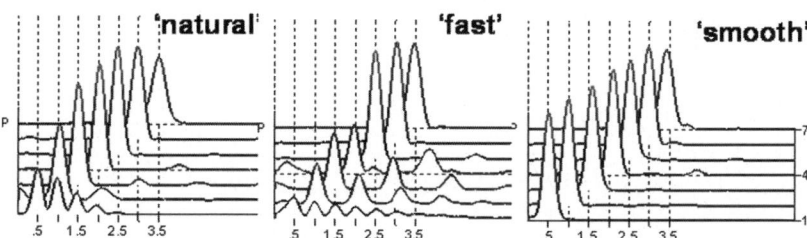

Fig. 4. Spectral density estimates for the 'natural' (no instruction), 'fast' and 'smooth' conditions (*left, middle* and *right panel*, respectively). In each panel, the frequency of the spectral estimates is depicted on the x-axis; the speed plateau on the y-axis; and the (normalized) spectral power P on the z-axis. Under the 'no instruction' and 'fast' instruction conditions, the contribution of the super-harmonics is very pronounced at the low speed plateaus. (The presence of the super-harmonics reflects the 'sharpness' of fast executed movements.) At the higher speed plateaus, the spectral distribution indicates that the trajectories had become close to sinusoidal under all instruction conditions

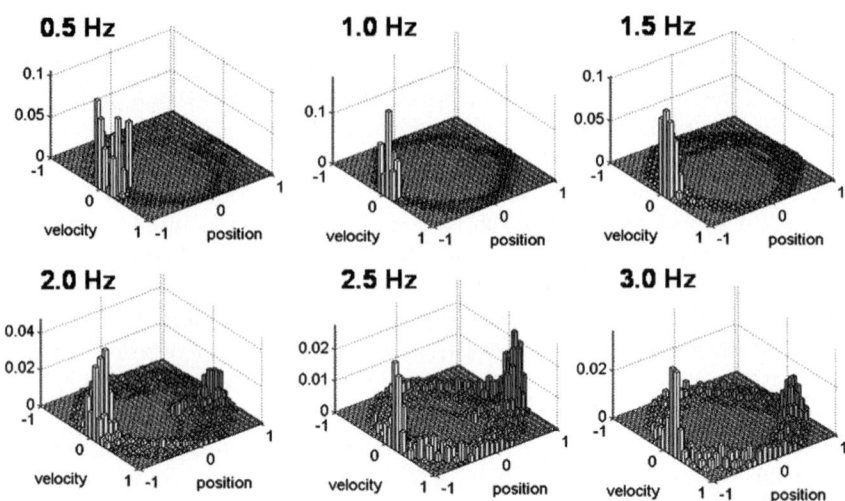

Fig. 5. Two-dimensional probability distributions (averaged across five trials) of the (normalized) position [x] and velocity [$\dot{x}$] trajectories in phase space for a single representative participant corresponding to the first up to the sixth speed plateau (i.e., at 0.5–3.0 Hz; step size 0.5 Hz for the upper left, middle and right, and lower left, middle and right panels, respectively). It can be clearly observed that at the lower frequencies (*upper row*) most of the time is spent around a single position with $\dot{x} \approx 0$ (indicative of a fixed point), while at the higher frequencies the time spent at the location corresponding to the movement trajectory is more evenly distributed (indicative of a limit cycle) even though some clustering around both positions with $\dot{x} \approx 0$ is present (a phenomenon known as 'anchoring'; cf. [3, 9]). (Notice the different scaling of the z-axis)

rhythmic movements [55, 58]. In other words, the role of the cerebellum in discrete movements relative to rhythmic movements remains to be elucidated.

4 Another Dynamic Possibility: Time Variant Phase Flows

So far, we have motivated the potential existence of qualitatively distinct classes of movements (i.e., discrete and rhythmic) based on phase flow topology and provided evidence testifying to their utilization by humans. The perspective outlined above furthermore allows for another class of movements, referred to as the bi-stable scenario. Whether this scenario is actually realized or, at least, provides a useful conceptualization for discrete point-to-point movements is still an open (empirical) question that remains to be answered. Regardless, for these two classes the phase flow remains invariant at the time scale of movement execution. Under the assumption of two state variables, and in view of the Poincaré–Bendixson theorem and the Hartman–Grobman theorem no other classes are possible for invariant phase flows. Under the release of the latter constraint, however, another class may be postulated in which the phase flow changes on the time scale of the movement. In phase space, this class is represented by a fixed point dynamics in which the fixed point is 'externally' driven and (either continuously or intermittently) moving even though the flow topology remains invariant under the impact of the external driving. We do not refer to this scenario as a normal form, however, because of the parametric change on the time scale of the movement.

This notion resembles a dynamic conception of equilibrium point models [1, 15, 16, 50, 51] as briefly discussed above [38]. In these models, the equilibrium point shifts due to the adjustment of a control parameter. Proponents of equilibrium point models have tried to estimate the duration of the control parameter adjustment that leads to the repositioning of the equilibrium (or fixed) point and argued that these adjustments occur 'long' before movement termination in fast point-to-point movements [22, 63]. We hypothesize, however, that the repositioning of the fixed point may be of arbitrary duration (although it obviously cannot be instantaneous). For instance, slow continuous movement of a fixed point could well underwrite tracking movements or locomotive-like behavior. In the former case, the fixed point is environmentally defined; in the latter case the fixed point exists virtually (it has a 'cognitive representation'). The duration of repositioning the fixed point, however, is irrelevant for our present taxonomy. What is pertinent is a (continuous) modification of a topologically invariant flow field, which may underpin a range of – at face value – different movements such as point-to-point movements, tracking, and locomotive-like movements.

The tracking movements of the eyes (i.e., smooth pursuit eye movements; cf. [8, 10, 45]) and hands (cf. [13, 52, 68]) are well studied. In general, these movements are described in terms of a gain factor (the amplitude of the end-effector trajectory relative to that of the stimulus) and a temporal (or phase) delay between the end-effector and the stimulus. (As an aside, Engel and colleagues [14] have shown strong similarities in the response of smooth pursuit eye movements and tracking movements of the hands to changes in stimulus direction (notwithstanding the inertial differences between both systems). While these authors suggested that these similarities hint at commonalities in the neural mechanisms underwriting both movements, they may also be taken to suggest that the same dynamical organization underwrites the tracking movements of both systems.) We do not believe that these kind of movements are governed by mono-stable (i.e., 'discrete') or limit cycle dynamics, even though a limit cycle description may successfully be applied to the manual tracking of a rhythmically moving stimulus [40]. In that regard, in all likelihood due to the strength of the evidence testifying to the organization of (some, if not most) rhythmic movements in terms of limit cycle dynamics (cf. [4, 35, 36]), it has become widespread practice to regard all repetitive motion as governed by limit cycles. However, while limit cycle dynamics are represented by a 'closed loop' in phase space, the observation thereof does not necessarily testify to the existence of a limit cycle (even under the assumption that the system is appropriately described by two state variables). For instance, Leist and colleagues [39] showed that when people visually track a sinusoidal moving stimulus with a low frequency (approximately up to 1 Hz) increases, their eye movements are of the smooth pursuit type. Representing the corresponding data in phase space would result in a circular-like structure. Upon increasing stimulus frequency, (ballistic) saccadic components (whose execution requires external triggering, cf. [11, 33]) set in. At higher frequencies (approximately 2 Hz and higher) eye movements are no longer executed. In other words, if one were to interpret the circular-like structure in phase space at low frequency as a limit cycle, then the latter dynamics would gave way to the ballistic-like saccadic movements with increasing frequency. This scenario is incompatible with the framework we outlined above in that the switch from externally triggered to limit cycle dynamics as a function of movement frequency would be reversed. In fact, at sufficiently low frequency a system in the limit cycle régime will undergo a Hopf bifurcation and adopt fixed point dynamics. We propose that, to keep on moving at a (very) slow pace, one has to adopt a dynamics governed by a moving fixed point.

5 Conclusion

We have proposed a phenomenological taxonomy of movement based on phase flow. The proposed classification is founded upon the notion of topological equivalence, which provides an unambiguous qualitative demarcation scheme to classify dynamical systems: two systems belong to the same class if, and only if, they are topologically equivalent. The existence of qualitatively distinct (movement) dynamics implies that the mechanisms underlying (human) trajectory formation are different. On the basis of these considerations, we argued for the utilization of two dynamic possibilities (for constant phase flows), to which we referred to as normal forms of movement. The phase flows corresponding to these two phase flows – represented by a separatrix and a stable fixed point and limit cycle in phase space – are invariant on the time scale of movement. We reported computational and experimental evidence testifying to the existence of these normal forms that underwrite discrete (ballistic) movements and rhythmic movements, respectively. In addition, we put forward the notion of another class of movements whose corresponding flow field is translated on a time scale similar to that of the executed movement. In other words, we propose to distinguish flow fields, which change on a time scale similar to the movement process from flow fields with a very slow time scale. The latter can be considered to be constant and are subject to our topological classification as outlined in Sects. 2 and 3. A summarizing representation of the proposed classification can be appreciated in Fig. 6.

What could be the functional meaning of phase flows that are invariant on the time scale of the movements? By speculation, biological solutions that involve (non-instantaneous) transitions from one flow topology to another may be less robust than those that are not dependent on such switches as

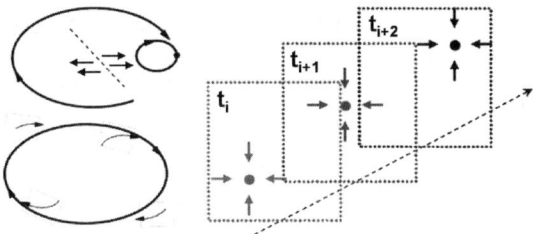

Fig. 6. Schematic representations of two normal forms (invariant phase flows) and time varying phase flow underlying human trajectory formation (see text); a fixed point with a separatrix (*upper left plot*), a limit cycle (*lower left plot*) and an externally driven fixed point (*right plot*)

the timing of switching flow topology becomes crucial to attain a certain task goal in the former scenario(s). For instance, in Schöner's conception a discrete movement involves the destruction of a fixed point in favor of a limit cycle and the creation of a new fixed point [56]. The ill-timed creation of the second fixed point could easily induce the execution of additional (half) cycles relative to the intended movement. Also, the annihilation and creation of flow fields will be time consuming. So far, it is not clear on which time scale flow field changes operate, but it does not seem far fetched to assume that constant flow fields are more adequate for fast movements.

References

1. Balasubramaniam R, Feldman AG (2004) Guiding movements without redundancy problems. In: Jirsa VK, Kelso JAS (eds) Coordination dynamics: Issues and trends. Springer, Berlin, pp. 155–176
2. Baratto L, Morasso P, Zaccaria R (1986) Complex motor patterns: Walking. In: Morasso P, Tagliasco V (eds) Advances in Psychology, 33, Human Movement Understanding. Elsevier Science Publishers B.V., North-Holland, pp. 61–81
3. Beek PJ (1989) Juggling dynamics. Free University Press, Amsterdam
4. Beek PJ, Schmidt RC, Morris AW, Sim MY, Turvey MT (1995) Linear and nonlinear stiffness and friction in biological rhythmic movements. Biol Cybern 73:499–507
5. Bizzi E, D'Avella A, Saltiel P, Tresch M (2002) Modular organization of spinal motor systems. Neuroscientist 8:437–442
6. Buchanan JJ, Park J-H, Ryu YU, Shea CH (2003) Discrete and cyclical units of action in a mixed target pair aiming task. Exp Brain Res 150:473–489
7. Buchanan JJ, Park J-H, Shea CH (2004) Systematic scaling of target width: dynamics, planning and feedback. Neurosci Lett 367:317–322
8. Burke MR, Barnes GR (2006) Quantitative differences in smooth pursuit and saccadic eye movements. Exp Brain Res 172:596–608
9. Byblow WD, Carson RG, Goodman D (1994) Expressions of asymmetries and anchoring in bimanual coordination. Hum Mov Sci 13:3–28

10. Carey MR, Lisberger SG (2004) Signals that modulate gain control for smooth pursuit eye movements in monkeys. J Neurophysiol 91:623–631
11. Carpenter RMS (1977) Movements of the eyes. Pion Limited, London
12. Daffertshofer A (1998) Effects of noise on the phase dynamics of nonlinear oscillators. Phys Rev E 58:327–338
13. Engel KC, Soechting JF (2000) Manual tracking in two dimensions. J Neurophysiol 83:34383–3496
14. Engel KC, Anderson JH Soechting JF (2000) Similarity in the response of smooth pursuit and manual tracking to a change in the direction of target motion. J Neurophysiol 84:1149–1156
15. Feldman AG (1980) Superposition of motor program – I. Rhythmic forearm movements in man. Neuroscience 5:81–90
16. Feldman AG (1986) Once more on the equilibrium-point hypothesis (model) for motor control. J Mot Behav 18:17–54
17. Fink P, Kelso JAS, Jirsa VJ (under review) Excitability is exogenous to discrete movements.
18. FitzHugh R (1961) Impulses and physiological states in theoretical models of nerve membrane. Biophys J 1:445–466
19. Flash T, Hochner B (2005) Motor primitives in vertebrates and invertebrates. Curr Opin Neurobiol 15:660–666
20. Ghahramani Z (2000) Building blocks of movement. Nature 407:682–683
21. Guckenheimer J, Holmes P (1983) Nonlinear Oscillations, Dynamical Systems, and Bifurcations of Vector Fields. Springer, New York
22. Ghafouri M, Feldman AG (2001) The timing of control signals underlying fast point-to-point arm movements. Exp Brain Res 137:411–423
23. Guiard Y (1993) On Fitts's and Hooke's laws: Simple harmonic movement in upper-limb cyclical aiming. Acta Psychol 82:139–159
24. Guiard Y (1997) Fitts' law in the discrete vs cyclical paradigm. Hum Mov Sci 16:97–131
25. Haken H (1983) Synergetics: An introduction: nonequilibrium phase transitions and self-organization in physics, chemistry, and biology, 3rd rev and enl ed. Springer, Berlin; New York
26. Haken H (1996) Principles of brain functioning. A synergetic approach to brain activity, behavior and cognition. Springer, Berlin, Heidelberg, New York
27. Huys R, Studenka BE, Rheaume N, Zelaznik NH, Jirsa VK (submitted) Discrete timing mechanisms produce discrete and continuous movements
28. Ivry R (1997) Cerebellar timing systems. Int Rev Neurobiol 41:555–73
29. Ivry RB, Spencer RM, Zelaznik HN, Diedrichsen J (2002) The cerebellum and event timing. Ann N Y Acad Sci 978:302–317
30. Jirsa VK (2004) Information processing in brain and behaviour displayed in large scale topologies such as EEG and MEG. Int J Bifurcat Chaos 14:679–692
31. Jirsa VK, Kelso JAS (2004) Coordination Dynamics: Issues and Trends Springer, Berlin Heidelberg
32. Jirsa VK, Kelso JAS (2005) The Excitator as a minimal model for the coordination dynamics of discrete and rhythmic movement generation. J Mot Behav 37:35–51
33. Joiner WM, Shelhamer M (2006) An internal clock generates repetitive predictive saccades. Exp Brain Res 175:305–320
34. Kay BA (1988) The dimensionality of movement trajectories and the degrees of freedom problem: A Tutorial. Hum Mov Sci 7:343–364

35. Kay BA, Kelso JAS, Saltzman EL, Schöner G (1987) Space-time behavior of single and bimanual rhythmical movements: data and limit cycle model. J Exp Psychol Hum Percept Perform 13:178–192
36. Kay BA, Saltzman EL, Kelso JAS (1991) Steady-state and perturbed rhythmical movements: a dynamical analysis. J Exp Psychol Hum Percept Perform 17:183–197
37. Kelso JAS (1995) Dynamic patterns: the self-organization of brain and behavior. MIT Press, Cambridge, MA
38. Kugler PN, Kelso JAS, Turvey MT (1980) Coordinative structures as dissipative structures I Theoretical lines of convergence. In: Stelmach GE, Requin J (eds) Tutorials in motor behaviour. Amsterdam, North Holland
39. Leist A, Freund HJ, Cohen B (1987) Comparative characteristics of predictive eye-hand tracking. Hum Neurobiol 6:19–26
40. Liao MJ, Jagacinski RJ (2000) A dynamical systems approach to manual tracking. J Mot Behav 32:361–378
41. Miall RC, Ivry R (2004) Moving to a different beat. Nat Neurosci 7:1025–1026
42. Mitra S, Riley MA, Turvey MT (1998) Intermediate motor learning as decreasing active (dynamical) degrees of freedom. Hum Mov Sci 17:17–65
43. Morasso P (1986) Understanding cursive script as a trajectory formation paradigm. In: Kao HSR, van Galen GP, Hoosain R (eds) Graphonomics: Contemporary research in handwriting. Elsevier Science Publishers BV, North-Holland pp. 137–167
44. Mottet D, Bootsma RJ (1999) The dynamics of goal-directed rhythmic aiming. Biol Cybern 80:235–245
45. Mrotek LA, Flanders M, Soechting JF (2006) Oculomotor responses to gradual changes in target direction. Exp Brain Res 172:175–192
46. Murray JD (1993) Mathematical Biology. New York: Springer
47. Mussa-Ivaldi FA, Bizzi E (2000) Motor learning through the combination of primitives. Philos Trans R Soc Lond B 355:1755–1769
48. Mussa-Ivaldi FA, Giszter SF, Bizzi E (1994) Linear combinations of primitives in vertebrate motor control. Proc Natl Acad Sci USA 91:7534–7538
49. Perko L (1991) Differential Equations and Dynamical Systems. Springer, New York
50. Polit A, Bizzi E (1978) Processes controlling arm movements in monkeys. Science 201:1235–1237
51. Polit A, Bizzi E (1979) Characteristics of motor programs underlying arm movements in monkeys. J Neurophysiol 42:183–194
52. Poulton EC (1974) Tracking skill and manual control. Academic Press Inc, New York
53. Salman MS (2002) The cerebellum: it's about time! But timing is not everything – new insights into the role of the cerebellum in timing motor and cognitive tasks. J Child Neurol 17:1–9
54. Schaal S, Sternad D (2001) Origins and violations of the 2/3 power law in rhythmic three-dimensional arm movements. Exp Brain Res 136:60–72
55. Schaal S, Sternad D, Osu R, Kawato M (2004) Rhythmic arm movement is not discrete. Nat Neurosci 7:1136–1143
56. Schöner G (1990) A dynamic theory of coordination of discrete movement. Biol Cybern 53:257–270

57. Spencer RMC, Zelaznik HN, Diedrichsen J, Ivry RB (2003) Disrupted timing of discontinous but not continuous movements by cerebellar lesions. Science 300:1437–1439
58. Sternad D (2008) Rhythmic and discrete movements – behavioral, modeling and imaging results. In: Fuchs A, Jirsa VK (eds), Coordination: Neural, Behavioral and Social Coordination. Springer (this volume)
59. Sternad D, Schaal S (1999) Segmentation of endpoint trajectories does not imply segmented control. Exp Brain Res 124:118–136
60. Sternad D, Dean WJ, Schaal S (2000) Interaction of rhythmic and discrete pattern generators in single-joint movements. Hum Mov Sci 19: 627–664
61. Sternad D, Wei K, Diedrichsen J, Ivry RB (2006) Intermanual interactions during initiation and production of rhythmic and discrete movements in individuals lacking a corpus callosum. Exp Brain Res 176:559–574
62. Steward I (1995) Concepts of modern mathematics. Dover Publications, New York
63. St-Onge N, Adamovich SV, Feldman AG (1997) Control movements underlying elbow flexion movements may be independent of kinematic and electromyographic patterns: Experimental study and modeling. Neuroscience 79:295–316
64. Strogatz SH (1994) Nonlinear dynamics and chaos With applications to physics, biology, chemistry, and engineering. Perseus Books Publishing LLC, Cambridge, MA
65. Throughman KA, Shadmehr R (2000) Learning of action through adaptive combination of motor primitives. Nature 407:742–747
66. Teulings H-L, Mullins PA, Stelmach GE (1986) The elementary units of programming in handwriting. In: Kao HSR, van Galen GP, Hoosain R (eds) Graphonomics: Contemporary research in handwriting. Elsevier Science Publishers BV, North-Holland pp. 21–32
67. van Mourik AM, Beek PJ (2004) Discrete and cyclical movements: Unified dynamics or separate control? Acta Psychol 117:121–138
68. Viviani P, Campadelli P, Mounoud P (1987) Visuo-manual tracking of human two-dimensional movements. J Exp Psychol Hum Percept Perform 13:62–78
69. Viviani P, Flash T (1995) Minimum-jerk, two-thirds power law, and isochrony: converging approaches to movement planning. J Exp Psychol Hum Percept Perform 21:32–53
70. Wei K, Wertman G, Sternad D (2003) Interactions between rhythmic and discrete components in a bimanual task. Motor Control 7:134–154
71. Wing AM, Kristofferson AB (1973) Response delays and the timing of discrete responses. Percept Psychophys 13:5–12
72. Woch A, Plamondon R (2004) Using the framework of the kinematic theory for the definition of a movement primitive. Motor Control 8:547–557
73. Zelaznik HN, Spencer RMC, Ivry RB (2002) Dissociation between explicit and implicit timing in repetitive tapping and drawing movements. J Exp Psychol Hum Percept Perform 28:575–588
74. Zelaznik HN, Spencer RMC, Ivry RB, Baria A, Bloom M, Dolansky L, Justice S, Patterson K, Whetter E (2005) Timing variability in circle drawing and tapping: Probing the relationship between event and emergent timing. J Mot Behav 37:395–403

Dynamical Systems and Internal Models

James R. Lackner[1] and Betty Tuller[2]

[1] Ashton Graybiel Spatial Orientation Laboratory, Brandeis University, Waltham, MA 02254-9110, USA
[2] Center for Complex Systems and Brain Sciences, Florida Atlantic University, Boca Raton, FL 33431, USA

1 Introduction

We are pleased to participate in honoring our friend and colleague, J.A.S. Kelso, on the occasion of his 60th birthday. For three decades, he has made seminal contributions to the understanding of movement control from both an experimental and a theoretical standpoint. He pioneered the dynamical systems approach to the analysis of movement which has unified the understanding of movement control across multiple domains from animal behavior to human social interaction.

An important conceptual development in movement control during the past 10 years has been the refinement of the concept of internal models. Internal models, forward and inverse, kinematic and dynamic, have become key explanatory concepts for understanding motor behavior and the patterns of adaptation and aftereffect following exposure to dynamic perturbations of movement trajectories. Our goal here is to integrate work on motor control and language carried out by us some years ago, with more recent work to show that internal models also serve as explanatory concepts in the sensory and cognitive domains. Internal model concepts have actually been an integral component of theories of speech comprehension and production for nearly half a century. We will concentrate on how internal models participate in determining how we perceive our external environment, our own bodies, and even our "willed" actions. We will then show how the dynamical models that Scott Kelso has been instrumental in investigating provide an alternative to internal model formulations and invigorate work on perceptual adaptation.

We continually face changes in our external environment as well as in ourselves. Our bodies are not constant; our cellular elements are continually being replaced. Hayflick estimates that after a cell has undergone mitosis 50 times it has reached its full lifespan and will die, a phenomenon known as "Hayflick's Limit". Cortical bone tissue turns over at a rate of about 5%/year, cancellous bone at 20%/year, and neurons at 1%/year. Despite such changes in physical makeup, the structural integrity of the body and of memory and cognition are retained except in the case of disease or injury or advanced age. The dimensions of our body and our weight also change greatly from birth to adulthood, and to a lesser extent and more gradually thereafter. Changes in body loading related to clothing and other external encumbrances can occur within seconds. Muscle efficiency also depends on temperature because muscle contraction is the consequence of a chemical reaction. The Q_{10} of muscle is 2.5, meaning that the velocity of muscle contraction more than doubles for each 10°C increase in temperature. Even relatively modest changes in muscle temperature affect contraction speed which is why on cold days bumble bees have to raise the temperature of their flight muscles by "shivering" before they are able to fly [24]. Presumably, the mammalian CNS in controlling body movements has some way of adjusting for muscle temperature to preserve accurate control.

Such rapid adaptive accommodations to external and internal conditions represent a common feature of animal as well as insect behavior. David Katz, perhaps the most creative of the gestalt psychologists, studied locomotory adaptation in creatures deprived of one or more appendages [11]. He found, for example, that dogs deprived of both limbs on one side of the body could still learn to run. Interestingly, he showed that a beetle deprived progressively of one and then another and another leg until all six legs were absent could still locomote. The creature would grab with its mandibles and pull its legless body along. This is a quintessential example of behavioral modifiability.

Temperature variations and leg removals represent perturbations to which the affected organism must accommodate. They represent transgressions of both the internal and external situations, to which accommodation must be made. Henderson [10] introduced the concept of the "Fitness of the Environment" to characterize the constituent features of the earth's environment that allow life forms to develop. For example, when water freezes, it expands and the resulting ice floats to the surface of streams and oceans allowing life beneath the ice layer to survive during cold periods. Similarly, adaptive capacities of animals can be thought of as reflecting the fitness of the organism, the ability to self-calibrate and adjust constantly to altered internal and external demands.

2 Role of Internal Models

The fundamentally adaptive nature of motor control has, over the past 10 years, been examined using internal models as powerful explanatory schemas. The notion is that an internal model enables the CNS to anticipate the consequences of motor commands for forthcoming body movements and for the sensory feedback resulting from the movements. Early forms of internal models were developed for the oculomotor system. Von Holst and Mittelstaedt [27] attributed the perceptual stability of the visual world during voluntary eye movements to a central comparator relating an efference copy of the neural command signal sent to the extraocular muscles with the sensory feedback ("reafference") from the retina contingent on the movement. In the event the commanded movement and the resulting movement coincided in magnitude and expected sign, the visual world would seem stable; otherwise it would be seen to jump during an eye movement. This simple model explains why the visual world appears to move when the eye is passively displaced: the retinal displacement is not associated with an efference copy signal. It also explains why an afterimage appears to move in the direction of an eye movement: the efference copy signal indicates the direction and magnitude of the change in eye position, and the unchanging retinal signal indicates that the stimulus is linked to eye position.

Two of our early studies point to the presence of a similar monitoring mechanism in speech production. One makes use of the "verbal transformation effect". When isolated speech sounds are played over and over again, the listener will begin to hear them undergo changes, or verbal transforms, such that they sound like a different speech sound. For example, the sound /bi/ repeated may be heard as /pi/ [21]. The voiced stop consonant is heard as unvoiced, a single distinctive feature change has occurred. However, if one repeats a speech sound aloud to oneself, perceptual changes will not be heard. By contrast, if one listens to a tape recording of the sounds one had actively spoken, then perceptual changes are commonplace [16]. Thus, in the case of speech production, there is an efferent representation of the intended speech sound that is compared with the returning speech sound, and the motor signal has priority in determining perceptual assignment.

In another study, we demonstrated that there are multiple levels of corollary discharge signals available in speech [23]. We had subjects simply repeat different sequences of speech sounds that were designed to make speech errors likely. The subject's task was to press a key each time he or she made a speech error. The important finding was that most speech errors were detected with latencies well below usual conscious reaction times, i.e., less than 200 ms. Moreover, many errors were detected with near zero and even negative times with respect to the onset of the acoustic signal representing the speech error.

These results have two implications. "Negative" detection latencies occur because neural conduction and muscle contraction take time. The negative latencies indicate that there can be a comparison of the intended speech sound's motor command representation with representations of actually issued motor commands prior to actual articulatory movements. This allows speakers to realize they are making a speech error just before or just after sound production begins. Error detections with reaction times between 0 and 200 ms after sound onset indicate that there can be a comparison of the expected sensory feedback from the issued command signals to the articulatory muscles with the proprioceptive and auditory feedback from the speech sound being produced. One of our experimental conditions used masking noise to obscure the auditory feedback from the self-produced speech sounds. Such masking actually enhanced the overall rate of error detection and lowered detection latencies.

Together these two experiments provide strong evidence for internal models in speech production, mechanisms which incorporate multiple levels of motor representations and comparisons of both motor–motor and motor–sensory signals. A key feature of speech motor control is the ability to accommodate to perturbations of the articulatory apparatus during the production of speech sounds. One of us with Kelso has shown a spontaneous ability for the articulatory system to make online adjustments to preserve the vocal quality of the intended sounds. For example, if the motion of the jaw is perturbed to disrupt the required closure of the lips (e.g., during the production of a bi-labial stop consonant, the jaw is unexpectedly prevented from moving upward), the closure will nevertheless be achieved by compensatory adjustments of the upper lip [13]. This is a remarkable example of the phenomenon of motor equivalence, the same external event can be associated with many patterns of neuromotor activation.

Ostry and colleagues have shown an interesting sidelight of such compensatory jaw movements [25]. In their experiments, a robot was used to apply a perturbation to the jaw, just as robots are commonly used to study limb movement control. The device delivered a jaw-velocity-dependent perturbation as the speaker either produced a speech sound or made the same movements of the jaw without producing a sound. For production of speech sounds, after repeated exposure to the perturbation, the trajectory of the jaw showed adaptive modifications and negative aftereffects on catch trials in which the perturbation was absent. By contrast, when subjects simply made the jaw movements, no adaptation occurred. This result is important because to adapt to perturbations of limb-reaching movements, we have found it is necessary to have a spatial goal, i.e., to point to a target, as opposed to simply repeating the same movement in the same way [15]. Speech motor control is a quintessential form of goal-directed behavior involving both a speaker and a perceiver. Thus, adaptive modifications in speech communication must be defined by a combined system determined by the speaker, the perceiver, and the task requirements.

3 Calibration of Body Dimensions

During everyday life, we constantly change the configuration of our limbs and rotate and bend our neck and body and move about. Goal-directed movements must be coordinated relative to the overall configuration of the body. It was long thought that perception of the spatial configuration of the body was based on signals from sensory endings within the joints, the joint receptors, acting as internal goniometers. However, Goodwin et al. [7] showed that muscle spindles influence conscious appreciation of limb position, and it was found that after total joint replacement (which destroys joint receptors) position sense is preserved. Goodwin et al. [7] found that if a limb muscle is vibrated to elicit a tonic vibration reflex and the limb is physically restrained from moving, illusory displacement will be experienced. For example, vibration of the biceps brachii leads to contraction of the biceps and when the forearm is prevented from flexing it will be perceived, even though stationary, to be moving into extension. In related experiments, we found that when subjects standing in the dark are restrained in position and their Achilles tendons are vibrated they will experience forward motion of their body pivoting about the ankles [22]. Most subjects also exhibit involuntary eye movements, a nystagmus, with the slow phase compensatory for their apparent motion. When given a stationary visual target to fixate in the otherwise dark experimental chamber, they will see it as moving with their body leading it slightly. Our studies thus demonstrate that muscle spindle signals contribute not only to the sense of body configuration but also to body orientation with respect to space and sensory localization.

In experiments making use of such vibration-induced illusions, we found that the apparent dimensions of the body are influenced by contact with the hands. The hands can be calibrated in terms of their dimensions by contact with external objects. But the hands can affect the perceptual calibration of body parts that do not normally come in contact with external objects. For example, we have found that if the hand is grasping the nose and a tonic vibration reflex is elicited in the biceps muscle, then as the stationary forearm is experienced as moving into extension, the nose is perceived as lengthening, like a Pinocchio nose [17]. If both hands are in contact with the waist while one is sitting with arms akimbo, and tonic vibration reflexes are elicited in the triceps muscles, then the forearms will be experienced as moving into flexion and the waist will seem to shrink. Thus, the apparent motion of the hands influences the perceived dimensions of the body. It seems reasonable that contact of the body with the hands is used as a way of updating the dimensions of the body. Similarly, if one is grasping a stick with both hands and the biceps muscles are vibrated, the stick will be perceived as lengthening as the arms seem to move into extension; the stick will seem to shorten during vibration of the triceps muscles.

We found that the hands can also influence the interpretation of retinal local signs. A simple paradigm demonstrates this [6]. A target light is attached to each of a subject's index fingers, and the target lights are viewed either monocularly or binocularly in an otherwise dark room. Then, the illusion of arm displacement is induced by simultaneously vibrating both biceps or triceps brachii. The two retinally fixed target lights will appear to separate spatially by many degrees during biceps vibration and during triceps vibration they will appear to move closer to one another. In other words, there is a virtually instant remapping of retinal local signs. When vibration ceases, the illusory displacements abate. It seems likely that such haptic information is used in the course of everyday life to calibrate the retinae. Such haptic-proprioceptive inputs are probably as important or more important than visual information in mechanisms of self-calibration because in sighted individuals it is necessary also to calibrate the visual sense, and this can only be achieved through some form of interaction with the environment involving motor systems and proprioception.

Interestingly, current internal models that have been proposed to account for control of body movements take for granted the dimensions of the body. They do not incorporate mechanisms for updating these dimensions which are of course key to accurate performance.

4 Calibration of Limb Movement Control

We have found that muscle spindle signals also participate in the calibration of motor control. An important feature of spindle signals is that they are interpreted in relation to ongoing patterns of afferent and efferent signals and information about body loading. The role of spindle signals in motor calibration becomes apparent during exposure to rotating "artificial gravity" environments. When a reaching movement is made in a rotating environment, the forward motion of the arm generates a lateral Coriolis force that acts in the direction opposite to body rotation. Such Coriolis forces are dependent on the velocity of rotation of the environment and the linear velocity of the arm in relation to that environment; consequently, they are absent at the beginning and the end of a movement. Because velocity profiles of normal arm movements are bell-shaped, the Coriolis force perturbation are bell-shaped as well. When one first makes reaching movements in a rotating environment, movement trajectories and endpoints are displaced in the direction of the transient Coriolis forces [18]. However, as additional reaches are made, trajectories gradually become straighter and endpoints more accurate until after 10 or 20 reaches, movements are again as straight and accurate as pre-rotation. Movements made post-rotation show a mirror image trajectory and endpoint pattern to that initially seen during rotation.

The presence of a mirror aftereffect implies that the nervous system is anticipating and compensating for the consequences of Coriolis forces on the impending arm movement, i.e., a feedforward compensation is implemented. An interesting feature of exposure to Coriolis forces is that when reaches initially are made during rotation, the Coriolis force is felt to deviate the path of the arm, but as additional reaches are made and adaptation proceeds, the Coriolis forces generated seem to be less and less intense until they seem to be absent. They are no longer perceived even though they are still present in full magnitude during reaches. Post-rotation Coriolis forces are no longer generated during reaches, but when a reach is first made, it feels as if a Coriolis force is present and deviating the arm. What is sensed as an external force is actually the self-initiated internal compensation for an anticipated but absent Coriolis force. These observations mean that perception of the forces acting on the body is also consistent with the action of an internal model.

Adaptation to Coriolis forces occurs even without visual feedback about arm position. This means that some combination of motor, proprioceptive, and somatosensory information must be used in the updating process. We have found that if we have subjects make repeated reaching movements in the dark to targets which are turned off as they start to reach and have them point just above the target, then they will adapt to the Coriolis forces generated but only in terms of trajectory curvature. That is, they again reach straight but will continue to miss the position of the target. By contrast, if they are allowed to touch down at the end of their movement, even though contact with the surface provides no direct information about movement accuracy with respect to the target, they adapt both their movement curvature and endpoint, and again reach straight and accurately. The reason for this pattern is that trajectory curvature information is derived from motor commands compared with proprioceptive feedback signals from the Golgi tendon organs and muscle spindles. When a reaching movement is made and the arm does not follow the path intended, the pattern of muscle spindle and tendon organ feedback received will be inappropriate for the planned movement. The nervous system uses this information to reprogram the movement to restore spindle feedback to normal.

Endpoint adaptation appears to be related to finger contact at the end of the reaching movement. We have found that when reaching movements are made to targets projected in different locations on a surface, the pattern of forces on the fingertip during the first 50 ms after contact with the surface provides directional information about the location of the fingertip in relation to the body [19, 20]. The three-dimensional reaction force vector generated by finger contact always points to a location on the shoulder of the pointing arm. This means that every time one contacts a surface in everyday life, information about the location of the hand can be updated on the basis of the shear force reaction vector present during the initial landing of the fingertip. Taken together, our experiments point to a common theme: the location of the hand with respect to the body is part of a constant calibration and tuning that takes place during everyday life. The same mechanisms are used in adjusting motor control to characteristics of the environment and in calibrating the dimensions of the body.

Similar mechanisms may be used in updating the control of speech. It is notable that individuals lacking somatosensory feedback from the vocal tract are unable to develop normal speech. Delayed auditory feedback also has a deleterious effect on the production of speech. Delays of the order of 200 ms or near average syllabic duration tend to be most disruptive causing people to block or repeat sounds. We have tried to adapt people to delayed auditory feedback by incrementally exposing them to progressively longer delays to reach 200 ms gradually [12]. However, we have failed to find significant adaptation. Instead, we find that various strategies are used. Some subjects speak as rapidly as they can, trying to produce the speech as pre-programmed chunks so that they will not be disrupted by the delay of their voice. Others try to ignore the delayed auditory feedback concentrating instead on articulatory control. The overall implication of our studies is that auditory feedback does not seem to play an important role in the ongoing production of speech but may serve as a source of confirmatory feedback which allows the speaker to know whether or not the intended speech sounds were produced. That is, the auditory feedback may serve much like shear forces do in the case of finger contact (indicating where the hand landed in relation to the body) and be used to update speech motor control.

5   Dynamic Systems Approach

The foregoing examples underscore the adaptive ability of the perceptuomotor system across different sensory and motor domains. It is possible to reconceptualize this adaptive flexibility as a dynamic system that evolves in response to the altered actor–environment interaction in a given goal-directed task rather than as the recalibration of an internal model. The theory of dynamic patterns that Scott Kelso has been exploring and extending since the 1980s emphasizes the fit, or complementarity between an individual's intrinsic dynamics, the available information, and the constraints imposed by a particular task or goal. In this view, perceptual adaptation corresponds to finding stable solutions (conceptualized as attractors) in the system dynamics. Adaptive behavior is thus a property of a system that includes the actor, the environment, and the task or goal. Since any of the system components can, and usually does, change over time, the layout of stable attractors also changes with time, as well as with the goal. Thus, it may be useful to conceptualize changes in calibration of the perceptuomotor system as a shift in the layout of stable attractors.

One area of perceptual adaptation that is readily modeled as a dynamical system is the verbal transformation effect (VTE) described above. Although most investigations of the VTE have focused on the phonemic relationship between the stimulus and the perceived transformations [21], Ditzinger et al. [5] explored the dynamics of switching among transformations over time. They found evidence that when the identity of the transforms is ignored, sustained oscillations between two transforms are far more frequent than expected on the basis of a random arrangement of alternatives. Moreover, these two-form alternations show a faster and more stable dynamic than observed when more than two perceptual forms are cycled. This suggests a coupling mechanism that underlies the VTE (much like that proposed for the alternation of visual reversible figures [2, 3]).

There are similarities between the idea of recalibration of perceptual and motor mechanisms and a dynamic model of the VTE. The model [4], based on the synergetic computer [8], incorporates an associative memory that enables it to identify an input pattern by calculating the overlap of the input pattern with each of a set of previously stored prototype patterns, much like the comparison of actual and intended consequences in an internal model. However, the role of this comparison is somewhat different in the two conceptualizations. Recalibration assumes a minimization of some quantity such as absolute error. In the dynamic model, the overlap is a measure of the strength of possible patterns or attractors that in part determines the current percept. There are also global effects on the attractor landscape (on the internal model) that are equally important. In reality, humans encounter a variety of patterns that continuously change in space and time so that their attraction changes as well. Individuals also have internal perceptual-motor biases that affect the attractor landscape and perceptual systems are themselves not static but evolve over time [14] and changing conditions [9]. The influence of the time dependence can be seen in the pattern of coupled oscillations of transforms in the VTE. Recall as well that repeating the same syllable or word over and over to oneself does not result in perceptual transformations. Previously, we suggested that this indicates a continuous calibration based on a high-priority motor-based signal. An alternative description is that the motor signal helps stabilize an otherwise changing perceptual dynamic.

A second relevant area in speech illustrates how a new calibration of acoustic signals into phonemic categories can also be considered as a dynamic system. Native speakers of a language consider a range of acoustic objects as being phonemically identical even though in another language the same acoustic range might span two or more phonemic categories. When asked to learn a category that is close to the native one, listeners must become attuned to distinctions that are not phonologically meaningful in their native language. This can be very difficult, sometimes impossible, for adult learners when the new sound is acoustically very similar to an existing sound in their native language. We have found that an individual's initial perceptual biases greatly affect the likelihood that an acoustic range will be perceived reliably as a new (non-native) speech sound. Moreover, learning the new category shifts the acoustic parameters of what is perceived as the best exemplar of the native category. Although this process might be described as recalibrating the acoustic signal and perceptual categories, an alternative approach, akin to that used to examine dynamics in verbal transforms, views category learning as a dynamical process that modifies perceptual space over time. For example, Case et al. [1] showed that when listeners initially perceive the non-native sound as "different" from the native one, the progressive stabilization of the sound to be learned is relatively fast. In other words, the rate of change of the perceptual landscape (the arrangement of attractors corresponding to phonologically meaningful distinctions, or the recalibration of the acoustic range into more than one phonemic category) depends on the initial conditions of the listener's perceptual/linguistic system.

Tuller et al. [26] observed that for listeners who learned a new phonological category, the strength of the attraction of the to-be-learned sound increased until a qualitative change (a bifurcation, or phase transition, marked by high variability) reflected the emergence of the new attractor. The bifurcation was most evident when a strong attractor, the native sound, preceded the non-native sound. In contrast, other subjects became increasingly sensitive to acoustic differences in the stimulus set until they could parse the stimuli into two groups using a well-calibrated labeling of the stimuli. No bifurcation was observed for these listeners during learning. The important point is that different modes of learning may differ in their reliance on recalibration processes. Just as the previously described work on motor control requires internal models to incorporate mechanisms for updating the dimensions of the body, learning to perceive a new sound categorically requires a mechanism for updating the dimensions and structure of perceptual space. The adaptive capacity of adult language users is hence reminiscent of the adaptive capabilities of a dog or beetle: they reflect the ability to self-calibrate constantly and to adjust flexibly and continuously to altered situational demands.

In conclusion, we have explored the possibility that the dynamical systems approach to human behavior that Scott Kelso has tirelessly and creatively advanced can be integrated with current conceptions of internal models. The parallels across the fields are theoretically deep and, we believe, will invigorate and re-cast work on perceptual adaptation as the normal moment-to-moment adjustments of humans to the ever-changing internal and external milieu.

Acknowledgments

NSF grant BCS-0414657, AFOSR grant FA9550-06-1-0102, and NIH grant R01 AR48546-01.

References

1. Case P, Tuller B, Kelso JAS (2003) The dynamics of learning to hear new speech sounds. Speech Pathol, November 17, 2003, retrieved from SpeechPathology.com Jan 1, 2007, http://www.speechpathology.com/articles/article_detail.asp?article_id=50
2. Ditzinger T, Haken H (1989) Oscillations in the perception of ambiguous patterns. Biol Cybern 61:279–287
3. Ditzinger T, Haken H (1990) The impact of fluctuations on the recognition of ambiguous patterns. Biol Cybern 63:453–456
4. Ditzinger T, Tuller B, Haken H, Kelso JAS (1997a) A synergetic model for the verbal transformation effect. Biol Cybern 77:31–40
5. Ditzinger T, Tuller B, Kelso JAS, Haken H (1997b) Temporal patterning in an auditory illusion: the verbal transformation effect. Biol Cybern 77:23–30
6. DiZio P, Lathan CE, Lackner JR (1993) The role of brachial muscle spindle signals in assignment of visual direction. J Neurophysiol 70:1578–1584

7. Goodwin GM, McCloskey DI, Matthews PBC (1972) Proprioceptive illusions induced by muscle vibration: contribution by muscle spindles to perception. Science 175:1382–1384
8. Haken H (1991) Synergetics, computer, and cognition. Springer-Verlag, Berlin
9. Helson H (1964) Adaptation level theory: an experimental and systematic approach to behavior. Harper & Row, New York
10. Henderson JL (1958) The fitness of the environment; an inquire into the biological significance of the properties of matter, Beacon Press, Boston, MA
11. Katz D (1953) Animals and men; studies in comparative psychology. Penguin Books, Baltimore, MD
12. Katz D, Lackner JR (1977) Adaptation to delayed auditory feedback. Percept Psychophys 22:476–486
13. Kelso JA, Tuller B, Vatikiotis-Bateson E, Fowler CA (1984) Functionally specific articulatory cooperation following jaw perturbations during speech: evidence for coordinative structures. J Exp Psychol Hum Percept Perform 10:812–832
14. Köhler W, Wallach H (1944) Figural after-effects: an investigation of visual processes. Proc Amer Philos Soc 88:296–357
15. Kurtzer I, DiZio P, Lackner JR (2003) Task-dependent motor learning. Exp Brain Res 153:128–132
16. Lackner JR (1974) Speech production: evidence for corollary-discharge stabilization of perceptual mechanisms. Percept Motor Skills 39:899–902
17. Lackner JR (1988) Some proprioceptive influences on the perceptual representation of body shape and orientation. Brain 111:281–297
18. Lackner JR, DiZio P (1994) Rapid adaptation to Coriolis force perturbations of arm trajectory. J Neurophysiol 72:299–313
19. Lackner JR, DiZio P (2000) Aspects of body self-calibration. Trends Cogn Sci 4:279–288
20. Lackner JR, DiZio P (2005) Motor control and learning in altered dynamic environments. Curr Opin Neurobiol 15:653–659
21. Lackner JR, Goldstein LM (1975) The psychological representation of speech sounds. Quart J Exp Psychol 27:173–185
22. Lackner JR, Levine MS (1979) Changes in apparent body orientation and sensory localization induced by vibration of postural muscles: vibratory myesthetic illusions. Aviat Space Environ Med 50(4):346–354
23. Lackner JR, Tuller B (1979) Role of efference monitoring in the detection of self produced speech errors. In: Walker ECT, Cooper WE (eds) Sentence Processing MIT Press, Cambridge, MA, pp. 281–294
24. Loli D, Bicudo JE (2005) Control and regulatory mechanisms associated with thermogenesis in flying insects and birds. Biosci Rep 25:149–180
25. Tremblay S, Shiller DM, Ostry DJ (2003) Somatosensory basis of speech production. Nature 423:866–869
26. Tuller B, Jantzen MG, Jirsa V (in press) A dynamical approach to speech categorization: two routes to learning. New Ideas in Psychol, Special issue on Dynamics & Psychology
27. von Holst E, Mittelstaedt H (1973) The reafference principle. In: Martin R (ed) The behavioural physiology of animals and man: the collected paper of Erich von Holst, Vol. 1. University of Miami Press, Coral Gables, FL, pp. 139–173

Towards a Unified Theory of Rhythmic and Discrete Movements – Behavioral, Modeling and Imaging Results

Dagmar Sternad

Departments of Kinesiology and Integrative Biosciences, Pennsylvania State University, University Park, PA 16802, USA

Summary. Since the seminal paper on phase transitions in bimanual rhythmic movements, research from the dynamical systems perspective has given primacy to rhythmic coordination. While rhythmic movements are a ubiquitous and fundamental expression in biological behavior, non-rhythmic or discrete movements are of similar importance. In fact, rhythmic and discrete movements are commonly intertwined in complex actions. This review traces our strategy of extending a dynamic systems account from rhythmic to non-rhythmic behavior. Behavioral and modeling work on uni- and bimanual, single- and multijoint coordination increasingly investigated more complex movement tasks consisting of rhythmic and discrete elements. The modeling work suggested a three-tiered architecture consisting of a biomechanical, internal and parameter level with different responsibilities for different components of movement generation. A core question raised in the modeling is what are the fundamental units and principles that are tuned to make up complex behavior. Are rhythmic pattern generators the primitives for generating both rhythmic and non-rhythmic behaviors? Alternatively, are discrete pattern generators fundamental, or are there two primitives of action? fMRI experiments compared brain activation in continuously rhythmic and discrete movements. Significantly more activation in discrete movements suggested that discrete movements have higher control demands and may be distinct primitives, different from rhythmic movements. This result corresponds to the modeling work that highlighted that discrete movements require more parameterization. Our behavioral, modeling and imaging research built on and extended the dynamical systems approach to rhythmic coordination with the goal to develop a comprehensive framework to address complex everyday actions in a principled manner.

1 Introduction

It is by now redundant to describe Kelso's seminal work on phase transitions in rhythmic interlimb movements [30, 31]. This experiment marks the beginning of the "dynamical systems perspective" to movement coordination and has entered virtually every introductory course and textbook in motor control.

In bimanual rhythmic movements there are two stable states, and changes in the control parameter frequency induce a transition from the less stable to the more stable state. The vocabulary of order and control parameters, relative phase and movement frequency, is in everybody's use. The paradigm has even received a nickname – "the finger wiggling experiment" – and has become a hands-on exercise in lecture rooms. Behind this nickname is a robust phenomenon that has provided deep insights into coupling, entrainment, phase transitions, symmetry and generally the non-linear processes that constrain and coordinate human behavior. This seminal experiment was followed by numerous studies that corroborated and elaborated the early findings and it has been discussed as an example for a paradigm shift in the sense of Kuhn [38].

Over 20 years of research on rhythmic interlimb coordination has produced a remarkably extensive and consistent body of literature on this topic. Held together by the HKB modeling framework two major strands of investigation performed an impressive series of studies extending the original observations: Kelso's group continued to investigate questions around the phenomenon of the phase transition; Turvey's group examined bimanual rhythmic coordination at steady state. The variety of specific issues examined is striking: from the asymmetry between coordinated limbs, interpersonal coordination, to intentional switching; from synchronization with an external timer, cognitive influences, to learning new coordinative patterns; from polyrhythmic coordination, attentional effects, to perceptual anchoring and multi-sensory influences. This list is by no means complete and the appropriate references would overwhelm the writing. Importantly, the research direction also spread to other labs and investigators: Beek and colleagues spearheaded investigations on juggling and polyrhythmic coordination; Carson and colleagues brought biomechanical and physiological issues to the fore to provide further insights into the constraints underlying the phase transitions. This experimental body of work was complemented by theoretical work: the initial HKB model was extended by a stochastic element, behavioral dynamics was introduced to address issues of learning and memory, limb asymmetry was expressed by a detuning parameter, and laterality necessitated an extension of the basic coupling terms. To repeat, this review can only be an incomplete list of key terms.

2 In the Shadow of Rhythmic Interlimb Coordination – Studies on Discrete Movements

Somewhat overshadowed but of equally important status than the discovery of phase transitions is an earlier study that also showed the overriding influence of coupling between two simultaneous but discrete arm movements. This study on bimanual reaching threw a critical light on the famous Fitts' law which states that movement time scales linearly with the log of movement amplitude over target width [9, 10, 34, 35]. However, as Kelso and colleagues convincingly demonstrated, this seemingly law-like phenomenon only holds for a controlled pointing task performed with a single effector. When a subject simultaneously reaches to two targets of different widths or distances, i.e., with different indices of difficulty, this index is no longer the only predictor for movement time. The movement time to the easier target is significantly slowed due to synchronization with the second simultaneous movement to the more difficult target. Although the trajectories of the two hands were still different in velocity, the peaks of the velocity and acceleration profiles were synchronized in time. This was a powerful demonstration of coupling effects in two discrete goal-oriented movements and essentially showed that coupling between two hands can override Fitts' law. Despite this demonstration of the limited generalization of Fitts' law, this study has not had as much influence as one might expect. Was the reason that the point was made? Or was it because the discovery of the transitions in rhythmic interlimb coordination, only shortly afterwards, deflected the research community's attention? Fact is that bimanual synergy effects were subsequently examined in rhythmic, not in discrete coordination, with numerous, almost endless experimental variations and theoretical extensions.

Another study of Kelso and colleagues of the early 1980s that is again less frequently cited but nevertheless extremely insightful is an experiment by Goodman and Kelso [19] who explored the effect of rhythmic onto discrete movements. More specifically, the authors examined constraints exerted by the underlying physiological tremor on the initiation of a discrete movement. Does the continuously present low-amplitude high-frequency rhythm influence the timing of the discrete movement's onset? Stimulated by the homeokinetic theory by Iberall and colleagues that postulated the pervasiveness of cycles in biological systems [26, 27] and with reference to an early report on pathological tremor by Travis [66], Goodman and Kelso performed fine-grained kinematic analyses on healthy subjects and demonstrated that the initiation of a discrete finger extension is phase-coupled to the ongoing background tremor. The discrete movement was constrained to occur in a limited phase window of the ongoing oscillation; most frequently it occurred at the peak of angular velocity in the tremor oscillation, which is equivalent to the moment of maximum momentum in the direction of the discrete movement. This particular moment represents the most advantageous time to initiate the voluntary movement as momentum can be exploited. This was observed for instructions that either emphasized speed (as fast as possible following a trigger signal) or that permitted a comfortable pace. The authors suggested that the perceptual-motor system is sensitive to its own dynamics and capitalizes on this opportune moment to initiate a voluntary discrete action.

These three selected studies were at the beginning of a long fruitful career of Kelso whose impact on the research community began with insightful experimental work that significantly extended our understanding of the principles of coordination. Kelso's impact grew further when theoretical work started to complement his empirical work and provided not only explanations but also more testable predictions for numerous follow-up studies. Most notably the theory of synergetics by Haken became influential and led to the formulation of the Haken–Kelso–Bunz model which became prominent in all research on rhythmic interlimb coordination for the past 20 years [21, 22, 23]. Kelso's scientific productivity further blossomed when he turned his attention to the functioning of the brain. Imaging studies combined with pioneering theoretical work on these new types of data aimed to unravel the neural substrate in the most complex part of the human dynamical system, the brain [12, 13, 14, 28]. EEG, fMRI and MEG data extended the understanding from behavior to brain activation, all centered around the rhythmic finger task [15, 43]. Kelso's prolific career demonstrated how the confluence of different levels of analyses, behavioral, modeling, and brain imaging, can heighten insights into a central problem: stability in the coordination of – rhythmic – movement patterns.

Returning to the pioneering papers from the early 1980s, one is taken back to discrete movements that somehow have fallen by the wayside. However, nobody would deny that both rhythmic and discrete movements exist and often co-exist, probably with equal importance. Walking is seldom a regular periodic pattern; more often it is interspersed with goal-directed steps, for example onto stairs, or combined with reaching out, for example to open a door. Piano playing, while paradigmatically rhythmic, would not be successful if it were not combined with reaches to the targeted keys. Goodman and Kelso's study on tremor raised awareness that both rhythmic and discrete movements are ubiquitous in everyday behavior and are commonly intertwined. And yet, despite these reflections and commonsense observations, in motor control the two movements are mostly examined in separate lines of work. Kelso's own research trajectory and the research community that further developed the dynamical systems account for movement coordination gave primacy to rhythmic movements [2, 32, 37, 68, 69].

Admittedly, there were few exceptions. Schöner [52] showed how a discrete movement could be understood as a fixed-point attractor co-existing with limit cycle attractors in a different parameter space. But this was only a theoretical demonstration using a suitable dynamic model that permitted analytical treatment. This interesting theoretical work however made little impact on experimental research. The equilibrium point or λ-hypothesis, developed for discrete single-joint movements, was close in spirit to a dynamical systems conceptualization of discrete movements but it was formulated at the neurophysiological level of analysis [40]. The movement endpoint was viewed as an equilibrium point set by the nervous system via the threshold in the reflex system of the muscles. A formalization in terms of dynamical systems was suggested only by Turvey and colleagues but was not pursued further [36, 63]. Discrete movements played a Cinderella existence in a dynamical systems account.

Goal-directed reaching movements, on the other hand, have been the favorite paradigm for neuroscience and computational research. Voluntary reaching in different directions, the so-called center-out task, has been the predominant task for examining motor cortical functions [17, 39, 45, 46]. In work that aims to explicate the function of the cortical and cerebellar structures in their role of instantiating internal models again goal-directed reaching has been prominent [53, 54, 73, 74, 75]. Rhythmic movements play a Cinderella existence in these quarters of motor control research. Does this difference of explanatory constructs for rhythmic and discrete movements reflect different control mechanisms of these behaviors or simply the theoretical backgrounds of different investigators? Are discrete movements fundamentally distinct from rhythmic movements or should one also develop a dynamical account for discrete movements?

3 A Dynamical Approach to Rhythmic and Discrete Movements: Asymmetric Bimanual Coordination

Coordination dynamics has characterized coordination patterns in terms of relative phase between two moving limbs. This is a rich framework but how can one relate this approach to discrete translatory movements? If rhythmic movements are to be understood as limit cycle attractors, should discrete movements be understood as fixed-point attractors? Or are discrete movements truncated limit cycles? How are discrete and rhythmic movements coordinated when they co-exist in more complex behavior? What are the fundamental units of actions – rhythmic movements, discrete movements or both? These were the guiding questions for research in our lab.

A first attempt to gradually move away from purely rhythmic interlimb coordination was a study by Sternad et al. [62] on a partly discretized bimanual task that examined coupling effects between two different movements. Subjects rhythmically moved one hand-held pendulum in continuous fashion in their right hand while they also performed a single discrete cycle with the left hand-held pendulum every fourth right-hand cycle. This combined uni- and bimanual pattern repeated itself every four cycles. Viewed differently, the task consisted of one cycle performed by one hand that was intermittently locked into an ongoing sequence of cycles in the other hand (Fig. 1A). Kinematic analyses revealed systematic coupling effects across the two hands: the relative phase measure demonstrated immediate synchronization of the "discrete" event directly after its onset, suggesting ongoing coupling between hands even when the discrete hand rested. The same coupling effect was signaled by the "ringing" after the termination of the full cycle, indicating continued phase and frequency locking in the damped oscillation. During the bimanual portion the amplitude of the continuous oscillations was enlarged indicating energy transfer across hands (Fig. 1A). When the two hand-pendulum systems had different eigenfrequencies, phase differences were observed that were dependent on the magnitude of eigenfrequency differences, consistent with observations in parallel rhythmic coordination (Fig. 1B). In sum, significant influences of coupling existed across rhythmic and non-rhythmic movements.

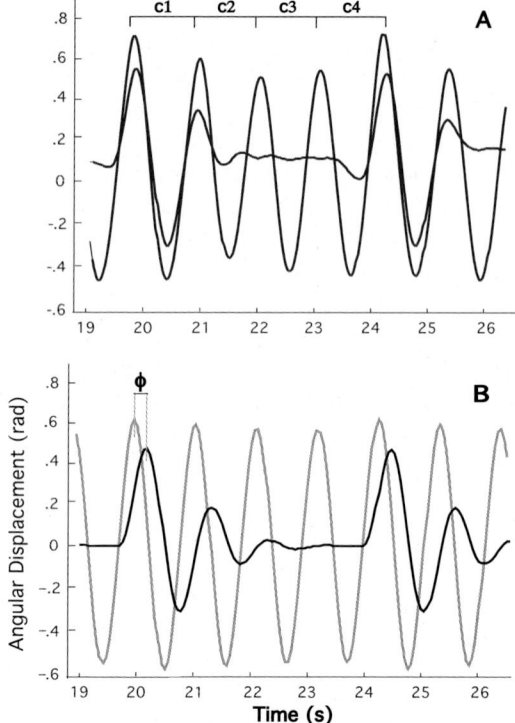

Fig. 1. Time series of the bimanual task. (**A**) Exemplary time series of one participant showing the left and right-hand for six cycles; the right hand performs continuous cycles, the left hand performs one cycle in synchrony with every fourth continuous cycle. (**B**) Simulated time series using different eigenfrequencies of the two pendulums. This results in a small phase difference between the peaks of the two simultaneous oscillations that scales with the magnitude of the eigenfrequency difference similar to the experimental results

The departure from an invariant spatiotemporal pattern posed new problems for modeling. Some mechanisms had to be introduced that initiated and terminated the single discrete event. Further, this new mechanism should not affect the relative phase as it results from the biomechanical differences between moving limbs. Hence, in line with task dynamics by Saltzman and Kelso [48], a new layered architecture was introduced to minimally account for these effects. Figure 2 schematically displays the three-tiered model structure. The biomechanical level expresses the known pendular limbs with their measurable eigenfrequencies. This level is separated from the internal level that schematically represents internal driving of the pendular effectors by self-sustained oscillations [37]. Each hand is governed by its own oscillator to allow for different movements in each hand. The specific representations of this internal driving are the hybrid oscillators as developed by Haken, Kay and colleagues [21, 29],

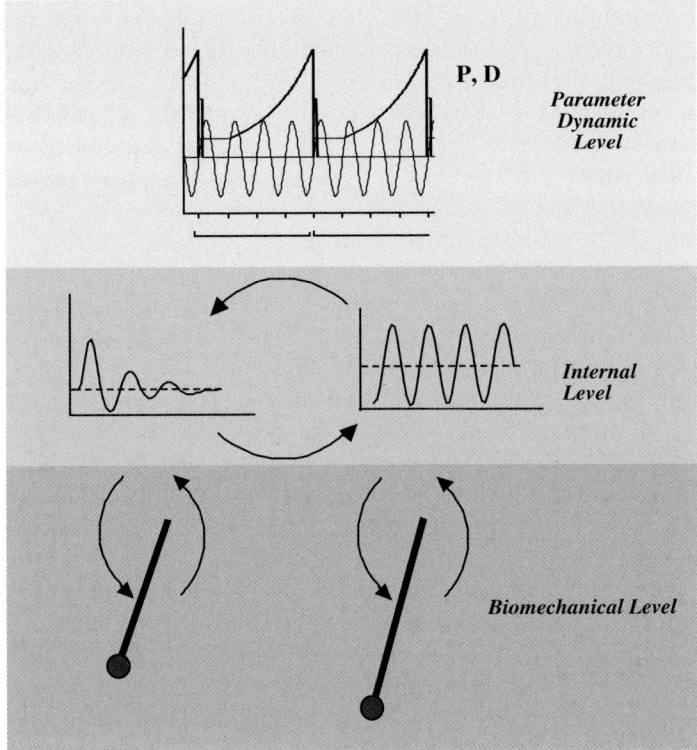

Fig. 2. Three-tiered model. The parameter dynamic level shows the schematic time series of the parameters stiffness (P) and damping (D) of the oscillatory units at the internal level. Stiffness P has a pulse increase of short duration, followed by an exponential increase of the damping parameter D from a rest level; D is reset after four cycles to its rest value. The timing of this dynamics is synchronized with the continuous oscillation. The internal level consists of two oscillators, one for each hand. The left hand's oscillation is initiated by the resetting of P and gradually terminated by the damping dynamics. The right-hand oscillator is at steady state. The two units are coupled symmetrically via the same coupling terms as the HKB model. The biomechanical level consists of two pendulums with the known nonlinear pendulum equations. Each pendulum is bidirectionally coupled to one oscillator at the model level. The two pendulums can have different eigenfrequencies giving rise to systematic phase differences. The driving from the oscillator to the pendulum is via position difference between the oscillator and the pendulum, an equilibrium point-type control, the feedback coupling is via velocity difference (for details see [62]). The system output is measured from the biomechanical pendulums

although alternative formulations for autonomous oscillators may be used. To address the new problem of initiation and termination of one cycle, a higher level has to be introduced at which the model parameters change at their own time scale [49]. To obtain a single discrete cycle, the parameters stiffness and damping of one oscillator are given a simple dynamics, a pulse onset and an exponential increase, that is phase-locked into the continuous oscillations.

Yet, the non-rhythmic task was still a cycle and was therefore best generated by an oscillator. As such, the step away from continuous rhythmic coordination was modest. Therefore, Wei et al. [71] conceived a different bimanual task where participants performed one translatory movement while the other arm performed a continuous rhythmic movement. Both arm movements were elbow rotations in the horizontal plane. The central difference to the task above was that the discrete displacement was triggered at all phases of the ongoing oscillation. A metronome produced a trigger signal upon which subjects had to initiate the discrete movement as fast as possible. The objective was to examine the nature of the constraints between arms when they performed different, i.e., rhythmic and discrete, movements. Can a discrete movement be initiated at all phases of the rhythmic movement? Or does the initiation phase-lock with the ongoing rhythmic movement? In the control task participants combined the ongoing oscillation with initiating a rhythmic movement, again triggered at all phases. Figures 3A and 3B illustrates the two tasks with exemplary time series. Kinematic and also electromyographic analyses extracted the mutual influences that the two movements exerted onto each other and contrasted the discrete–rhythmic with the rhythmic–rhythmic bimanual coupling. The central result was that initiation of a rhythmic movement was constrained to occur inphase with the ongoing rhythmic movement and the two limbs remained coupled in an inphase mode across the duration of the trial. In contrast, a discrete movement could be initiated at any arbitrary phase of the ongoing rhythmic movement. Figures 4A and 4B shows histograms of the phases of initiation (onset phase) pooled over all subjects for the discrete and the rhythmic initiation. Although the phase of the trigger signal was uniformly distributed across all phases, the rhythmic initiation clearly has a non-uniform distribution and favors onset phases around zero to $\pi/4$ rad. However, the discrete movement was not completely uncoupled from the rhythmic movement as seen by a scaling of the peak velocity with the oscillatory period (four periods were examined).

These results gave further support that the coupling between two rhythmic movements is strong, even before the secondary movement is initiated. As in the pendular task, initiation of a secondary rhythmic movement is constrained to be inphase which is indicative of coupling effects that exist at some higher level prior to initiation. Discrete movements showed different and less strong signs of coupling with the ongoing rhythmic movements. This difference points to the hypothesis that discrete and rhythmic actions are controlled differently, a question that will be further pursued in brain imaging experiments below.

To further unravel the neural substrate of the coupling between bimanual movements, the same experimental task was tested with three individuals lacking a corpus callosum and three age-matched healthy controls [64]. In contrast to the control participants the acallosal individuals failed to show phase entrainment at the initiation of the secondary rhythmic movements

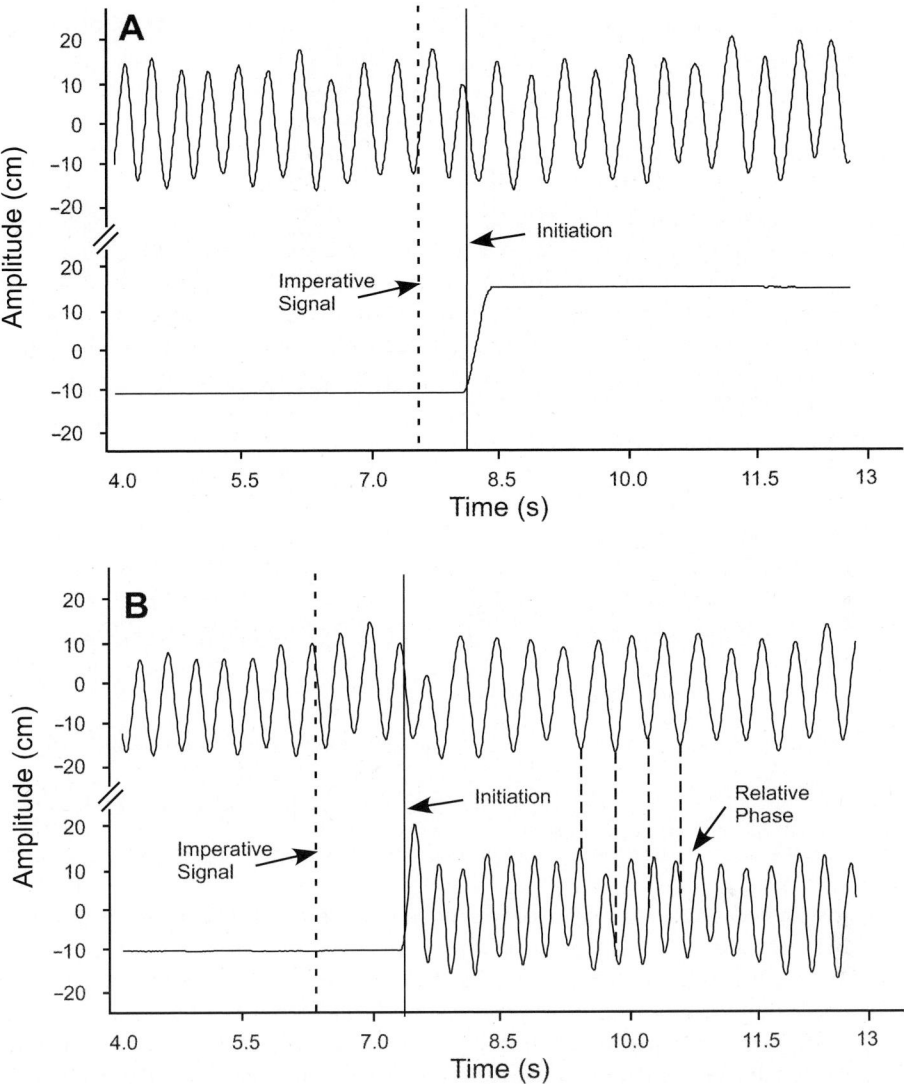

Fig. 3. Time series of position of the two hands performing one continuous oscillation and one discrete movement. (**A**) Exemplary time series of a healthy individual initiating a discrete movement as fast as possible following an imperative signal at an arbitrary phase; the instruction also emphasized to maintain the continuous oscillation as best as possible. (**B**) Exemplary time series of two hands of one individual without a corpus callosum. The secondary rhythmic movement was similarly triggered to occur at an arbitrary phase. For this acallosal individual the rhythmic movement was initiated not inphase with the ongoing rhythmic movement and also drifted with respect to the ongoing movement during the bimanual continuation

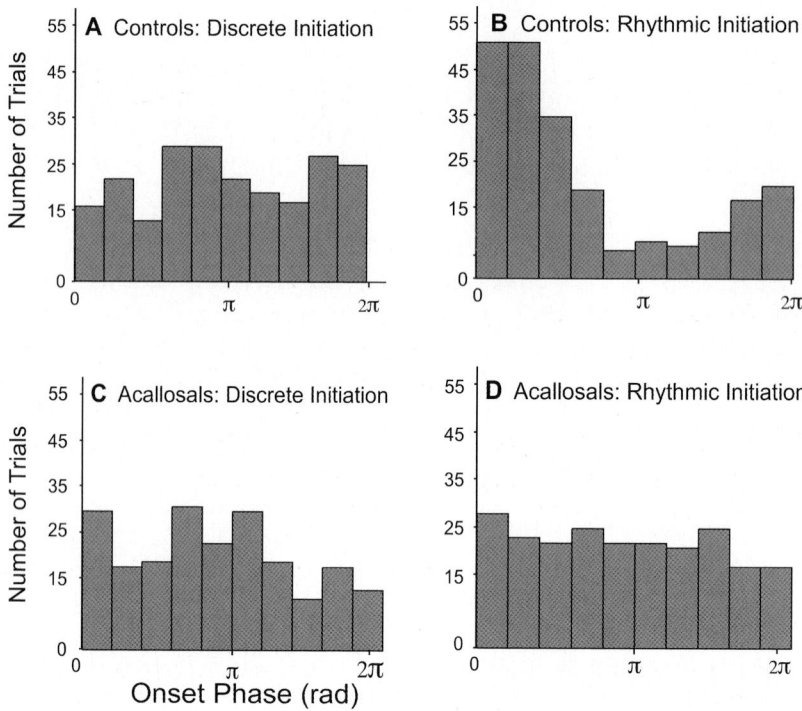

Fig. 4. Histograms of onset phases of the secondary movement pooled over all subjects. (**A, B**) Distributions of onset phases for healthy individuals in discrete and rhythmic initiation. (**C, D**) Distributions of onset phases for acallosal individuals in discrete and rhythmic initiation

and exhibited weak coupling between the rhythmically moving limbs. Figure 3B shows such an exemplary time series of an acallosal individual that shows no entrainment, neither at the moment of initiation, nor at the continued bimanual oscillation. Figures 4C and 4D shows the histograms of the onset phases of the acallosal individuals for both tasks. Unlike the healthy individuals the distributions for both tasks are approximately uniform just like the imperative signal. Additional analyses of the perturbations of the continuous hand revealed that in acallosal participants the continuous movement was perturbed in both secondary movement conditions, apparently indicating interhemispheric coupling at some level lower than the corpus callosum. These results are consistent with the hypothesis that intermanual interactions arise from various levels of control, and that these are distinct for discrete and rhythmic movements. Temporal coupling during rhythmic movements arises in large part from transcallosal interactions between the two hemispheres. The imposition of a secondary movement may transiently disrupt an ongoing rhythmic movement even in the absence of the corpus callosum. This may reflect subcortical interactions associated with response initiation.

These results did not agree with results by Tuller and Kelso [67] who reported similar stable coordination of inphase and antiphase patterns in one split-brain individual compared to a healthy control participant in continuous rhythmic coordination. Besides different severity of the lesion, a reason for this difference to our results may be that Tuller and Kelso's subject was paced by a metronome while the subjects in our two studies were only initially paced to control the frequency. Further, Tuller and Kelso used tapping movements where the haptic contact may have provided additional stabilization. This fact that perceptual information, both auditory and haptic, stabilizes rhythmic synchronization has been demonstrated later by Kelso and colleagues [33]. This work showed that transitions occurred for higher frequencies when both haptic and auditory information was synchronized with flexion. With a different focus of interest Ivry and colleagues reported that tapping movements with intermittent contacts have different control demands than continuous rhythmic movements without such contact events [56, 57]. This is concluded from the data from cerebellar patients who show similar variability in tapping movements than controls, while variability in continuous rhythmic movements is higher than in healthy subjects. They attribute this observation to the fact that haptic contacts provide temporal goals that the cerebellum utilizes in movement coordination. Taken together, these results highlight some of the potential neural substrates that are involved in rhythmic bimanual coordination. Further, seemingly small features of a movement, such as contacts or stops, may enhance the coupling by engaging additional neural circuits.

4 From Interlimb to Intralimb Coordination: Combination of Discrete and Rhythmic Elements in Single-Joint Movements

Another step away from purely rhythmic bimanual coordination was taken in a study by Sternad et al. [60]. This experiment investigated a single-joint elbow rotation that combined a discrete and a rhythmic component with the objective to investigate the interaction between discrete and rhythmic movement elements. Participants performed elbow flexions and extensions in the horizontal plane, oscillating at a prescribed frequency and amplitude around one target; upon a trigger signal they shifted the oscillation to a second target as fast as possible, without stopping the oscillation. As in the experiments above, the trigger signal occurred at all phases with equal probability across trials, leading to the expectation that the discrete shifts would also be initiated at all phases with equal probability. Figure 5 illustrates the task with exemplary kinematic and electromyographic data. Analyses focused on extracting the mutual influences of the rhythmic and the discrete components, very much in the spirit of Goodman and Kelso [19]. Only, this task involved voluntary rhythmic movements with larger amplitudes and lower frequencies than tremor. Yet, as seen for tremor, the results showed that the rhythmic

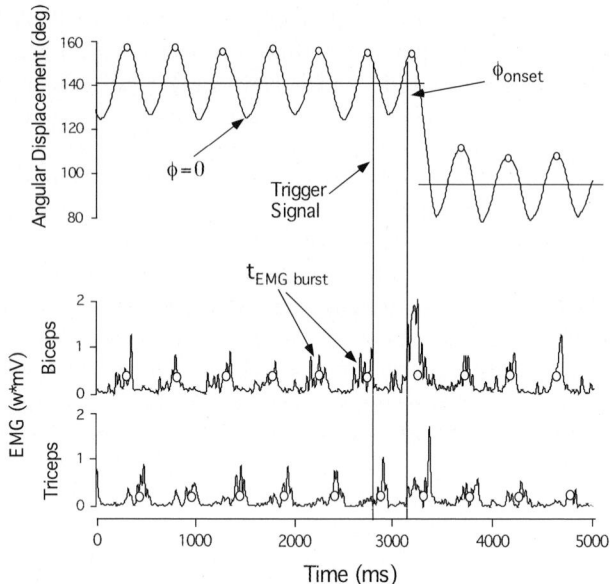

Fig. 5. Exemplary data showing the kinematic signal of the elbow movements and the electromyographic recordings of biceps and triceps. The time of the trigger signal precedes the onset of the discrete movement. The phase of the discrete movement's onset is calculated from supra-threshold activity of the EMG signal of the agonist (biceps) and then projected onto the kinematic signal. In alternative calculations the onset phase was determined within the EMG signal, where phase zero was defined at the moment of the center of mass of each EMG burst (indicated by white circles)

cycle acted as a constraint for the onset of the discrete movement and the initiation was confined to a limited phase window. In addition, our analysis of the discrete movement revealed that its duration was influenced by the period of oscillation; further, the oscillation was "perturbed" by the discrete movement as indicated by phase resetting. Analysis of the timing of the EMG bursts showed that the discrete muscle activation was closely "synchronized" with the muscle activation of the rhythmic movement in the same direction. These results are comparable to those of related studies although their interpretation and modeling differs [1, 58, 72]. Our interpretation of these findings is that constraints reside at the muscular or higher neural level in the form of mutual inhibition across antagonistic muscles.

This interpretation was expressed in a model that consisted of two pattern generators, one for the discrete and one for the rhythmic movement. To introduce the notion of inhibition a neurophysiologically motivated oscillator model was used, the half-center oscillator of Matsuoka, which mimics two mutually inhibiting neural units [41, 42]. For the discrete displacement a point attractor dynamics was developed close to the VITE model [6]. The two units were coupled to mutually inhibit each other. This meant that, for example, when the extensor muscle is active, the flexor is inhibited and cannot be activated for the discrete flexion movement. The beginning of the discrete movement is suppressed until the agonist is activated as part of the rhythmic movement. Simulation results replicated the kinematic effects recorded in human performance.

These results and interpretations, however, highlighted different mechanisms than Goodman and Kelso who had argued that the motor system explicitly takes advantage of the dynamics of the system utilizing the moment of maximum momentum for initiation. One possible reason for this difference was that the oscillation frequencies were different; another reason was that the different studies based their analyses either on kinematic or on electromyographic signals. Not only are the calculation details different for these two kinds of data, but also the landmarks in these two signals are phase-shifted with respect to each other, with different relative phases for different frequencies. To address these issues, we conducted the same experiment again where the rhythmic movements were paced at five relatively fast frequencies [61]. Analyses based on EMG activity revealed that discrete initiation is constrained to occur at or slightly advanced to the burst of the ongoing rhythmic activity, in accord with previous studies. Analyses of kinematic data show further details of the rhythmic–discrete interaction: duration, peak velocity and the overshoot of the discrete movement varied systematically with the frequency of the rhythmic movement. Effects of the discrete onto the rhythmic component were seen in a phase resetting of the oscillation and a systematic acceleration after the discrete movement, which also varied as a function of the oscillation frequency.

The results on the discrete movement's onset were again consistent with an inhibitory mechanism that suppressed muscular activity of the discrete movement until it was synchronized with a rhythmic burst. Determining the corresponding onset phase in the kinematic signal, the increased EMG activity began half a cycle before the time of maximum momentum. Although this result appears incompatible with Goodman and Kelso's hypothesis, this conclusion needs to be taken with a grain of salt: While the initiation of the discrete movement may not be coincident with maximum momentum, the unfolding of the discrete movement itself may still utilize this momentum.

In continuation of this line of experiments de Rugy and Sternad [8] extended the behavioral task with three variations in order to further test the hypothesis that two pattern generators, one discrete and one rhythmic, underlie more complex movement sequences. Three combinations of discrete and rhythmic elements were designed (Fig. 6): (1) a discrete movement that displaces the midpoint of oscillations (MID), identical to the task above; (2) a discrete change that increases the amplitude of oscillation without changing the midpoint of oscillation (AMP); (3) a combination of MID and AMP where the center of oscillation is displaced together with an increase in amplitude (MID+AMP). In the experiment subjects performed single-joint elbow oscillatory movements and initiated, upon a randomly timed signal, one of the three

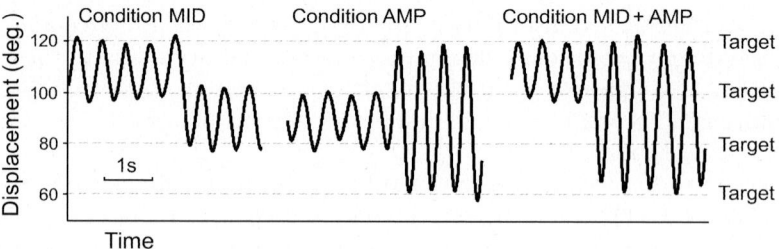

Fig. 6. Three task variations that combined rhythmic and discrete elements in different ways

tasks. The discrete movements were performed either in a reaction time or in a self-paced fashion. Figure 7 summarizes the phases of onset in histograms for the three tasks under both instructions. The preference to initiate inphase, i.e., around 0 or 2π rad, with the ongoing rhythm is again visible. In fact, when subjects were not forced to respond as fast as possible, it is evident that they avoided a wide range of phases around π rad of the ongoing rhythmic movement. In replication of earlier results, EMG burst synchronization was found similarly in all three tasks. In the self-paced instruction the synchronization was even more pronounced. Reaction times were longest for the combined task (MID + AMP), indicating higher control demands due to a combination of two types of changes.

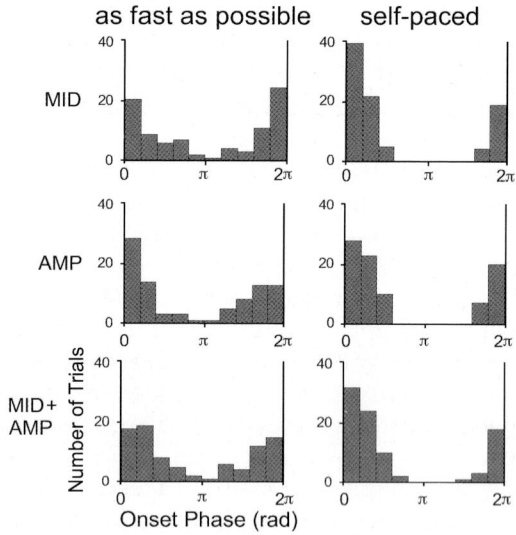

Fig. 7. Histograms of onset phases for the three different tasks (MID, AMP and MID + AMP) and two instructions that set different timing constraints for the initiation: "as fast as possible" and "self-paced"

To simulate both the recorded kinematic and electromyographic activity for the three task variations a three-tiered model was used consisting of an internal level, a biomechanical level and a parameter dynamic level oscillator similar to the first model for bimanual coordination (Figs. 2 and 8). Including elements from the model of Sternad et al. [60], the pattern generator at the internal level was now the Matsuoka half-center oscillator as it captures the alternating activation of flexor and extensor muscle groups in rhythmic behavior (Fig. 8A). The output of the oscillating unit is governed by a simple dynamic of the parameters, similar to what was developed for the bimanual task earlier. A continuous oscillation was generated by a step change of the activation parameter to a constant magnitude; a discrete activation was initiated by a pulse onset of the same parameter followed by an exponential decay to achieve an appropriately timed termination (Fig. 8B, top panel). Rhythmic activation produced a sequence of alternating flexor and extensor bursts; discrete output shows three bursts similar to the commonly observed triphasic EMG activation (Fig. 8B, middle panel). This Matsuoka output was then interpreted as muscle activations or torques driving a simple single-joint mechanical limb. The final output of the pattern generator was either a continuous oscillation or a discrete output that resembled the typical rhythmic and discrete trajectories (Fig. 8B, bottom panel).

The three types of tasks were simulated by a simple unweighted summation at the parameter level; the sum of the discrete and rhythmic activation signals then drives the Matsuoka oscillator; the output of the Matsuoka oscillator drives the limb. Figure 9 illustrates the basic output for the three parameterizations. The simulated trajectories resemble the human trajectories for the three tasks, although some caveats apply (for discussion see [8]). Yet, analysis of the onset constraints in the simulated data reproduced the major results from the human data. Most centrally, the constraint of the phase of initiation of the discrete movement was reproduced. Note that, different to above, the notion of two task elements was realized as two components at the parameter level, while only one pattern generator produced the output for the single-limb task. Hence, no explicit inhibitory coupling between the discrete and rhythmic components was present. Instead, the two state variables of the Matsuoka oscillator itself instantiated the notion of mutual inhibition and the combined input from the parameter level was subject to these constraints.

These experiments and simulations emphasized the advantages of the layered model structure: The internal level represents basic pattern generators that are invariant and produce stable steady state output depending on their parameterization. By the separation between internal and biomechanical level subtle mechanical effects such as phase lags or filtering can be relegated to this passive dynamics without requiring explicit specification at the internal or parameter level. Initiation, timing and modulation of these

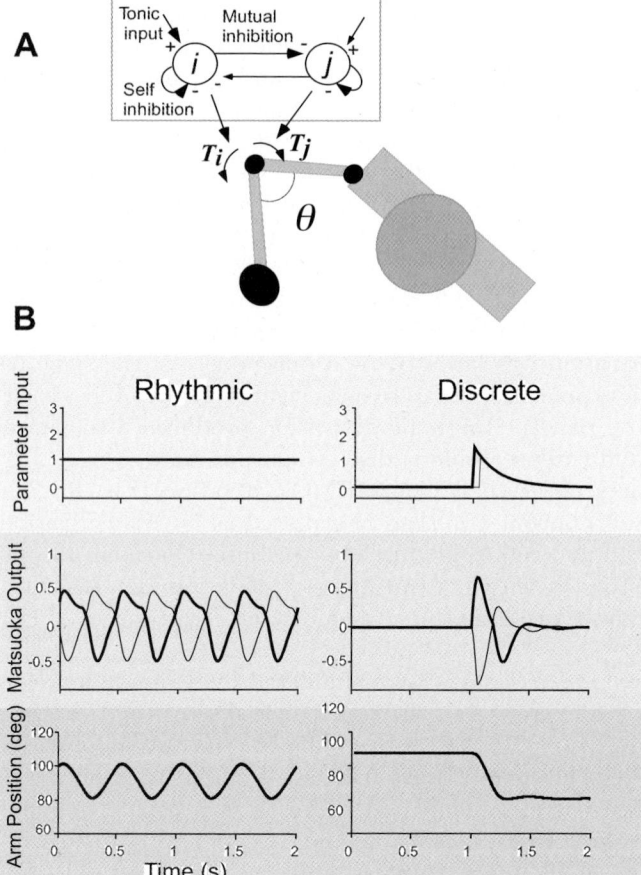

Fig. 8. Three-tiered model. (**A**) Illustration of the arm model as driven by the half-center oscillator, consisting of antagonistic units i and j. The outputs of this Matsuoka oscillator serves as torques, T_i and T_j, driving the agonist and antagonist of the elbow movement, θ. (**B**) Three-tiered model structure consisting of the level with parameter dynamics, the internal level with pattern generators (Matsuoka oscillators) and the kinematic output produced by the torque-driven arm model. The parameter input for the rhythmic unit is constant, leading to an alternation of activation in the Matsuoka units i and j. The torques T_i and T_j produce the trajectories of the model arm. For the discrete pattern generator, the parameter input consists of a step change followed by an exponential decrease leading to a triphasic burst pattern of the units i and j. Using these as torque input to the model arm a discrete displacement of the arm is generated

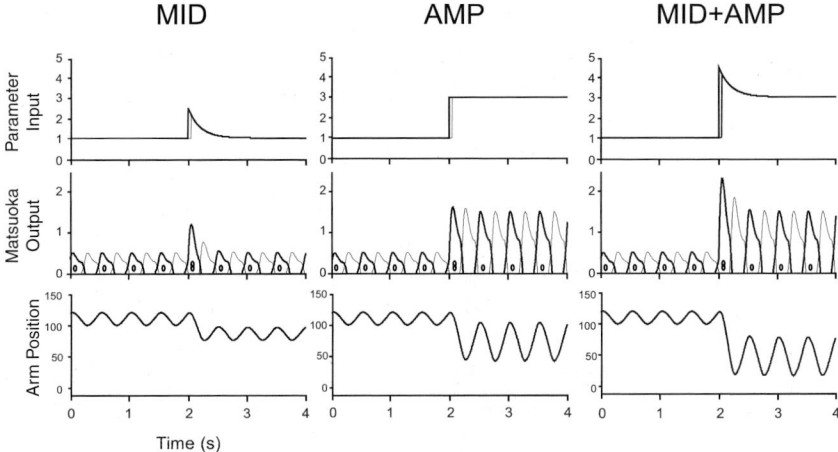

Fig. 9. Simulation of the three tasks using unweighted superposition of the two types of parameterization: The task MID is produced by the superposition of a constant activation with the discrete movement's pulse activation leading to a change in the midpoint of oscillation without effect on amplitude; the task AMP is produced by a step change from one constant activation level to a higher one, leading to a change in amplitude; the task MID + AMP is produced by the superposition of the step change in constant activation at the same time as the pulse activation for the discrete change of the midpoint of oscillation

pattern generators are governed by a simple dynamics in the parameters. The explicit specification of these parameters is the domain of higher levels of control.

5 The Role of Neural and Biomechanical Constraints in Multijoint Intralimb Coordination

With the objective to further explore the interplay of biomechanical and neural/internal constraints Sternad and Dean [59] investigated a two-joint task that combined discrete and rhythmic elements. The new challenge in two- or, generally, multijoint coordination is the presence of interaction torques exerted by one limb upon the other. Following Bernstein's insights several studies demonstrated that skilled actors utilize interaction torques and shape their endpoint trajectories accordingly [3, 47, 51]. However, these demonstrations were for relatively simple, usually discrete movements. One open question is whether such interaction torques still pose the predominant constraints for more complex movements? Are physiological constraints such as those identified for discrete and rhythmic movements subordinate to these biomechanical constraints? Alternatively, if phase constraints are observed, are they due to the exploitation of interaction torques? This latter conjecture is close in spirit to what Goodman and Kelso speculated for single-joint movements [19].

The task consisted of an elbow oscillation in the plane that was to be merged with a fast discrete adduction or abduction in the shoulder triggered by an auditory signal. Put into more ecological terms, subjects were instructed to "rub" a spot at the table surface with a small amplitude at a given frequency, as in cleaning the table. Upon a trigger signal, they were asked to change the focus of the oscillation to another spot marked on the table. The translatory movement between the two target "spots" required a change in shoulder angle of 40°. The rubbing oscillations were not restricted in their direction, a feature that made the task redundant. Typically, though, the oscillations were confined to the elbow joint. Two hypotheses were tested: (1) The observed kinematic constraints between discrete and rhythmic elements are identical to those documented for the single-joint task; hence, they arise at a neural level. (2) The initiation of the discrete shoulder movement is constrained by biomechanical factors; specifically, intersegmental torques exerted from the forearm onto the upper arm are utilized to perform the task.

The results provided support for Hypothesis 1. Without explicit instruction the oscillatory and translatory components were separated into the shoulder and elbow joints, respectively. In replication of the single-joint results the initiation of the discrete shoulder action was constrained to a preferred phase of the ongoing elbow, i.e., shoulder oscillation (Fig. 10A). Additionally, the rhythmic movement showed a systematic phase resetting during the discrete shift even though these elements occurred in different joints. Reaction times of the discrete movement were relatively long and peak velocities were slower than reported for isolated discrete movements, due to the simultaneous presence of the oscillation. Finally and importantly, interaction torques acting onto the shoulder joint prior to onset and during the accelerating segment of the discrete movement did not assist the movement. This conclusion was derived from the comparative analyses of onset phases and the prevalent torques acting on the shoulder at that phase. Figure 10B illustrates the three torque components, interaction torque, muscle torque and net torque acting upon the shoulder during one elbow cycle (zero phase is maximum extension). If interaction torques were utilized, as predicted by Hypothesis 2, then the onset phase ϕ for discrete shoulder abduction should favor values between $\pi/2$ and $3\pi/2$ rad (Fig. 10B) as the torque values are positive in the direction of abduction. Given that the modal distribution in the histogram of onset phases is around $3\pi/2$ rad, ranging between π and 2π rad, this expectation was only marginally supported. During half of the range of the preferred onset phases, the interaction torques worked against the direction of the discrete abduction movement. However, Hypothesis 2 is supported indirectly. While the shoulder angle showed only negligible angular displacements, the EMG activity of the prime movers showed rhythmic co-activation, counteracting the torques exerted by the elbow oscillations upon the shoulder. This rhythmic muscle activation then became the constraining EMG activity that limited

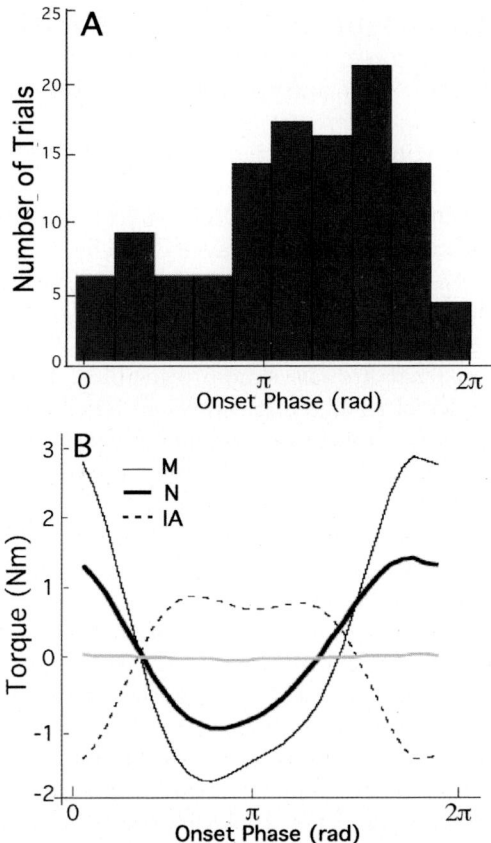

Fig. 10. Results of the kinematic and kinetic analyses of discrete onset of shoulder movement during elbow oscillations. The reference phase was taken from the elbow joint cycle where 0 rad corresponds to maximum extension, π rad refers to maximum flexion. (**A**) Histogram of the onset phase of the discrete shoulder movement, determined as a function of the phase of the elbow joint rotations. (**B**) Mean torque profiles during one cycle of the elbow oscillation onto the shoulder prior to the discrete movement in the shoulder. IA symbolizes interaction torques, M muscle torque and N net torques

the onset of the discrete movement in the same fashion as observed in the previous studies.

Overall, the results were not consistent with the notion that the nervous system makes use of interaction torques to facilitate movements. Rather, in more complex tasks additional constraints arise from different task elements or control signals that interact at a neural level. It is unclear at this stage whether these interactions occur at the internal or the parameter dynamic level. More experiments are needed to detail the type and locus of neural interactions.

6 From Complex Behaviors to the Primitives of Action

Thus far, the behavioral experiments explored interactions between rhythmic and discrete elements in different task contexts, with the aim to detail the kinds and possible origins of these interactions. The more general objective was to incrementally move beyond the widespread focus on purely rhythmic or purely discrete movements and thereby create experimental tasks that come closer to the complexity seen in everyday skills. The premise was to keep the explanatory units as simple as possible and introduce only simple control elements at higher levels, i.e., the parameter dynamics. This route of modeling directly connects to another core question: what are the basic units or primitives of action and what are the simple principles? If such primitives were found, basic principles of their interactions can be explored, leaving minimal demands to executive control. Given the fundamental nature of this question a number of suggestions have been made: Inferring from animal experiments, for example Giszter and colleagues have suggested force fields generated by spinal cord compartments to serve as primitives [4, 18, 24]. Relatedly, synergies have been proposed as primitives by numerous researchers who inferred these from electromyographic data [7, 65].

These propositions for primitives are grounded in a physiological level of analysis and tie their notion of primitive to some physiological substrate. A dynamical systems perspective suggests another more abstract approach: a primitive defined by an organizing principle. One evident candidate for such a fundamental unit of organization is an oscillation. It was Iberall and colleagues who started to posit cyclicities at all temporal and spatial scales, as the core piece of their theorizing about biological systems [26, 27]. Ample support has been provided for the notion that the nervous system assembles stable limit cycle attractors to generate rhythmic movements. However, as illustrated above, there are many behaviors that may be difficult to generate by assuming oscillations as the one and only primitive. From the same nonlinear dynamics perspective, it is straightforward to suggest fixed-point attractors as another type of pattern generator for goal-oriented discrete movements, parallel to and complementing limit cycle attractors. In fact, this notion is at the core of the equilibrium point hypothesis which has attracted much attention and discussion. Are point attractors and limit cycle attractors two distinct primitives from which complex actions are built? Alternatively, some of the modeling above implied that discrete movements are damped oscillations, parameterized by an additional parameter dynamics, but generated by the same type of pattern generator. Yet, a third option, not elaborated above, is that rhythmic and discrete movements are composed of the same discrete elements, where continuous rhythmic movements are generated by concatenating discrete elements. This latter hypothesis has been implied or explicitly stated in cognitive or computational points of view that concentrated their efforts to discrete reaching movements [1, 11, 16].

7 Rhythmic and/or Discrete Primitives: Brain Imaging Results

These three alternative hypotheses were tested in a functional neuroimaging experiment where participants performed rhythmic and discrete wrist movements in a fMRI scanner [50]. If rhythmic and discrete movements are independent primitives, then each type of action should have cerebral structures dedicated exclusively to them. Of course, some areas like the primary motor areas, should be active in all movements, but discrete and rhythmic movements should have their own independent circuitry. Miall and Ivry [44] schematized this pattern of brain activation by a Venn diagram with two sets with only partial overlap. Second, if discrete movements (D) are truncated cyclic movements (R), then discrete activation should be equal to or a superset of rhythmic activation: $R \subseteq D$. Third, if rhythmic movements are a special case of rhythmic movement such as a concatenation of discrete movements, then brain activity in rhythmic movements should be equal to, or a superset of discrete movements due to the extra demands of sequencing: $R \supseteq D$. In the Venn diagram discrete activation should be a subset or superset of rhythmic activation, respectively.

The experiments compared the cerebral activation for single-joint wrist movements that were either performed in a continuous rhythmic fashion or initiated at random times as discrete single flexions or extensions. Using a block design participants performed the rhythmic wrist movements at their self-chosen comfortable frequency for the duration of 30 s. For the discrete movements subjects initiated and terminated wrist flexions and extensions at random self-selected times, again for 30 s. The results identified two different networks in cortical and subcortical regions for the two types of action (Fig. 11): rhythmic movements mainly activated contralateral areas in the sensorimotor cortex, the supplementary motor area, the caudal cingulate cortex and the ipsilateral cerebellum. In contrast, discrete movements induced additional neural activity in the prefrontal and parietal cortices; especially the dorsal premotor cortex, BA7 and BA47 were significantly activated only in discrete movements. The activations in the cerebrum and the cerebellum were generally more bilateral for discrete movements.

Given the extensive activation areas during discrete movements, hypothesis $R \supseteq D$, that discrete movements are fundamental and rhythmic movements are a concatenation of discrete strokes, was ruled out. The finding that rhythmic movements activated regions that were largely a subset of those needed for discrete movements was consistent with the second hypothesis, $R \subseteq D$. However, there was also exclusive activation of caudal cingulate cortex and the supplementary motor area that supported the first hypothesis that rhythmic movements constitute a separate class of movements with its own dedicated network. Taken together, the one message is that Hypothesis 3 can be ruled out. Whether discrete movements can be considered truncated

Axial Slice at z = +48 mm

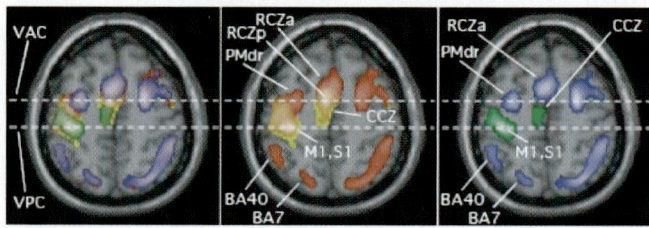

Fig. 11. Brain activity in rhythmic and discrete performance superimposed on a Montreal Neurological Institute (MNI) brain template (SPM). Four different statistical comparisons are displayed in different colors. *Rhythmic-Rest* (yellow) and *Discrete-Rest* (red) in the middle plot demonstrate the main effects of brain activity during *Rhythmic* and *Discrete* movement conditions. *Rhythmic-Discrete* (green) shows brain areas where rhythmic movements have stronger activity than discrete movements (only areas that were active in *Rhythmic-Rest* were included in this contrast). *Discrete-Rhythmic* (blue) displays areas that showed significantly more activation than rhythmic movement, using *Discrete-Rest* as a mask. The right plot shows only the *Rhythmic-Discrete* and *Discrete-Rhythmic*. The left plot superimposes the activities from the other plots to allow an easy comparison of activation locations. Abbreviations: RCZa and RCZp: anterior and posterior rostral cingulate zone; CCZ: caudal cingulate zone; PMdr: dorsal premotor area; M1: primary motor cortex; S1: primary sensory cortex; BA: Brodmann area; VAC: vertical line perpendicular to the AC-PC, passing through AC; VPC: vertical line perpendicular to the AC-PC, passing through PC

rhythmic movements or rather a second primitive could not yet be answered conclusively.

A follow-up experiment aimed to replicate these findings and conducted additional control conditions to disambiguate the remaining two hypotheses. To this end finger movements were performed to auditory trigger signals at two periods (500 ms, 1100 ms) presented either in a periodic fashion or with pseudo-randomized inter-stimulus intervals [76]. The results were generally consistent with the previous findings: rhythmic movements were associated with contralateral primary motor areas, and non-rhythmically paced movements implicated additional areas, corroborating the significant differences between rhythmic and discrete control structures. In contrast to previous results, however, activation in the pre-SMA and the anterior cingulate area was specific to non-rhythmic performance. Moreover, there was very little activation exclusive to rhythmic movements, suggesting that rhythmic movements have similar control demands as discrete movements with the latter having additional requirements of initiation and termination. This conclusion is consistent with the modeling above. However, there was also indication of activation of the lower midbrain or brainstem in voluntary rhythmic movements. While thought-provoking, these results need to be interpreted with care, as resolution and recording artifacts may interfere at these lower levels of the brain. More experimentation is required to address these issues.

One more experiment tested the idea that it is not periodicity but the starting and stopping that differentiates the two types of movements. Parkinson patients and age-matched controls performed thumb pressing movements paced by a metronome set to a relatively fast and a very slow pace [77]. Without any explicit instructions participants performed the slow movements with relatively long stops separating successive movements into aperiodic discrete movements, in contrast to the faster movements. For healthy subjects the slow movements elicited the same circuitry of activation as the randomly initiated discrete movements. This study on elderly subjects corroborated the previous findings.

Across all these experiments, one message is clear: discrete movements activate significantly larger areas than rhythmic movements. But what aspects of such seemingly simple movements require this significantly more extensive and different activation? Is it the starting and the timed stopping? The involvement of parietal and prefrontal areas signals that more "executive" control is required as these areas are attributed to be the seat of higher cognitive functions. With a view to the suggested three-tiered model (Figs. 2 and 8), the level of parameter dynamics is the correlate of such higher functions. For both model variants, the parameterization for the single discrete cycle and for the discrete output of the Matsuoka oscillator required more complex and accurately timed changes in the parameters. This modeling is therefore consistent with the finding that more cortical areas are active.

Again, what differentiates rhythmic from discrete movements? What if initiation and termination recur in a periodic fashion? Is a sequence of cycles, separated by pauses still a rhythmic movement? What if these pause events become extremely short as in tapping or drumming? Is periodicity the hallmark of rhythmic movements or is it the deviation from strict periodicity that makes movement rhythmic? This plethora of questions and the associated confusion of terms was the motivation for Hogan and Sternad [25] to attempt a formal classification of rhythmic and discrete movements. Adopting the level of the observable movement outcome as the basis for their mathematical classification, Hogan and Sternad define discrete movements as those that are bounded by postures, implying that all derivatives go to zero. Rhythmic movements, on the other hand, have as their minimum requirement recurrence. Given that there are different degrees of departure from strict periodicity, "rhythmic" is suggested as an umbrella term for several subclasses. Although many movements can thereby unambiguously be classified as either rhythmic or discrete, there are also many variants in natural behaviors where discreteness and rhythmicity shade into each other. To evaluate such behaviors a smoothness criterion is introduced that permits quantification of discreteness and rhythmicity. One important outcome of this analysis is that some rhythmic movements cannot be concatenations of back-to-back discrete movements. This is consistent with inferences drawn from empirical studies [5, 20, 55, 70].

8 Towards a Theoretical Framework for Complex Actions

Natural behavior manifests itself as a continuous flow of actions where these actions may be rhythmic, discrete, or a combination of both. Fundamental locomotory actions in both legged and also non-legged animals are rhythmic; manual skills such as reaching and grasping are typically goal-directed and discrete with a beginning and an end. These two are the well-studied classes of behavior with a lot of insightful modeling work. For more complex and richer types of behaviors principled theoretical approaches are still missing. This realization has been the motivation for our research program that adopted a dynamical systems framework extending from the rich theoretical framework developed for rhythmic interlimb coordination. Originating in the seminal work of Kelso and Turvey, our empirical and theoretical strategy aimed to extend and complement this scientific body of research to address non-rhythmic behavior. Similar to Kelso's research program we adopted multiple levels of analysis, including kinematic, kinetic, muscular and fMRI data complemented by modeling work, with the objective to gain insights into the principles of movement generation in biological systems.

A series of behavioral studies highlighted that the co-existence of rhythmic and discrete movement elements is subject to different constraints compared to purely rhythmic coordination. Inhibitory coupling between discrete and rhythmic components at the neural/muscular level was indicated in a series of uni- and bi-manual single- and multi-joint coordination. These behavioral observations pointed to the fact that rhythmic and discrete movements are different and have different demands on the central nervous system. It seems likely that discrete movement cannot be reduced to rhythmic movements and vice versa. Several modeling propositions made explicit how such a co-existence of two behavioral elements and their interaction can be understood. Common to the proposed models is that a minimal account for such asymmetrical tasks required two additional features compared to the HKB model architecture: First, a biomechanical level should be separated from an internal level; second, a parameter dynamic level to represent higher-level timing and modulation functions that generate initiation and termination.

A deep issue directly following from this empirical and modeling work is whether all the pattern generators postulated for the internal level should be reduced to one primitive or whether there should be two or more. To obtain an answer fMRI studies compared brain activation in rhythmic and discrete movements. The results are unambiguous and consistent with the modeling: activations for the two behavioral tasks are strikingly different, giving support for the notion of two primitives. The more extensive activation elicited by discrete movements speaks to the increased control demands that were highlighted in the modeling of discrete elements.

Given the infinite variety of natural behaviors a theoretical framework is needed that permits a principled foundation to examine them in their variety. Such framework should enable researchers to investigate arbitrary tasks that go beyond simple controlled experimental tasks, and approach them in a principled fashion. We hope that our research stimulates the development of such a framework.

Acknowledgments

This research was supported by grants from the National Science Foundation (SBR 97-10312, BCS-0096543 and PAC-0450218), the Tobacco Settlement Funds 4100020604 and the National Institutes of Health (RO1-HD045639). I would also like to take this opportunity to thank all my collaborators that inspired me to develop different angles onto these topics and my students William Dean, Aymar de Rugy, Hong Yu and Kunlin Wei who developed this line of research with me. Finally and importantly, I would also express my sincere thanks to my advisors Michael Turvey and Elliot Saltzman who attracted and guided me towards this fascinating line of research.

References

1. Adamovich SV, Levin MF, Feldman AG (1994) Merging different motor patterns: coordination between rhythmical and discrete single-joint movements. Exp Brain Res 99(2):325–337
2. Beek PJ (1989) Juggling dynamics. Unpublished Doctoral Dissertation, Free University Press, Amsterdam
3. Bernstein N (1967) The coordination and regulation of movement. Pergamon Press, London
4. Bizzi E, Mussa-Ivaldi SA, Giszter SF (1991) Computations underlying the execution of movement: a biological perspective. Science 253:287–291
5. Buchanan JJ, Park JH, Shea CH (2006) Target width scaling in a repetitive aiming task: switching between cyclical and discrete units of action. Experimental Brain Research epub ahead of print.
6. Bullock D, Grossberg S (1991) Adaptive neural networks for control of movement trajectories invariant under speed and force rescaling. Hum Movement Sci 10:3–53
7. d'Avella A, Bizzi E (2005) Shared and specific muscle synergies in natural motor behaviors. Proc Natl Acad Sci 102(8):3076–3081
8. de Rugy A, Sternad D (2003) Interaction between discrete and rhythmic movements: reaction time and phase of discrete movement initiation against oscillatory movements. Brain Res 994:160–174

9. Fitts PM (1954) The information capacity of the human motor system in controlling the amplitude of movement. J Exp Psychol 47:381–391
10. Fitts PM, Peterson JR (1964) Information capacity of discrete motor responses. J Exp Psychol 67:103–112
11. Flash T, Hogan N (1985) The coordination of arm movements: an experimentally confirmed mathematical model. J Neurosci 5(7):1688–1703
12. Fuchs A, Jirsa VK, Kelso JAS (2000a) Issues in the coordination of human brain activity and motor behavior. Neuroimage 11(5):375–377
13. Fuchs A, Jirsa VK, Kelso JAS (2000b) Theory of the relation between human brain activity (MEG) and hand movements. Neouroimage 11(5):359–369
14. Fuchs A, Kelso JAS, Haken H (1992) Phase transitions in the human brain: spatial mode analysis. Int J Bifurcat Chaos 2:917–939
15. Fuchs A, Mayville JM, Cheyne D, Weinberg H, Deecke L, Kelso JAS (2000) Spatiotemporal analysis of neuromagnetic events underlying the emergence of coordinative instabilities. Neuroimage 12(1):71–84
16. Georgopoulos AP (1991) Higher order motor control. Annu Rev Neurosci 14:361–377
17. Georgopoulos AP, Kettner RE, Schwartz AB (1986) Neuronal population coding of movement direction. Science 233:1416–1419
18. Giszter SF, Mussa-Ivaldi FA, Bizzi E (1993) Convergent force fields organized in the frog's spinal cord. J Neurosci 13(2):467
19. Goodman D, Kelso JAS (1983) Exploring the functional significance of physiological tremor: a biospectroscopic approach. Exp Brain Res 49:419–431
20. Guiard Y (1993) On Fitts and Hooke's law: simple harmonic movements in upper-limb cyclical aiming. Acta Psychologica 82:139–159
21. Haken H (1983) Synergetics. An introduction. Springer, Berlin
22. Haken H (1987) Advanced synergetics. Springer, Berlin
23. Haken H, Kelso JAS, Bunz H (1985) A theoretical model of phase transition in human hand movements. Biol Cybern 51:347–356
24. Hart CB, Giszter SF (2004) Modular premotor drives and unit bursts as primitives for frog motor behaviors. J Neurosci 24(22):5269–5282
25. Hogan N, Sternad D (in press) On rhythmic and discrete movements: Reflections, definitions and implications for motor control. Exp Brain Res 181:13–30
26. Iberall AS (1972) Toward a general science of viable systems. McGraw-Hill, New York
27. Iberall AS, Soodak H (1987) A physics of complex systems. In: Yates FE (ed) Self-organizing systems: the emergence of order. Plenum Press, New York, pp. 158–173
28. Jirsa VK, Fuchs A, Kelso JAS (1998) Connecting cortical and behavioral dynamics: bimanual coordination. Neural Comput 10(8):2019–2045
29. Kay BA, Kelso JAS, Saltzman EL, Schöner G (1987) Space-time behavior of single and bimanual rhythmical movements: data and limit cycle model. J Exp Psychol Hum Percept Perform 13(2):178–192
30. Kelso JAS (1981) On the oscillatory basis of movement. Bull Psychonomic Soc 18(63):49–70
31. Kelso JAS(1984) Phase transitions and critical behavior in human bimanual coordination. Am J Physiol Regul Integrative and Comparative Physiology 15:R1000–R1004
32. Kelso JAS (1995) Dynamic patterns: the self-organization of brain and behavior. MIT Press, Cambridge, MA

33. Kelso JAS, Fink PW, DeLaplain CR, Carson RG (2001) Haptic information stabilizes and destabilizes coordination dynamics. Proc Biol Sci 268(1472):1207–1213

34. Kelso JAS, Southard DL, Goodman D (1979a) On the coordination of two-handed movements. J Exp Psychol Hum Percept Perform 5:229–238

35. Kelso JAS, Southard DL, Goodman D (1979b) On the nature of human inter-limb coordination. Science 203(4384):1029–1031

36. Kugler PN, Kelso JAS, Turvey MT (1980) On the concept of coordinative structures as dissipative structures: I. Theoretical lines of convergence. In: Stelmach GE, Requin J(eds.) Tutorials in motor behavior. North Holland, New York, pp. 3–47

37. Kugler PN, Turvey MT (1987) Information, natural law, and the self-assembly of rhythmic movement. Erlbaum, Hillsdale, NJ

38. Kuhn T (1962) The structure of scientific revolutions. University of Chicago Press, Chicago

39. Kurtzer IL, Herter TM, Scott SH (2006) Nonuniform distribution of reach-related and torque-related activity in upper arm muscles and neurons of primary motor cortex. J Neurophysiol 96(6):3220–3230

40. Latash ML (1993) Control of human movement. Human Kinetics Publisher, Champaign, IL

41. Matsuoka K (1985) Sustained oscillations generated by mutually inhibiting neurons with adaptations. Biol Cybern 52:367–376

42. Matsuoka K (1987) Mechanisms of frequency and pattern control in the neural rhythm generators. Biol Cybern 56:345–353

43. Mayville JM, Fuchs A, Ding M, Cheyne D, Deecke L, Kelso JAS (2001) Event-related changes in neuromagnetic activity associated with syncopation and synchronization timing tasks. Hum Brain Mapp 14(2):65–80

44. Miall RC, Ivry RB (2004) Moving to a different beat. Nat Neurosci 7(10):1025–1026

45. Naselaris T, Merchant H, Amirikian B, Georgopoulos AP (2006a) Large-scale organization of preferred directions in the motor cortex I: motor cortical hyperacuity for forward reaching. J Neurosci 96(6):3231–3236

46. Naselaris T, Merchant H, Amirikian B, Georgopoulos AP (2006b) Large-scale organization of preferred directions in the motor cortex II: Analysis of local distributions. J Neurophysiol 96(6):3231–3236

47. Sainburg RL, Ghez C, Kalakanis D (1999) Intersegmental dynamics are control by sequential anticipatory, error correction, and postural mechanisms. J Neurophysiol 81:1045–1056

48. Saltzman EL, Kelso JAS (1987) Skilled actions: a task-dynamic approach. Psychol Rev 94(1):84–106

49. Saltzman EL, Munhall KG (1992) Skill acquisition and development: the roles of state-, parameter-, and graph-dynamics. J Motor Behav 24(1):84–106

50. Schaal S, Sternad D, Osu R, Kawato M (2004) Rhythmic arm movement is not discrete. Nat Neurosci 7(10):1136–1143

51. Schneider K, Zernicke RF, Schmidt RA, Hart TJ (1987) Intersegmental dynamics during the learning of a rapid arm movement. J Biomech 20:816

52. Schöner G (1990) A dynamic theory of coordination of discrete movement. Biol Cybern 63:257–270

53. Shadmehr R, Brashers-Krug T, Mussa-Ivaldi FA (1995) Interference in learning internal models of inverse dynamics in humans. In: Tesauro G, Touretzky DS Leen KT (eds.) Advances in neural information processing systems 7, Morgan Kaufmann, San Mateo, CA, pp. 1117–1124

54. Shadmehr R, Wise SP (2005) Computational neurobiology of reaching and pointing: a foundation for motor learning. MIT Press, Cambridge, MA

55. Smits-Engelman BCM, Van Galen GP, Duysens J (2002) The breakdown of Fitts' law in rapid, reciprocal aiming movements. Exp Brain Res 145:222–230

56. Spencer RM, Ivry RB, Zelaznik HN (2005) Role of cerebellum in movements: control of timing or movement transitions? Exp Brain Res 161(3):383–396

57. Spencer RM, Zelaznik HN, Diedrichsen J, Ivry RB (2003) Disrupted timing of discontinuous but not continuous movements by cerebellar lesions. Science 300(5624):1437–1439

58. Staude G, Dengler R, Wolf W (2002) The discontinuous nature of motor execution II. Merging discrete and rhythmic movements in a single-joint system – the phase entrainment effect. Biol Cybern 86(6):427–443

59. Sternad D, Dean WJ (2003) Rhythmic and discrete elements in multijoint coordination. Brain Res 989:151–172

60. Sternad D, Dean WJ, Schaal S (2000) Interaction of rhythmic and discrete pattern generators in single-joint movements. Hum Movement Sci 19:627–665

61. Sternad D, de Rugy A, Pataky T, Dean WJ (2002) Interactions of discrete and rhythmic movements over a wide range of periods. Exp Brain Res 147:162–174

62. Sternad D, Saltzman EL, Turvey MT (1998) Interlimb coordination in a simple serial behavior: a task dynamic approach. Hum Movement Sci 17:393–433

63. Sternad D, Turvey MT (1995) Control parameters, equilibria, and coordination dynamics. Behav Brain Sci 18:780

64. Sternad D, Wei K, Diedrichsen J, Ivry RB (2006) Intermanual interactions during initiation and production of rhythmic and discrete movements in individuals lacking a corpus callosum. Exp Brain Res 76:559–574

65. Ting LH, Macpherson JM (2005) A limited set of muscle synergies for force control during a postural task. J Neurophysiol 93(1):609–613

66. Travis LE (1929) The relation of voluntary movement to tremors. J Exp Psychol 12:515–524

67. Tuller B, Kelso JAS (1989) Environmentally-elicited pattern of movement coordination in normal and split-brain subjects. Exp Brain Res 75:306–316

68. Turvey MT (1990) Coordination. Am Psychol 45:938–953

69. Turvey MT, Carello C (1995) Dynamics of Bernstein's level of synergies. In: Latash M, Turvey MT (eds.) On dexterity and its development. Erlbaum, Hillsdale, NJ, pp. 339–376

70. van Mourik A, Beek PJ (2004) Discrete and cyclical movements: unified dynamics or separate control? Acta Psychol 117(2):121–138

71. Wei K, Wertman G, Sternad D (2003) Discrete and rhythmic components in bimanual actions. Motor Control 7(2):134–155

72. Wierzbicka MM, Staude G, Wolf W, Dengler R (1993) Relationship between tremor and the onset of rapid voluntary contraction in Parkinson disease. J Neurol Neurosur Psychia 56:782–787

73. Wolpert DM, Ghahramani Z, Jordan MI (1995) An internal model for sensorimotor integration. Science 269:1880–1882

74. Wolpert DM, Kawato M (1998) Multiple paired forward and inverse models for motor control. Neural Netw 11(7–8):1317–1329
75. Wolpert DM, Miall RC, Kawato, M (1998) Internal models in the cerebellum. Trends Cogn Sci 2(9):338–347
76. Yu H (2005) Rhythmic timing in human movements: behavioral data, a model and fMRI studies. Pennsylvania State University, University Park, PA
77. Yu H, Vaillancourt DE, Corcos D, Sternad D (2007) Hyper- and hypo-activation of cortical and subcortical structures during rhythmic movements in Parkinson's disease. Neuroimage 35:222–233

Part II

Neural Dynamics

Imaging the Neural Control of Voluntary Movement using MEG

Douglas Cheyne

Program in Neurosciences and Mental Health, Hospital for Sick Children Research Institute, Toronto, Ontario M5G 1X8 Canada and Department of Medical Imaging, University of Toronto, Toronto, Ontario M5S 3E2 Canada

1 Introduction

Throughout the history of behavioral neuroscience there has been a great interest in how the human brain processes sensory information and the neural mechanisms that underlie our perception of the outside world. Indeed, much of experimental psychology has its roots in the study of sensation and perception. Similarly, the study of human perceptual-motor skills has largely focused on our ability to perform a variety of tasks that involve the processing of sensory input and the mapping of that input onto motor responses. The development of neuroimaging techniques such as PET, functional MRI, high-resolution EEG, and magnetoencephalography (MEG) has contributed greatly to our understanding of the neural processes that perform such sensory-motor transformations. However, the neural mechanisms underlying internally generated movements – movements that are performed in the absence of a sensory cue – are less well understood, due in part to the difficulty in measuring those processes that guide our spontaneous actions and even the simplest of motor skills that we use on a daily basis. In 1965, Kornhuber and Deecke published their observation of electrical brain activity preceding a self-initiated movement after striking upon the idea of backwards averaging of brain potentials recorded during voluntary finger movements [49]. This pioneering discovery of the so-called Bereitschaftspotential (readiness potential) led to an increased interest in using electrophysiological techniques to study not only the brain's response to external events as reflected in the processing of sensory stimuli, but also to gain insights into neural activity that precedes the generation of voluntary movements, perhaps even elucidating the organizational principles by which spontaneous, self-willed actions arise. In this regard, neuroimaging of human brain activity during self-paced movements offers the unique opportunity to study those neural mechanisms that govern the planning, initiation, and control of volitional behavior.

Modern neuroimaging techniques such as functional MRI provide us with powerful methods with which to localize brain areas involved in various aspects of motor control. However, methods based on brain hemodynamics lack the temporal resolution required to investigate the precise sequence of neural events as they unfold over the course of a few hundred milliseconds during the execution of a single voluntary movement. High-resolution EEG and MEG studies utilizing state-of-the-art source modeling techniques, on the other hand, may hold the greatest promise to determine the time-dependent neural processes underlying motor control in humans. The following chapter provides a brief overview of the use of MEG to study the cortical control of voluntary movements and also describes the development of new MEG source reconstruction techniques that offer the ability to non-invasively image human brain activity during motor tasks with unparalleled spatial and temporal resolution.

2 Movement-Related Magnetic Fields

The first recordings of magnetic fields accompanying simple finger movements were reported by Deecke and colleagues [26] who observed slow magnetic field changes in recordings over sensorimotor areas of the brain preceding voluntary movements of the digits using a single channel magnetometer. These early recordings revealed that slow magnetic field shifts began several hundred milliseconds prior to movement and appeared to be the magnetic counterpart of the electrically recorded readiness potential. Subsequent studies using more extensive spatial sampling showed that these movement-related magnetic fields (MRMFs) were widely distributed over the scalp and indicated that pre-movement MEG activity could not be explained by a single generator in the contralateral primary motor area [21]. In addition, a number of transient field reversals were observed following the onset of electromyographic activity in the involved muscles, presumably reflecting cortical activity involved in movement execution and the processing of proprioceptive feedback, and were termed movement-evoked fields [21, 51]. Early attempts to model the generators of movement-related fields as equivalent current dipoles showed that at least two sources were necessary to account for pre-movement magnetic field patterns, even during unilateral movements, due to consistently observed activity in the ipsilateral hemisphere [21, 51]. The observation of bilateral motor cortex activation for unilateral movements was somewhat contrary to conventional models of motor control at the time; however, this finding is consistent with recent evidence of bilateral cortical activity during unilateral motor tasks.

The introduction of whole-head MEG systems in the early 1990s and recent advances in source reconstruction methods now offer the possibility of constructing more detailed models of distributed brain activity accompanying movement. Initial whole-head MEG studies of voluntary finger movements combined with dipole modeling have confirmed earlier findings regarding primary motor area activation, including the presence of ipsilateral MI activation [22, 63]. Fig. 1 shows the time course and magnetic field patterns accompanying

Movement-related magnetic fields (right index finger)

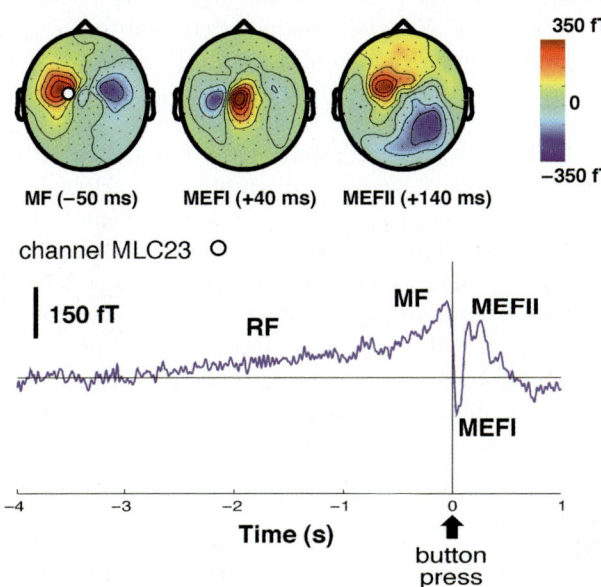

Fig. 1. Movement-related magnetic fields accompanying a self-paced movement of the right index finger, recorded from 151 first-order axial gradiometers covering the whole-head (VSM Medtech Ltd, Coquitlam, BC, Canada) and averaged over 80 trials. Color maps of magnetic field amplitude (red = ingoing, blue = outgoing) show the topography of the three main peaks in the time trace (below) for a sensor located over the contralateral hemisphere (white dot). RF = readiness field, MF = motor field, MEF = movement-evoked fields. fT = femtoTesla. Modified from [17]

a self-paced movement of the right index finger in a right-handed subject using a 151-channel whole-head magnetometer system. The averaged waveform from a sensor overlying the contralateral motor cortex shows a typical slow pre-movement readiness field (RF) with a rapid increase in amplitude beginning approximately 600 ms prior to movement onset and reaching peak amplitude (motor field – MF) about 50 ms prior to movement onset, roughly coincident with onset of EMG activity in the active muscles [21]. Movement-evoked fields can be seen during the execution phase of the movement, with the largest component (MEFI) reaching maximum amplitude at a latency of 40 ms (approximately 100 ms following the onset of EMG activity) followed by a second field reversal at about 140 ms (MEFII). One can also observe in the example shown in Fig. 1, the typical bilateral topography of the MF for a unilateral movement, with larger amplitude over the hemisphere contralateral to the side of movement [51]. Although this pattern has the appearance of a single midline source, it can be shown to reflect bilateral activation of the sensorimotor areas (bimanual movements produce an almost identical pattern) with cancellation of oppositely oriented magnetic fields over the central scalp, thus complicating reconstruction of the underlying sources. The degree of hemispheric asymmetry of the MF tends to vary across subjects and with side of movement, with greater degree of asymmetry for movements of the dominant hand in comparison to the non-dominant hand in right-handed subjects [51]. This functional asymmetry of motor cortex activation was confirmed in a whole-head MEG study [93] and additionally found that left-handed subjects showed a mixed pattern of asymmetry compared to right handers.

In contrast to the complex patterns of pre-movement brain activity, the first movement-evoked field (MEFI) at a latency of 30 or 40 ms following the button press (Fig. 1, middle map) shows a much more lateralized pattern of activity that can be modeled as a single source in the contralateral hemisphere [51]. This MEFI occurs about 100 ms after onset of EMG activity in the agonist muscle, around the same latency as single unit activity recorded from primary somatosensory cortex [6, 64]. Sources for the MEFI have been localized to the postcentral gyrus and show a homuncular distribution for movements of different body parts [55] suggesting that this component reflects somatotopically organized activation of neuronal populations in SI. Studies have shown evidence that the MEFI is produced by muscle activity in the periphery since it is delayed by cooling of the peripheral nerves [18], modulated by transient deafferentation [53] and localizes to portions of the postcentral gyrus that receive tactile and proprioceptive afferents [54, 69].

The exact role of peripheral feedback in the generation of the MEFI is not clear. The above studies suggest that such feedback may involve both afferent input from muscle spindle receptors monitoring changes in muscle length as well as sensory organs in joints and tendons, although a recent study found that MEFI latency was more time-locked to onset of muscle stretch than physical movement of the finger joint [70]. The hypothesis that the MEFI may reflect encoding of kinematic variables associated with muscle stretch is also consistent with the findings of Kelso and colleagues [43] who showed that MEG activity overlying the contralateral sensorimotor cortex during a continuous motor task correlated most highly with movement velocity, independently of explicit task requirements (flexion version extension). Interestingly, in a recent EEG study, the electrical counterpart to the MEFI was shown to be absent in a deafferented patient who could still perform flexions of their index finger without any sensation of movement [50]. The authors speculated that the MEFI might therefore reflect feedback from the periphery to the sensorimotor cortex necessary for the "perceptual awareness" of movement. Movement-evoked fields at latencies greater than 100 ms (e.g., MEFII and MEFIII) have rather complex topographies and are not amenable to modeling as simple configurations of dipole sources. The MEFIII component, occurring at latencies of up to 300 ms after EMG onset, may be associated with movement termination or subsequent muscular contractions during complex movements as it is more strongly detected during flexion–extension movements than for flexion only [39].

2.1 Rhythmic Brain Activity Accompanying Voluntary Movements

Recently there has been renewed interest amongst neuroscientists in the role of synchronized rhythmic activity in the brain, particularly as it pertains to the coupling or phase-locking of different neuronal networks that may under-lie cognitive processes such as object recognition or perceptual 'binding' of multiple sensory inputs [91, 96]. Similarly, there has been an increased in-terest in the role of neural oscillations in motor control [7, 60]. It has long been known that movements elicit frequency-specific changes in the EEG, particularly within the β (15–30 Hz) and μ (8–14 Hz) frequency bands. These changes in ongoing or *induced* oscillations are similar to those observed during cognitive tasks in that they tend to be only coarsely time-locked to specific components of the movement. Within the EEG literature, these increases and decreases in spectral power have been traditionally termed event-related synchronization (ERS) and desynchronization (ERD), respectively, and have been shown to be elicited during and following preparation and performance of voluntary movements [72, 73, 80], passive movements [11], and even dur-ing imagined movements [74, 83]. Using MEG, Jensen and colleagues [40] have demonstrated that augmentation of β-band activity by benzodiazepines was localized to regions of the sensorimotor cortex with a concomitant shift in peak frequency. The authors compared their findings with a neural net-work model showing that the frequency specificity of cortical β-rhythms may reflect the inherent time-constant of neural networks involving both long-range thalamocortical connections, as well as local intracortical connectivity between inhibitory interneurons and excitatory pyramidal cells. Higher fre-quency (>30 Hz) γ-band oscillations have also been reported in motor corti-cal areas during a variety of visuomotor tasks and appear to be more related to aspects of motor planning, such as movement trajectory [15, 71] or atten-tional aspects of the motor task [8] and likely arise from premotor and parietal regions.

The functional role of μ and β rhythmic activity in the sensorimotor cor-tex is not well understood. One interpretation is that they simply reflect background or 'idling' rhythms that are disrupted during movement and are enhanced (rebound) during periods of cortical inhibition following movement onset [14]. The observation that β-band rebound exhibits frequency specificity related to the body part being moved has led to speculation that the frequency of oscillation reflects the extent of the cortical area activated, and in turn the dynamics of a local cortical oscillatory network [73]. Others have speculated that these oscillations may reflect perceptual "binding" in the motor system analogous to that found for gamma band oscillations in the visual system [9]. Alternatively, sensorimotor oscillations have been suggested to reflect a type of *sensorimotor sampling* [60] in which sensory input necessary for the guidance of movements can be more efficiently combined with synchronously firing neuronal populations involved in motor output.

2.2 Corticomuscular Coherence

Using a single channel magnetometer, Conway and colleagues [25] made the interesting observation of increased coherence (correlation in the frequency domain) between the surface electromyogram activity in a contracting muscle and MEG recordings made over the contralateral motor areas. This has led to a large number of studies on MEG–EMG or *corticomuscular* coherence (CMC) and speculation on the functional relationship between spontaneous cortical rhythms and EMG activity during movement [4, 9]. Changes in the frequency of coherence varies with the strength of muscular contraction [10] and MEG studies have shown changes in CMC frequency in patients with Parkinson's disease [79] leading to hypotheses regarding the role of CMC in maintaining efficient cortical 'drive' to the peripheral motor neurons during normal muscular contraction. The observed constant phase lag between cortex and muscle during CMC suggests that it is produced by fast corticomotoneuronal pathways, although reduction of CMC in a deafferented patient indicates some dependence on sensory afferents [46]. Patterns of corticomuscular coherence are best observed during extended periods of isometric force, as relatively long sampling periods are required to compute reliable estimates of coherence. Hence the relationship of CMC to specific phases of discrete voluntary movements is not clearly understood. However, recent MEG studies have shown CMC was related to task-specific aspects of movement [44] and was reduced during the phasic part of a movement relative to a hold phase [45]. In addition, there are periods of increased CMC following voluntary movements [29] that have been shown to correlate with the level of attention to motor performance [52] and CMC has been correlated with readiness to respond in a reaction time task [84]. This suggests that corticomuscular coherence may not only function as a mechanism for stabilizing corticospinal output to the muscles during periods of sustained muscular force but may also subserve a higher-order role in motor control.

3 Methodological Issues

In order to use MEG to model the rather complex patterns of cortical activity observed during even simple motor tasks, as well as concomitant changes in cortical oscillatory activity, we require more sophisticated approaches to magnetic source reconstruction than provided by the dipole modeling approach. Fortunately, more advanced MEG source reconstruction methods have been recently developed, capable of localizing both time-locked and induced oscillatory activity from multiple brain regions, thereby allowing us to take full advantage of the temporal information in the MEG signal. As described in the following section, these new approaches offer the promise of a true *spatiotemporal* brain imaging method that can disentangle the complex nature of ongoing and evoked neural activity during motor control.

3.1 Beamforming: A New Approach to Neuromagnetic Source Reconstruction

Neuromagnetic inverse solutions involve determining the distribution of electrical activity within the brain that contributes to the magnetic field recorded by sensors outside of the head. Such inverse solutions are non-unique (i.e., there are many possible combinations of sources that may produce a given pattern at the sensors) and often highly underdetermined (i.e., there may be many more sources than sensors). Simple dipole fitting or scanning methods are limited by the fact that the existence of unknown sources can result in erroneous solutions due to the superposition of signals across the sensor array. General linear inverse methods, such as those based on the minimum-norm solution [36], are thus preferred, as they can model distributed activity without assuming a fixed number of active sources. However, these methods must account for an arbitrary number of distributed sources by incorporating the magnetic field patterns (*lead fields*) of all possible sources in the solution. This requires minimizing the dimensions of the resulting inverse operator, usually by restricting solutions to the cortical surface and using regularization techniques to stabilize the solutions. An alternative source reconstruction approach involves a signal processing technique known as beamforming – a method introduced in the 1950s in the field of radio communications using multiple antenna arrays. These algorithms were developed to optimally reduce interference (cross-talk) between signals arriving from multiple directions at the antenna array. Signal processing based on beamforming has been adapted to the neuromagnetic inverse problem only in the last few years, as it relies on correlations across multiple detectors and is thus suited to modern MEG systems that consist of arrays of 100 or more sensors.

Various types of beamforming algorithms have been recently adapted to neuromagnetic measurements [34, 77, 87], the most common of which is the minimum-variance beamformer [95]. Beamforming methods are also described as *spatial filters*, since source activity at any brain location is derived by passing the measured data through a set of coefficients or weights that suppress signals coming from other sources. Compared to model-based linear inverse methods, the beamformer approach is a data-based or adaptive inverse solution, where the forward model for all possible sources is replaced by the estimated data covariance. Thus beamforming methods work best when the data (i.e., source) covariance is well estimated by including a sufficient amount of sample data. Beamformer solutions also have the added advantage of inherently attenuating artifacts, even those of non-brain origin (e.g., environmental noise, eye movements) as these signals are included in the data covariance, whereas minimum-norm-based solutions only model brain sources, and thus require relatively artifact-free data. In addition, since beamformer solutions are constructed volumetrically throughout the brain they are readily amenable to averaging across groups of subjects using spatial normalization techniques [12, 88].

3.2 Imaging Cortical Oscillations Using the SAM Beamformer

A beamforming technique known as *synthetic aperture magnetometry* (SAM) was introduced by Robinson and colleagues [77] that derived a single optimal current orientation at each voxel, based on the assumption that MEG signals are dominated by intracellular pyramidal cell currents flowing perpendicular to the cortical surface. This method is computationally simpler, and also produces slightly higher spatial resolution than multidimensional or vector beamformers [86] that model current flow in multiple orthogonal directions. Most importantly, however, the SAM beamformer was designed to work with single trial data and provided a single time series of source activity at each brain location, making the method ideal for time–frequency analysis of source activity. As a result, this method has been used primarily as a frequency-based imaging technique to measure induced changes in oscillatory neural activity. This is achieved by integrating the output from a lattice of SAM spatial filters constructed throughout the brain volume within specific frequency bands and measuring the changes in source power for selected time windows, relative to a pre-stimulus or pre-movement baseline. This approach has been successfully used to localize induced cortical oscillatory activity associated with visual [28], auditory [38] and somatosensory [33, 85] stimulation as well as oscillatory changes accompanying self-paced movements [19, 92].

In a recent study [42] we used the SAM algorithm to localize changes in the μ and β frequency bands during self-paced finger movements. As shown in Fig. 2, using this method one can construct robust images of frequency-specific source activity in the sensorimotor cortex, even within single subjects. When averaging images across subjects, we found that suppression of μ-band (8–15 Hz) activity preceded movement onset and was localized mainly to regions of the postcentral gyrus. In contrast, power in the β-band (15–30 Hz) also decreased prior to movement onset but showed a marked sudden increase beginning approximately 300 ms after termination of EMG activity in the active muscle (first dorsal interosseus) and lasting for over 500 ms. This *post-movement β rebound* (PMBR) localizes to bilateral regions of the precentral gyrus, with larger amplitude changes in the contralateral hemisphere, confirming previous reports that PMBR is generated by neural populations in the primary motor cortex [81]. Interestingly, PMBR is coincident with a period of decreased cortical excitability in MI as shown by transcranial magnetic stimulation [13]. We also found that PMBR was time-locked to the cessation, rather than the onset of movement [42] and other MEG studies have shown similar patterns of β rebound in motor cortex following median nerve [80] and tactile stimulation [33]. Thus, augmentation of β-band oscillatory activity appears to constitute an *off-response* to the termination of sustained afferent or reafferent input to the primary motor cortex.

Beamforming methods have also been adapted to model coherence between different brain sources or even corticomuscular coherence. A frequency domain beamforming method termed *dynamic imaging of coherence sources* developed

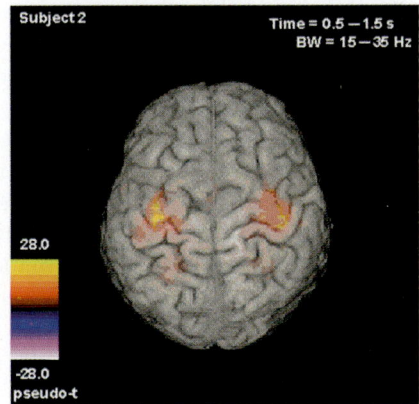

Fig. 2. Localization of induced cortical oscillations during a voluntary right finger movement in a single subject using the SAM beamformer source reconstruction algorithm. Differential source power (pseudo-t) decreases (blue) and increases (red) are superimposed on the subject's MRI. Left: Decreased power in the mu (8–15 Hz) frequency band during the pre-movement period is seen in the contralateral post-central gyrus and parietal areas (left). Right: Increases in the β (15–30 Hz) band begin approximately 0.5 s after movement onset and are localized to bilaterally to the primary motor cortex

by Gross and colleagues [34] has been recently used to image task-dependent oscillations during continuous motor tasks [35] localizing the generators of corticomuscular coherence in the β and γ frequency bands to regions of the primary motor cortex and supplementary motor area. This method has also been used to image coherence between multiple cortical regions during Parkinsonian tremor [94].

3.3 Spatiotemporal Imaging of Movement-Related Brain Activity

Beamformer localization methods based on coherence or band-limited power differences have provided a powerful new method for the localization of oscillatory brain activity, however, this approach sacrifices temporal resolution by integrating power over relatively long (e.g., 200 ms) time windows and is therefore not applicable to imaging transient motor and movement-evoked fields. We recently introduced the use of an "event-related" beamformer (ERB) algorithm adapted for the localization of instantaneous or evoked brain responses [17]. Similar to the SAM method, a beamformer filter is derived from the single trial MEG data with a single optimal current direction at each voxel. The output of the spatial filter for each voxel is then averaged with respect to stimulus or movement onset to measure the event-related source activity. Three-dimensional ERB images are constructed by mapping the spatial distribution of the averaged response at selected latencies, to produce millisecond-by-millisecond three-dimensional images of brain activity.

We used the ERB algorithm to localize the generators of movement-related fields in a study of self-paced movements (button press with right or left index finger) in eight right-handed adults. Selected time frames from the resulting spatiotemporal 'movie' of source activity are shown in Fig. 3 for one subject (left-sided movements) revealing a rapid sequential activation of regions of the contralateral precentral and postcentral cortices. Three peaks of maximal source activity are observed, the first occurring in the hand region of the precentral gyrus, reaching peak magnitude at the latency of the MF (-50 ms before button press). This is followed by a brief activation of the postcentral gyrus at around 40 ms and a second activation of the precentral gyrus at approximately 140 ms, corresponding to the latencies of the MEFI and MEFII,

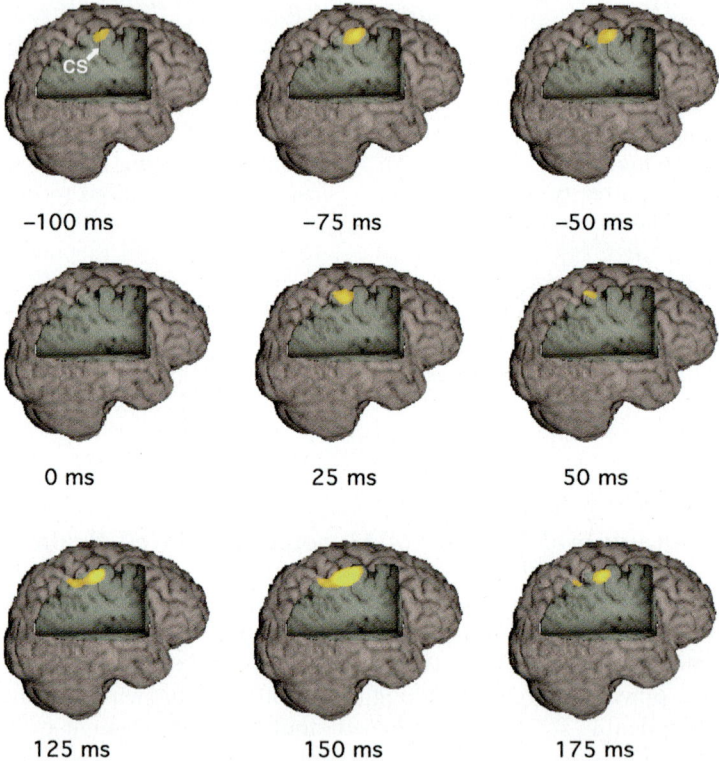

Fig. 3. Snapshots from a spatiotemporal "movie" of activity in the contralateral sensorimotor cortex of a subject performing self-paced movements of their left index finger. Images of instantaneous source power were computed using the event-related beamformer algorithm. Latencies are relative to the time of the button press, and show sequential activation of the pre- and postcentral gyrus during movement onset. The central sulcus (CS) is indicated by the white arrow

respectively. This sequence was observed for both left and right finger movements across subjects, and demonstrated the ability of the ERB algorithm to image movement-related brain activity with sufficient spatial and temporal resolution to resolve multiple generators within regions of the sensorimotor cortex only several millimeters apart.

We proposed a model of the putative generators of the MF, MEFI and MEFII [17] based on a sequential activation of MI and SI around the period of movement onset, as revealed by the event-related beamformer analysis. As shown in Fig. 4, this involves an initial activation of an anterior portion of the precentral gyrus that gives rise to the motor field (MF). The involvement of both the anterior bank of the central sulcus and the crown of the precentral gyrus may account for the somewhat variable amplitude of the MF (since MEG is less sensitive to radially oriented currents) as has been suggested in EEG modeling studies [63]. This also implies activation of neurons in Brodmann's area 6, rather than the somatotopically organized corticospinal output from area 4. This is followed by the MEFI component occurring about 100 ms later, representing a brief and highly time-locked activation of the postcentral gyrus due to reafferent feedback, involving both proprioceptive and tactile input to areas 3a and 3b, respectively. Finally, during the MEFII, activity shifts back to regions of the precentral gyrus. This activity was observed to be typically greater in strength, and slightly inferior in location to MF activity, suggesting the activation of a functionally different region of MI.

Prior to these new findings, the generator of the MEFII component was unknown, and assumed to be primarily sensory in nature since it occurs shortly after movement onset. However, these new results suggest that the MEFII reflects activity of neural populations in MI, possibly related to ongoing motor control such as onset of the antagonist burst (braking function)

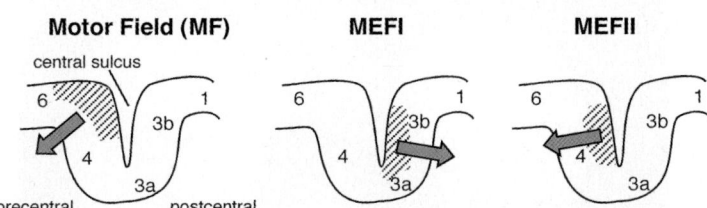

Fig. 4. A model of the generators of movement-related magnetic fields during voluntary finger movement. The pre-movement motor field (MF) reaches maximal amplitude at the time of movement onset and reflects activation of anterior portions of primary motor cortex in the anterior bank of the central sulcus and/or the crown of the precentral gyrus (shaded area). The movement-evoked field I (MEFI) is generated by activity in the posterior wall of the central sulcus due to feedback to regions of SI in the postcentral gyrus. The MEFII appears as a second activation of the precentral gyrus, slightly posterior and inferior to the location of the MF. Approximate location of Brodmann areas are shown relative to the central sulcus. Arrows indicate the direction of net intracellular current flow. Modified from [17]

or proprioceptive feedback to MI from the periphery via other cortical areas [57]. Interestingly, TMS studies have shown that latency of the MEFII corresponds to a transient period of increased excitability of motor cortex [13]. We also found that MEFII latency was remarkably consistent across subjects who showed variable patterns of EMG activity suggesting that the MEFII reflects neural activity in MI involved in the feed-forward control of the movement once it has been initiated [82] and is likely independent of movement duration or the precise timing of sensory feedback.

3.4 Ipsilateral Motor Fields

In addition to the robust source activations in the contralateral sensorimotor area during movement onset, significant activity was also observed in the ipsilateral precentral gyrus during both right and left finger movements in almost all subjects. Figure 5 shows MF activity superimposed on the MRI in one subject, demonstrating bilateral activation in the hand region of the precentral gyrus for both right and left index finger movements. The averaged time course of source activity at these locations is shown below. Bilateral MI activity can be seen beginning as early as 1 s prior to movement that returns to baseline shortly after movement onset. A sharp deflection corresponding to the MEFI just after movement onset is seen only in the contralateral MI source waveform (due to pick-up of source activity in the adjacent postcentral gyrus). These results provide further evidence that movement-related brain activity preceding voluntary movements involves very early activation of the primary

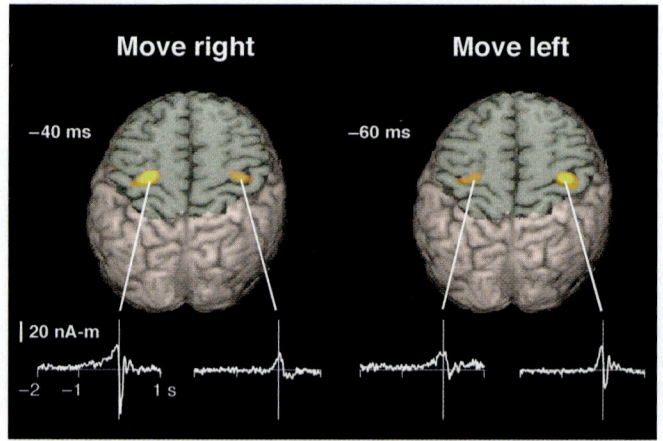

Fig. 5. Bilateral activation of the primary motor cortex during unilateral movements. Averaged source activity at the peak latencies of the motor field (MF) is shown for a subject performing right and left index finger movements. Time traces show the averaged output of the spatial filter in units of source strength (nanoAmpere-meters) at the location of the MF source. Modified from [17]

motor cortex of both hemispheres. In the example shown in Fig. 5, one can also observe an earlier onset of the pre-movement activity in the left MI in comparison to the right MI for both contralateral and ipsilateral movements. Interestingly, this earlier onset of the motor field in the right hemisphere motor cortex, independent of the side of movement, was consistent across the group of subjects. This may be related to handedness (all subjects were right handed) or may reflect a left hemisphere specialization for movement preparation.

3.5 Localization of Non-primary Motor Areas Using MEG

Neuroimaging studies of voluntary movement tasks using functional magnetic resonance imaging (fMRI) or positron emission tomography (PET) typically show activation of multiple secondary motor areas, including premotor areas of the frontal lobes thought to be critical in the initiation of movements [32, 41]. In addition, EEG studies have successfully modeled activation of the supplementary motor area (SMA) during the pre-movement period [3, 5]. MEG studies to date have only reported weak activation of the frontal midline [27] presumably due to cancellation of tangential currents in the opposing hemispheres [21, 56]. However, in the above study we were able to detect source activity in both left and right dorsal premotor regions and parietal cortex, as well as weak activation in the SMA when averaging ERB source images across subjects [17]. This suggests that MEG combined with beamformer source reconstruction methods can be used to localize activity in non-primary motor areas not previously reported by MEG studies of voluntary movements.

4 Theoretical Considerations

4.1 What is the Significance of Ipsilateral MI Activity?

The neural sources contributing to movement-related magnetic fields have been debated in the literature, particularly with respect to the role of the ipsilateral motor cortex. Based on our initial observations of ipsilateral MEG activity during unilateral movement, we speculated that this was evidence of an intrinsically bilateral organization of movement control within the motor system, demonstrated by the occasional presence of associated or mirror movements that must be somehow suppressed during precise independent and unilateral movements of the digits [16, 21]. Evidence of the existence of inhibitory transcallosal pathways between the primary motor cortices has been provided by a number of transcranial magnetic stimulation (TMS) studies, implying the need for inhibitory control over the ipsilateral motor cortex during movement [30, 48, 66]. While some fMRI studies have reported ipsilateral activation of the sensorimotor cortex during motor tasks [47], other studies have shown only weak or inconsistent activation of the ipsilateral precentral gyrus for simple movements [1], with dependence on complexity of the motor task [68] or handedness [48, 59]. Interestingly, fMRI studies have also reported *decreased* BOLD responses in MI during ipsilateral movements that appear to reflect neuronal deactivation [2, 89]. Based on their observation of BOLD co-activation of homologous regions of MI, Stippich and colleagues [90] suggested that unilateral movements were accompanied by a "global" activation of MI prior to movement, with an increasing ratio of excitation to inhibition in the area of MI contralateral to the generated movement.

MEG measures of pre-movement brain activity, taken together with results from fMRI and TMS studies, support the hypothesis that the ipsilateral motor cortex is actively inhibited during unilateral movements, most likely through transcallosal excitatory projections from the contralateral motor cortex onto local inhibitory interneurons [68, 89]. This mechanism alone would predict a net decrease or inhibitory pattern of neural activity in the ipsilateral motor cortex during movement; however, MEG source analysis shows that pre-movement activity in MI is similar in both location and direction of intracellular current flow for both ipsilateral and contralateral movements – indeed, pre-movement field patterns during non-dominant finger movements may be almost indistinguishable from those for bilateral movements [51]. This suggests that during the preparation of both bilateral and unilateral finger movements MI receives bilateral input from premotor brain structures (e.g., supplementary motor areas and dorsal premotor areas). During unilateral movements, MI of the non-moving hand also receives inhibitory input via transcallosal projections from the contralateral MI, presumably to suppress the occurrence of mirror movements, resulting in reduction of activity in the ipsilateral MI. TMS studies have shown that this contra- to ipsilateral MI inhibition is asymmetric [65] with stronger inhibitory connections from the dominant to non-dominant hemisphere. This would predict greater reduction of ipsilateral MI activity for movements of the dominant hand and is consistent with our observation of greater asymmetry of the MF for dominant hand movements [51, 93]. Recently it has been proposed that this interhemispheric inhibition may even play a role in motor deficits and subsequent recovery in unilateral stroke due to abnormal inhibition of the lesioned primary motor cortex by the intact (contralesional) motor cortex [62].

Bilateral preparation of unilateral movements may provide a mechanism by which the motor system can rapidly switch between different outputs during skilled motor tasks. Alternatively, this may reflect the ontogeny of the human motor system and the transition from bilateral control of the proximal musculature to the relative independent control of movements of the distal musculature [58]. TMS studies have shown that inhibitory transcallosal pathways between motor areas develop with age as demonstrated by the presence of ipsilateral motor evoked potentials prior to the age of 10 that are presumably suppressed in adults once these inhibitory pathways are fully developed [67]. This may underlie similar age-related changes in the development of fine motor skills, along with the disappearance of mirror movements at around the same age [61]. ERP studies [24] have shown changes in the topography and polarity of the readiness potential with age, with early pre-movement activity emerging after the age of 6 years as positive potential, and developing into the negative slow shift observed in adults only after 9 or 10 years of age [23]. Although MEG measurements in younger children, particularly during voluntary movement tasks, are technically challenging, studies of motor cortex function during development may provide an ideal avenue for the investigation of these issues.

4.2 What is the Functional Role of the Sensorimotor Cortex?

The activation of the sensorimotor cortex in the absence of overt movement raises questions regarding the neuronal populations involved in the generation of the motor field and their functional role in the production of movement. The presence of activity in both the contralateral and ipsilateral primary motor cortex during the pre-movement period, prior to any detectable increases in EMG activity, precludes the activation of direct monosynaptic excitatory inputs to spinal motoneurons, even though these *corticomotoneuronal* pathways are well developed in humans and thought to play a relatively important role in the independent control of the digits [58]. However, much of the output of the descending corticospinal tract converges indirectly on spinal motoneurons through polysynaptic pathways. Thus, although the primary motor cortex has a direct role in the efferent control of movement, it exerts both excitatory and inhibitory effects on the segmental spinal motoneurons via both direct and indirect connections, indicative of a complex and diverse role in movement control [58]. This is supported by the observation of activity of the primary motor cortex during a variety of tasks that appear to engage the sensorimotor system but do not involve motor output, such as motor imagery [83] or action observation [37]. Similarly, we observed highly similar patterns of oscillatory activity in the primary motor cortex during both tactile stimulation of the index finger and while the subject observed another individual's finger being stimulated [20]. This suggests a close interrelationship between neuronal networks in the same cortical area activated by both somatosensory stimulation and visual monitoring of events in the environment associated with either sensory input or movement. Thus, it is clear that activity of neural populations in MI can be modulated in a variety of situations that do not necessarily involve movement production or control, and reflect their role in more general aspects of sensorimotor coordination, or even motor learning [75].

Interestingly, cortical oscillations preceding and following unilateral voluntary movements are also bilaterally distributed. This might at first suggest that they simply mirror the activation of the sensorimotor cortex as described above for movement-related fields. However, the transition from desynchrony to synchrony does not appear to correspond to onset or offset of electromyographic activity in the involved muscles, or specific components of movement-related fields. This is demonstrated in Fig. 6, which shows the relationship between induced oscillatory changes (spectral power integrated over single trials) and event-related activity (source activity averaged over trials) in the contralateral precentral gyrus during a self-paced button press. The results, averaged over a group of eight subjects, show the very different time course of oscillatory activity in comparison to the movement-evoked field activity in the same cortical area, with onset of suppression in the μ and β frequency bands appearing more than 1 s prior to movement and continuing after EMG activity has terminated. In contrast, the averaged event-related responses show highly time-locked changes immediately around the period of the button press with no concomitant change in oscillatory power in these higher frequency bands, but rather appears as a transient increase in power at lower ($<10\,\text{Hz}$) frequencies. Fetz and colleagues [31] also noted the lack of a direct correspondence between oscillatory activity in the sensorimotor cortex and EMG activity during intracellular recordings in awake monkeys and suggested that oscillatory activity may be an indication of global facilitation of synaptic interactions related to changes in levels of attention or arousal.

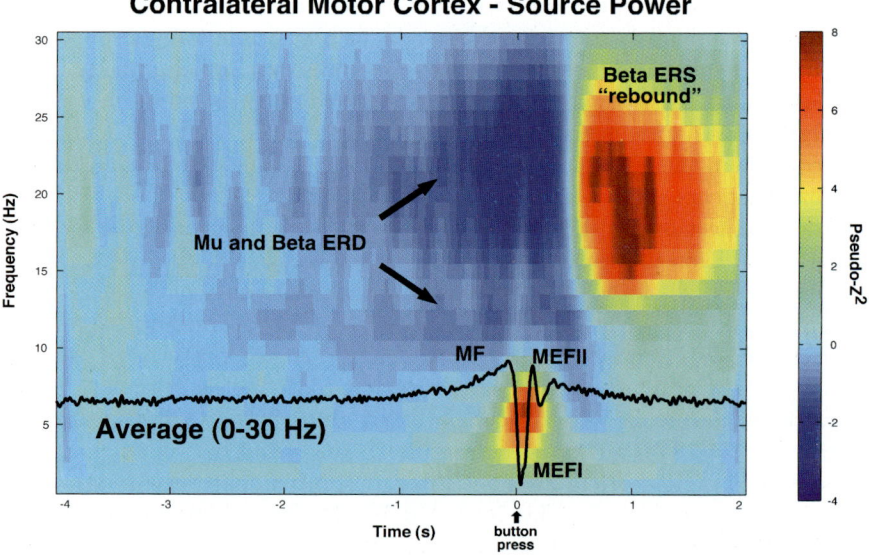

Fig. 6. Comparison of oscillatory and event-related activities in the contralateral motor cortex for a right index finger movement, averaged over eight subjects. Increases (red) and decreases (blue) in source power relative to pre-movement baseline (-4 to $-3\,$s) are shown as a time–frequency plot using a Morlet wavelet transform of the beamformer output, integrated over 80 movements. The same data filtered between 0 and $30\,\text{Hz}$ and averaged over all trials is superimposed on the same plot (black trace). MF = motor field, MEFI and MEFII = movement-evoked field I and II. ERD = event-related desynchronization, ERS = event-related synchronization

Taken together, these results suggest that induced rhythms and motor-evoked fields reflect the activation of different functional systems.

The fact that pre-movement μ-band activity arises from regions of the postcentral gyrus is also of interest, indicating that the primary somatosensory cortex (SI) is active during movement preparation in the absence of sensory input. However, a significant portion of the corticospinal tract originates from regions of the postcentral gyrus [78]. Moreover, these neurons project largely to the dorsal horn of the spinal cord and are thought to be involved in the descending control of sensory inputs or even 'gating' of proprioceptive input during movement [58]. Riddle and Baker [76] recently emphasized the role of sensorimotor integration in the generation of motor cortex oscillations and speculated that they may reflect a type of *recalibration* or interrogation of the 'system-state' after each movement. This favors the sensorimotor sampling theory of cortical oscillations [60] and the idea that oscillatory activity in the sensorimotor cortex reflects the monitoring of various parameters of the motor system (hand position, muscle length), even prior to feed-forward control of simple ballistic movements. In addition, in many studies of self-paced movements, including our own MEG experiments, movement tasks often require orientation of the hand to a button or similar triggering device and likely involves, to some degree, the integration of haptic information needed to perform the movement accurately from trial to trial.

Converging evidence from MEG studies of voluntary movement and other imaging and electrophysiological techniques points to the role of the sensorimotor cortex as a system for sensorimotor integration. However, this brain region has also evolved in humans to exert fast, feed-forward control of fractionated movements of the digits, as reflected by greater direct connectivity to the segmental neuronal pool. For example, the severe motor deficits resulting from damage to the primary motor cortex in humans in comparison to modest deficits in other species including non-human primates is thought to be due, in part, to increased direct monosynaptic terminations onto motorneurons in the ventral horn of the spinal cord, which also provides a possible mechanism for more dexterous control of the digits [58]. Superimposed on this lower level of input–output control of movement, however, are oscillatory changes in the same cortical areas that appear to have a more variable and indirect relationship to specific motor parameters. In their recent review, Bressler and Kelso [7] emphasized the need for flexibility (*metastability*) within distributed cortical networks that would allow rapid transitions from one functional coupling between networks to another, thereby allowing information to be combined across different neural networks in real time. It has been proposed that synchronization between cortical networks may underlie such large-scale integration in the brain [96]. Thus, the complex patterns of cortical oscillations that we observe during self-paced voluntary movements likely reflect the recruitment of additional cortical and subcortical networks involved in integration of interoceptive and exteroceptive information needed to rapidly adapt our movements to changes in the environment. The apparent tendency of these systems to become engaged, even when performing relatively simple self-paced finger movements, perhaps demonstrates the self-organizing nature of the human motor system in which the control of individual movements is embedded within larger scale neural networks that govern the cognitive control of our actions, plans and goals.

5 Conclusions

MEG studies of voluntary movement have contributed greatly to our knowledge of the function of cortical motor areas in humans. The results from studies carried out to date, taken together with results from animal studies and other imaging modalities, support the concept of distributed activation of the sensorimotor cortex during the preparation and performance of skilled voluntary finger movements. This activity reflects not only preparation for, and the subsequent efferent control of movement, but also the integration of sensory inputs needed to adjust parameters within the context of ongoing behavior, as well as the processing of proprioceptive feedback that may be important in motor learning. As shown in this review, the introduction of MEG instruments capable of recording magnetic fields simultaneously from many brain regions using large arrays of sensors covering the head, combined with recent advances in neuromagnetic source reconstruction techniques\offers the possibility of using MEG to study both evoked and induced cortical activity during voluntary movement tasks with exquisite spatial and temporal resolution. These techniques can thus provide unique insights into the spatiotemporal organization of brain activity controlling the preparation and performance of both volitional movements and the dynamics of sensorimotor coordination.

Acknowledgments

This work was supported by grants to the author from the Canadian Institutes of Health Research and the Natural Sciences and Engineering Council of Canada. I also wish to thank my many colleagues and collaborators who contributed to the studies described in this review.

References

1. Alkadhi H, Crelier GR, Boendermaker SH, Hepp-Reymond MC, Kollias SS (2002) Somatotopy in the ipsilateral primary motor cortex. Neuroreport 13:2065–2070
2. Allison JD, Meador KJ, Loring DW, Figueroa RE, Wright JC (2000) Functional MRI cerebral activation and deactivation during finger movement. Neurology 54:135–142

3. Babiloni F, Carducci F, Cincotti F, Del Gratta C, Pizzella V, Romani GL, Rossini PM, Tecchio F, Babiloni C (2001) Linear inverse source estimate of combined EEG and MEG data related to voluntary movements. Hum Brain Mapp 14:197–209

4. Baker SN, Kilner JM, Pinches EM, and Lemon RN (1999) The role of synchrony and oscillations in the motor output. Exp Brain Res 128:109–117

5. Ball T, Schreiber A, Feige B, Wagner M, Lucking CH, Kristeva-Feige R (1999) The role of higher-order motor areas in voluntary movement as revealed by high-resolution EEG and fMRI. Neuroimage 10:682–694

6. Bioulac B, Lamarre Y (1979) Activity of postcentral cortical neurons of the monkey during conditioned movements of a deafferented limb. Brain Res 172:427–437

7. Bressler SL, Kelso JA (2001) Cortical coordination dynamics and cognition. Trends Cogn Sci 5:26–36

8. Brovelli A, Lachaux JP, Kahane P, Boussaoud D (2005) High gamma frequency oscillatory activity dissociates attention from intention in the human premotor cortex. Neuroimage 28:154–164

9. Brown P (2000) Cortical drives to human muscle: the Piper and related rhythms. Prog Neurobiol 60:97–108

10. Brown P, Salenius S, Rothwell JC, Hari R (1998) Cortical correlate of the Piper rhythm in humans. J Neurophysiol 80:2911–2917

11. Cassim F, Monica C, Szurhaj W, Bourriez JL, Defebvre L, Derambure P, Guieu J-D (2001) Does post-movement beta synchronization reflect an idling motor cortex? NeuroReport 17:3859–3863

12. Chau W, McIntosh AR, Robinson SE, Schulz M, Pantev C (2004) Improving permutation test power for group analysis of spatially filtered MEG data. Neuroimage 23:983–996

13. Chen R, Hallett M (1999) The time course of changes in motor cortex excitability associated with voluntary movement. Can J Neurol Sci 26:163–169

14. Chen R, Yaseen Z, Cohen LG, Hallett M (1998) Time course of corticospinal excitability in reaction time and self-paced movements. Ann Neurol 44:317–325

15. Chen Y, Ding M, Kelso JAS (2003) Task-related power and coherence changes in neuromagnetic activity during visuomotor coordination. Exp Brain Res 148:105–116

16. Cheyne D (1988) Magnetic and electric field measurements of brain activity preceding voluntary movements: implications for supplementary motor area function. PhD Thesis: Simon Fraser University: Burnaby, BC, Canada.

17. Cheyne D, Bakhtazad L, Gaetz W (2006) Spatiotemporal mapping of cortical activity accompanying voluntary movements using an event-related beamforming approach. Hum Brain Mapp 27:213–229

18. Cheyne D, Endo H, Takeda T, Weinberg H (1997) Sensory feedback contributes to early movement-evoked fields during voluntary finger movements in humans. Brain Res 771:196–202

19. Cheyne D, Gaetz W (2003) Neuromagnetic localization of oscillatory brain activity associated with voluntary finger and toe movements. NeuroImage 19 (Suppl):1061

20. Cheyne D, Gaetz W, Garnero L, Lachaux JP, Ducorps A, Schwartz D, Varela FJ (2003) Neuromagnetic imaging of cortical oscillations accompanying tactile stimulation. Brain Res Cogn Brain Res 17:599–611

21. Cheyne D, Weinberg H (1989) Neuromagnetic fields accompanying unilateral finger movements: pre-movement and movement-evoked fields. Exp Brain Res 78:604–612

22. Cheyne D, Weinberg H, Gaetz W, Jantzen KJ (1995) Motor cortex activity and predicting side of movement: neural network and dipole analysis of pre-movement magnetic fields. Neurosci Lett 188:81–84

23. Chiarenza GA, Villa M, Vasile G (1995) Developmental aspects of bere-itschaftspotential in children during goal-directed behaviour. Int J Psychophysiol 19:149–176

24. Chisholm RC, Karrer R (1988) Movement-related potentials and control of associated movements. Int J Neurosci 42:131–148

25. Conway BA, Halliday DM, Farmer SF, Shahani U, Maas P, Weir AI, Rosenberg JR (1995) Synchronization between motor cortex and spinal motoneuronal pool during the performance of a maintained motor task in man. J Physiol 489 (Pt 3):917–924

26. Deecke L, Weinberg H, Brickett P (1982) Magnetic fields of the human brain accompanying voluntary movement: Bereitschaftsmagnetfeld. Exp Brain Res 48:144–148

27. Erdler M, Beisteiner R, Mayer D, Kaindl T, Edward V, Windischberger C, Lindinger G, Deecke L (2000) Supplementary motor area activation preceding voluntary movement is detectable with a whole-scalp magnetoencephalography system. Neuroimage 11:697–707

28. Fawcett IP, Barnes GR, Hillebrand A, Singh KD (2004) The temporal frequency tuning of human visual cortex investigated using synthetic aperture magnetometry. Neuroimage 21:1542–1553

29. Feige B, Aertsen A, Kristeva-Feige R (2000) Dynamic synchronization between multiple cortical motor areas and muscle activity in phasic voluntary movements. J Neurophysiol 84:2622–2629

30. Ferbert A, Priori A, Rothwell JC, Day BL, Colebatch JG, Marsden CD (1992) Interhemispheric inhibition of the human motor cortex. J Physiol 453:525–546

31. Fetz EE, Chen D, Murthy VN, Matsumura M (2000) Synaptic interactions mediating synchrony and oscillations in primate sensorimotor cortex. J Physiol Paris 94:323–331

32. Fink GR, Frackowiak RS, Pietrzyk U, Passingham RE (1997) Multiple nonprimary motor areas in the human cortex. J Neurophysiol 77:2164–2174

33. Gaetz W, Cheyne D (2006) Localization of sensorimotor cortical rhythms induced by tactile stimulation using spatially filtered MEG. Neuroimage 30:899–908

34. Gross J, Kujala J, Hamalainen M, Timmermann L, Schnitzler A, Salmelin R (2001) Dynamic imaging of coherent sources: studying neural interactions in the human brain. Proc Natl Acad Sci USA 98:694–699

35. Gross J, Pollok B, Dirks M, Timmermann L, Butz M, Schnitzler A (2005) Task-dependent oscillations during unimanual and bimanual movements in the human primary motor cortex and SMA studied with magnetoencephalography. Neuroimage 26:91–98

36. Hamalainen MS, Ilmoniemi RJ (1998) Interpreting measured magnetic fields of the brain: Estimates of current distribution. Report TKK-F-A559. Helsinki University of Technology: Espoo, Finland

37. Hari R, Forss N, Avikainen S, Kirveskari E, Salenius S, Rizzolatti G (1998) Activation of human primary motor cortex during action observation: a neuromagnetic study. Proc Natl Acad Sci USA 95:15061–15065

38. Herdman AT, Wollbrink A, Chau W, Ishii R, Pantev C (2004) Localization of transient and steady-state auditory evoked responses using synthetic aperture magnetometry. Brain Cogn 54:149–151

39. Holroyd T, Endo H, Kelso JAS, Takeda T (1999) Dynamics of the MEG recorded during rhythmic index-finger extension and flexion, in Recent Advances in Biomagnetism: Proceedings of the 11th International Conference on Biomagnetism, Yoshimoto T, Kotani M, Kuriki S, Nakasato N, and Karibe H, Editors. Tohoku University Press, pp. 446–449

40. Jensen O, Goel P, Kopell N, Pohja M, Hari R, Ermentrout B (2005) On the human sensorimotor-cortex beta rhythm: sources and modeling. Neuroimage 26:347–355

41. Joliot M, Papathanassiou D, Mellet E, Quinton O, Mazoyer N, Courtheoux P, Mazoyer B (1999) FMRI and PET of self-paced finger movement: comparison of intersubject stereotaxic averaged data. Neuroimage 10:430–447

42. Jurkiewicz MT, Gaetz WC, Bostan AC, Cheyne D (2006) Post-movement beta rebound is generated in motor cortex: evidence from neuromagnetic recordings. Neuroimage 32:1281–1289

43. Kelso JAS, Fuchs A, Lancaster R, Holroyd T, Cheyne D, Weinberg H (1998) Dynamic cortical activity in the human brain reveals motor equivalence. Nature 392:814–818

44. Kilner JM, Baker SN, Salenius S, Hari R, Lemon RN (2000) Human cortical muscle coherence is directly related to specific motor parameters. J Neurosci 20:8838–8845

45. Kilner JM, Baker SN, Salenius S, Jousmaki V, Hari R, Lemon RN (1999) Task-dependent modulation of 15–30 Hz coherence between rectified EMGs from human hand and forearm muscles. J Physiol 516(Pt 2):559–570

46. Kilner JM, Fisher RJ, Lemon RN (2004) Coupling of oscillatory activity between muscles is strikingly reduced in a deafferented subject compared with normal controls. J Neurophysiol 92:790–796

47. Kim SG, Ashe J, Georgopoulos AP, Merkle H, Ellermann JM, Menon RS, Ogawa S, Ugurbil K (1993) Functional imaging of human motor cortex at high magnetic field. J Neurophysiol 69:297–302

48. Kobayashi M, Hutchinson S, Schlaug G, Pascual-Leone A (2003) Ipsilateral motor cortex activation on functional magnetic resonance imaging during unilateral hand movements is related to interhemispheric interactions. Neuroimage 20:2259–2270

49. Kornhuber HH, Deecke L (1965) Changes in the brain potential in voluntary movements and passive movements in Man: Readiness Potential and Reafferent Potentials. Pflugers Arch Gesamte Physiol Menschen Tiere 284:1–17

50. Kristeva R, Chakarov V, Wagner M, Schulte-Monting J, Hepp-Reymond MC (2006) Is the movement-evoked potential mandatory for movement execution? A high-resolution EEG study in a deafferented patient. Neuroimage 31:677–685

51. Kristeva R, Cheyne D, Deecke L (1991) Neuromagnetic fields accompanying unilateral and bilateral voluntary movements: topography and analysis of cortical sources. Electroencephalogr Clin Neurophysiol 81:284–298

52. Kristeva-Feige R, Fritsch C, Timmer J, Lucking CH (2002) Effects of attention and precision of exerted force on beta range EEG-EMG synchronization during a maintained motor contraction task. Clinical Neurophysiology 113:124–131

53. Kristeva-Feige R, Rossi S, Pizzella V, Sabato A, Tecchio F, Feige B, Romani GL, Edrich J, Rossini PM (1996) Changes in movement-related brain activity during transient deafferentation: a neuromagnetic study. Brain Res 714:201–208

54. Kristeva-Feige R, Rossi S, Pizzella V, Tecchio F, Romani GL, Erne S, Edrich J, Orlacchio A, Rossini PM (1995) Neuromagnetic fields of the brain evoked by voluntary movement and electrical stimulation of the index finger. Brain Res 682:22–28

55. Kristeva-Feige R, Walter H, Lutkenhoner B, Hampson S, Ross B, Knorr U, Steinmetz H, Cheyne D (1994) A neuromagnetic study of the functional organization of the sensorimotor cortex. Eur J Neurosci 6:632–639

56. Lang W, Cheyne D, Kristeva R, Beisteiner R, Lindinger G, Deecke L (1991) Three-dimensional localization of SMA activity preceding voluntary movement. A study of electric and magnetic fields in a patient with infarction of the right supplementary motor area. Exp Brain Res 87:688–695

57. Lemon RN (1979) Short-latency peripheral inputs to the motor cortex in conscious monkeys. Brain Res 161:150–155

58. Lemon RN, Griffiths J (2005) Comparing the function of the corticospinal system in different species: organizational differences for motor specialization. Muscle Nerve 32:261–279

59. Li A, Yetkin FZ, Cox R, Haughton VM (1996) Ipsilateral hemisphere activation during motor and sensory tasks. AJNR Am J Neuroradiol 17:651–655

60. MacKay WA (1997) Synchronized neuronal oscillations and their role in motor processes. Trends Cogn Sci 1:176–181

61. Mayston MJ, Harrison LM, Stephens JA (1999) A neurophysiological study of mirror movements in adults and children. Ann Neurol 45:583–594

62. Murase N, Duque J, Mazzocchio R, Cohen LG (2004) Influence of interhemispheric interactions on motor function in chronic stroke. Ann Neurol 55:400–409

63. Nagamine T, Kajola M, Salmelin R, Shibasaki H, Hari R (1996) Movement-related slow cortical magnetic fields and changes of spontaneous MEG- and EEG-brain rhythms. Electroencephalogr Clin Neurophysiol 99:274–286

64. Neshige R, Luders H, Shibasaki H (1988) Recording of movement-related potentials from scalp and cortex in man. Brain 111(Pt 3):719–736

65. Netz J (1999) Asymmetry in transcallosal inhibition. Electroencephalogr Clin Neurophysiol Suppl 51:137–144

66. Netz J, Ziemann U, Homberg V (1995) Hemispheric asymmetry of transcallosal inhibition in man. Exp Brain Res 104:527–533

67. Nezu A, Kimura S, Uehara S, Kobayashi T, Tanaka M, Saito K (1997) Magnetic stimulation of motor cortex in children: maturity of corticospinal pathway and problem of clinical application. Brain Dev 19:176–180

68. Nirkko AC, Ozdoba C, Redmond SM, Burki M, Schroth G, Hess CW, Wiesendanger M (2001) Different ipsilateral representations for distal and proximal movements in the sensorimotor cortex: activation and deactivation patterns. Neuroimage 13:825–835

69. Oishi M, Kameyama S, Fukuda M, Tsuchiya K, Kondo T (2004) Cortical activation in area 3b related to finger movement: an MEG study. NeuroReport 15:57–62

70. Onishi H, Soma T, Kameyama S, Oishi M, Fuijmoto A, Oyama M, Furusawa AA, Kurokawa Y (2006) Cortical neuromagnetic activation accompanying two types of voluntary finger extension. Brain Res 1123:112–118

71. Pesaran B, Pezaris JS, Sahani M, Mitra PP, Andersen RA (2002) Temporal structure in neuronal activity during working memory in macaque parietal cortex. Nat Neurosci 5:805–811

72. Pfurtscheller G, Aranibar A (1977) Event-related cortical desynchronization detected by power measurements of scalp EEG. Electroencephalogr Clin Neurophysiol 42:817–826

73. Pfurtscheller G, Lopes da Silva FH (1999) Event-related EEG/MEG synchronization and desynchronization: basic principles. Clin Neurophysiol 110:1842–1857

74. Pfurtscheller G, Neuper C, Brunner C, da Silva FL (2005) Beta rebound after different types of motor imagery in man. Neurosci Lett 378:156–159

75. Richardson AG, Overduin SA, Valero-Cabre A, Padoa-Schioppa C, Pascual-Leone A, Bizzi E, Press DZ (2006) Disruption of primary motor cortex before learning impairs memory of movement dynamics. J Neurosci 26:12466–12470

76. Riddle CN, Baker SN (2006) Digit displacement, not object compliance, underlies task dependent modulations in human corticomuscular coherence. Neuroimage 33:618–627

77. Robinson SE, Vrba J (1999) Functional neuroimaging by synthetic aperture magnetometry, in Recent Advances in Biomagnetism, T. Y, Kotani M, Kuriki S, H. K, Nakasato N, Editors. Tohoku University Press, Sendai, pp. 302–305

78. Rothwell JC (1994) Control of human voluntary movement, 2nd edn. Chapman and Hall, London

79. Salenius S, Avikainen S, Kaakkola S, Hari R, Brown P (2002) Defective cortical drive to muscle in Parkinson's disease and its improvement with levodopa. Brain 125:491–500

80. Salenius S, Schnitzler A, Salmelin R, Jousmaki V, Hari R (1997) Modulation of human cortical rolandic rhythms during natural sensorimotor tasks. Neuroimage 5:221–228

81. Salmelin R, Hari R (1994) Spatiotemporal characteristics of sensorimotor neuromagnetic rhythms related to thumb movement. Neuroscience 60:537–550

82. Sanes JN, Jennings VA (1984) Centrally programmed patterns of muscle activity in voluntary motor behavior of humans. Exp Brain Res 54:23–32

83. Schnitzler A, Salenius S, Salmelin R, Jousmaki V, Hari R (1997) Involvement of primary motor cortex in motor imagery: a neuromagnetic study. Neuroimage 6:201–208

84. Schoffelen JM, Oostenveld R, Fries P (2005) Neuronal coherence as a mechanism of effective corticospinal interaction. Science 308:111–113

85. Schulz M, Chau W, Graham SJ, McIntosh AR, Ross B, Ishii R, Pantev C (2004) An integrative MEG-fMRI study of the primary somatosensory cortex using cross-modal correspondence analysis. Neuroimage 22:120–133

86. Sekihara K, Nagarajan SS, Poeppel D, Marantz A (2004) Asymptotic SNR of scalar and vector minimum-variance beamformers for neuromagnetic source reconstruction. IEEE Trans Biomed Eng 51:1726–1734

87. Sekihara K, Nagarajan SS, Poeppel D, Marantz A, Miyashita Y (2001) Reconstructing spatio-temporal activities of neural sources using an MEG vector beamformer technique. IEEE Trans Biomed Eng 48:760–771

88. Singh KD, Barnes GR, Hillebrand A (2003) Group imaging of task-related changes in cortical synchronisation using nonparametric permutation testing. Neuroimage 19:1589–1601
89. Stefanovic B, Warnking JM, Pike GB (2004) Hemodynamic and metabolic responses to neuronal inhibition. Neuroimage 22:771–778
90. Stippich C, Blatow M, Durst A, Dreyhaupt J, Sartor K (2006) Global activation of primary motor cortex during voluntary movements in man. Neuroimage 34: 1227–1237
91. Tallon-Baudry C, Bertrand O (1999) Oscillatory gamma activity in humans and its role in object representation. Trends Cogn Sci 3:151–162
92. Taniguchi M, Kato A, Fujita N, Hirata M, Tanaka H, Kihara T, Ninomiya H, Hirabuki N, Nakamura H, Robinson SE, Cheyne D, Yoshimine T (2000) Movement-related desynchronization of the cerebral cortex studied with spatially filtered magnetoencephalography. NeuroImage 12:298–306
93. Taniguchi M, Yoshimine T, Cheyne D, Kato A, Kihara T, Ninomiya H, Hirata M, Hirabuki N, Nakamura H, Hayakawa T (1998) Neuromagnetic fields preceding unilateral movements in dextrals and sinistrals. NeuroReport 9:1497–1502
94. Timmermann L, Gross J, Dirks M, Volkmann J, Freund HJ, Schnitzler A (2003) The cerebral oscillatory network of parkinsonian resting tremor. Brain 126: 199–212
95. Van Veen BD, van Drongelen W, Yuchtman M, Suzuki A (1997) Localization of brain electrical activity via linearly constrained minimum variance spatial filtering. IEEE Transactions on Biomedical Engineering 44:867–880
96. Varela F, Lachaux JP, Rodriguez E, and Martinerie J (2001) The brainweb: phase synchronization and large-scale integration. Nat Rev Neurosci 2:229–239

Nonlinear and Cooperative Dynamics in the Human Brain: Evidence from Multimodal Neuroimaging

Andreas Meyer-Lindenberg and Danielle S. Bassett

[1] Central Institute of Mental Health, J5. D-68159 Mannheim, Germany
A.Meyer-Lindenberg@zi-mannheim.de
[2] Unit for Systems Neuroscience in Psychiatry and Neuroimaging Core Facility, Genes, Cognition and Psychosis Program, National Institute for Mental Health, NIH, DHHS, 10-3C101, 9000 Rockville Pike, Bethesda, MD 20892-1365, USA
bassettda@mail.nih.gov

1 Nonlinear Dynamics and Human Brain Function

Even a cursory review makes it clear that the operation of the brain critically depends on a complex interaction of spatially segregated neural systems. An adequate description of these interactions and an understanding of their nature are therefore an important challenge for neuroscience. While this applies to normal and abnormal brain functions, a study of the nature of cortico-cortical interactions will be needed most of all in the study of diseases and conditions in which an alteration of connectivity is assumed to play a prominent role. A case in point is schizophrenia, in which convergent evidence from neuroanatomical, neurophysiological, pharmacological and theoretical studies suggests that a disturbance of cortical connections may play an important role in producing a functionally devastating and characteristic syndrome based on a pathology that is (comparatively) subtle and possibly diffuse [17, 22, 43, 70].

If this is accepted, then a fundamental challenge to any theory of brain function will be a description of the general nature of these cortical interactions. Controversy exists ([73], and discussion) as to whether these phenomena may be adequately described as linear or whether they have an important nonlinear, or even chaotic, component. The importance of this question arises from the fact that nonlinear/chaotic systems may exhibit a variety of phenomena such as unpredictability, adaptability, critical phase transitions that make intuitive sense with regard to mental functioning but are essentially absent in linear systems [8, 28]. Again, this is especially obvious in schizophrenia, a disease in which sudden and unpredictable transitions in psychopathology are common [34]. Consequently, alterations of nonlinear dynamics in schizophrenia have been described at the level of psychopathology [13], overt behavior [47] and electrophysiology [33, 39] and are regarded by several authors as a critical feature of the alteration of connectivity in this disorder [14, 21, 22].

In this setting, the pioneering work of Kelso has provided a guiding light for our own inquiries into the nature of neural interactions. It has not only provided us with well-validated and extensively studied paradigms from the domain of motion control [32] but most of all inspired an attempt to combine qualitative insights into the functional repertoire supported by human brain function with rigorous quantitative analysis. The latter facet becomes especially important as our field of psychiatric neuroscience moves into the characterization of the effects of biological determinants such as genetic variation, which are predicted to have subtle and often diffuse effects on order parameters of brain function that will require quantitative methods to detect and – even more importantly – understand their influence on neural dynamics.

In this contribution, we will review three lines of inquiry into the topic of nonlinear dynamics in brain function encompassing multimodal neuroimaging techniques ranging from EEG, MEG and PET to transcranial magnetic stimulation. In a first group of studies, we used measures from nonlinear dynamics to describe neural activity in brain development, healthy humans and subjects with schizophrenia both untreated and treated with neuroleptic drugs [38, 39, 40, 41]. This uncovered evidence for the utility of quantitative descriptors of nonlinear dynamics (such as Lyapunov exponents or correlation dimension) to classify neural activity. Since these parameters presuppose, but cannot prove, the presence of nonlinear dynamics in the human, these studies prompted an experiment designed to directly demonstrate a nonlinear phase transition in human brain [44]. For this, we employed the bimanual motor coordination paradigm studied by Kelso [32], used PET to image brain regions differentially affected by movement instability and then directly perturbed the brain at these sites using transcranial magnetic stimulation producing a behavioral switch from unstable to stable movement pattern. Finally, we have recently extended the characterization of neural dynamics in human brain by using graph theory [6]. In this work, the global pattern of connectivity is measured in successive temporal scales using a wavelet decomposition approach and then characterized by methods from graph theory. This work reveals self-similar and small-world properties of brain connectivity that link into nonlinear dynamics and shows that the dynamical consequences of these connectivity patterns place human brain function on the "edge of criticality", linking up topology of connectivity across temporal and spatial scales with mechanisms for emergent and adaptive behavior.

2 EEG Studies of Nonlinear Dynamics in Healthy Humans and Patients with Schizophrenia

Since a first report by Babloyantz et al. [3], the analysis of the human EEG using methods from nonlinear dynamics has emerged as an important advance in understanding the collective behavior of cortical neurons. The presence of nonlinear or chaotic behavior in various physiological (e.g., sleep [3]) and pathological states (e.g., epilepsy [18]) has been not only postulated but also disputed [61].

In our studies, we investigated several parameters derived from nonlinear dynamics. A general overview of these parameters with respect to their application to schizophrenia is provided in the recent excellent review by Breakspear [8]. Our study began with the parameter dimensional complexity. It is calculated by constructing, using the method of time-phase delays [60], a geometric equivalent of the examined EEG time series, called the attractor. The dimension of this structure is then estimated (see [48, 51] for basic concepts). The dimension will usually be noninteger in a system displaying nonlinear dynamics (the attractor is a fractal). The important property of this quantity is that it may be regarded as an estimate of the number of variables minimally needed to describe the system's behavior and thus a measure of its complexity. In addition, since parameters characterizing nonlinear dynamic behavior can be shown to be fundamentally independent from parameters derived from spectral analysis (e.g., EEG power spectra), the information contained in these measures represents a new line of study that is complementary to "classical" EEG analysis.

Central to the analysis of nonlinear dynamics is the concept of the phase space [48, 51]. In a hypothetical system governed by n variables, the phase space is n-dimensional. Each state of the system corresponds to a point in phase space whose n coordinates are the values assumed by the governing variables for this specific state. If the system is observed for a period of time, the sequence of points in phase space forms a trajectory. This trajectory fills a subspace of the phase space called the system's attractor. A hallmark of nonlinear systems exhibiting complex dynamics is the sensitive dependence on initial conditions. This property makes short-term predictions about future states impossible even if the systems dynamics are completely known (as in some instances of low-dimensional deterministic chaos). This important property can be measured using Lyapunov exponents. This measure is defined as follows: two points close to one another in phase space are chosen and their evolution over time determined. In a chaotic system, the two trajectories will diverge rapidly due to the sensitive dependence on initial condition. The Lyapunov exponents are the mean exponential divergences of such trajectories calculated for each of the n dimensions of the phase space. They are ordered by descending size. If the first (biggest) Lyapunov coefficient is positive, the system exhibits, in theory, chaotic behavior [48]. In another perspective, the Lyapunov exponent may be viewed as a measure of how quickly information about the system's state is lost, and therefore, as a measure of the "unpredictability" of the underlying dynamics.

Importantly, detection of a finite attractor dimension is in itself not sufficient to deduce a system's nonlinear or chaotic behavior [46]. For this reason, methods have been devised to allow for statistical testing for nonlinear behavior. We employed a method, further described below, that constructs random time series with the same power spectrum as the EEG examined [62]. Vectors in phase space were computed by a time-delay method as described by Takens [60].

For the calculation of the correlation dimension, we used a modified correlation integral algorithm (different from that first described by Grassberger and Procaccia [26]) as found in [67]. Briefly, the correlation dimension was calculated as the mean of 200 analysis runs. In each run, 30 000 data point pairs (x_i, y_i) were randomly selected from the epoch. To correct for autocorrelation, selected point pairs were required to be separated by at least the first local minimum of the autocorrelation function. The value

$$D^2(r) = \sum_{i=1}^{30\ 000} \theta(r- \parallel x_i - y_i \parallel) \tag{1}$$

was then calculated for a range of r, where $\theta(x)$ is the heaviside function, defined as 1 if $x > 0$ and 0 if $x \leq 0$, and $\parallel \ \parallel$ denotes Euclidean norm. Thus, $D^2(r)$ counts the number of point pairs whose distance is smaller than r. Since by the definition of the correlation dimension the relation $D^2(r) \propto r^c$ holds for small c, where c is the correlation dimension, a plot of $\ln(D^2)$ versus $\ln(r)$ yields an approximately linear region of constant slope c. The linear region was numerically determined as the subset of no less than 60 plot points, with $r < 0.7$ of maximal distance, that exhibits the highest linear correlation (at least $r = 0.97$). The correlation dimension is estimated by a least-squares-fit of that linear region. This approach avoids the problem of selecting appropriate reference points on the attractor present in the original algorithm.

In conventional, spectral analysis of the EEG, coherence is a well-known measure for the amount of variability shared by the time series recorded at two electrodes. In the nonlinear domain, a similar quantity can be derived that is called mutual dimension [11]. To estimate this measure, the dimensional complexity of the two electrode sites is first calculated separately, as described above. Then, an attractor is constructed from the combined values obtained at the two sites as follows: Using an embedding dimension that is twice the maximum of the dimension used to reconstruct the original attractors (M), points in phase space are formed by taking the first M values from a time-lag embedding of the first time series and the second M values from the corresponding measurements of the second series. The time lag used for the embedding of the combined attractor in this study was the maximum of the time lags used for the reconstruction of the individual attractors. Previous analyses had shown that this corresponds well to the first minimum of the autocorrelation function of the combined time series. Denoting the dimensional complexity of the combined attractor with D_C, the dimensions of the original attractors at electrode sites A and B with D_A and D_B, respectively, the mutual dimension D_m is then defined as $D_m = D_A + D_B - D_C$. Like coherence in the linear domain, mutual dimension is a measure of variability (or information) shared, higher numerical values of the parameter corresponding to a larger proportion of information that is common at the two sites.

The first Lyapunov coefficient was calculated using an algorithm by Frank et al. [18]. An arbitrary data point is chosen. In each step, another data point is sought whose distance to the reference point is less than 10% of maximum distance, up to which dynamics are assumed to be linear, but bigger than a minimal distance (160 μV) below which noise is assumed to perturb data. The trajectories of the data points are followed for a fixed time span (0.48 s) and the logarithm of the ratio of the distance of the evolved points to the original points' distance calculated. A replacement point is then sought that is associated with minimum orientation change relative to the distance vector, and the procedure is repeated until the data is exhausted. The sum of the logarithms of the distances, divided by the whole time elapsed, is the estimate of the first Lyapunov exponent.

EEG data were statistically tested for nonlinearity by using the method of surrogate data as described by Theiler et al. [62] and Prichard and Theiler [52]. The Fourier transform of the evaluated EEG epochs was calculated. From this, surrogate time series were calculated by randomizing the phases of the complex amplitude at each frequency, accomplished by multiplying each complex amplitude by $e^{i\phi}$, where ϕ is independently chosen for each frequency from the interval $[0, 2\pi]$, symmetrizing phases by observing $\phi(f) = -\phi(-f)$, and taking the inverse Fourier transform. By construction, these time series have the same power spectrum as the original data (and thus the same mean frequency, etc.), but nonlinear relationships have been destroyed by randomization. The attractor dimension is then calculated for the surrogate data sets, as described above. The results of 80 surrogate time series are compared to those obtained with the original EEG by independent t-test. If the EEG has nonlinear properties, its attractor dimension should differ significantly from those of the surrogate (randomized) data sets. This method, thus, allows for inferential statistical testing of a time series' nonlinearity at the price of increasing computation time by a factor of > 80.

The following 18 electrode pairs were selected for the calculation of mutual dimensions: interhemispheric Fp1–Fp2, F3–F4, C3–C4, P3–P4, O1–O2, F7–F8, T3–T4, T5–T6, parietal O1–P3, O2–P4, C3–P3, C4–P4, T5–P3, T6–P4, F3–P3, F4–P4 and fronto-temporal T3–F3, T4–F4. The parietal group was selected to reflect connections of heteromodal association cortex [58], the fronto-temporal interaction because of its implication in the pathophysiology of schizophrenia [23]. For comparison, absolute spectral power and coherence were calculated from the same data using standard methodology [35]. Seven EEG bands were used as defined for pharmacoelectroencephalographical studies by the CINP [65]. Brain map images were calculated by the method of Perrin et al. [50] using surface spline interpolation on a spherical head model.

2.1 Human Brain Development

In a study of normal human development, we investigated the correlation dimension and the first Lyapunov exponent in resting EEGs of 54 healthy children (newborns – 14 years) and 12 normal adults. Analysis of variance (ANOVA) was performed with probands grouped by age. The subgroup of children older than 1 year was further examined by regression analysis.

In all analyzed epochs, Lyapunov coefficients were significantly positive ($p < 0.0001$, t-test). The presence of nonlinear dynamics was asserted statistically in 64–76% of examined epochs. A highly significant increase in correlation dimension with age was found in all examined leads ($p < 0.0001$, ANOVA). In all age groups, marked differences in correlation dimension in different brain regions became evident ($p < 0.01 - 0.0001$, ANOVA), see Fig. 1.

We concluded that evidence for the presence of nonlinearity could be found even in newborns. Brain maturation was reflected in a marked and highly significant increase in correlation dimension (complexity). Our work indicates that nonlinear dynamics analysis is suitable for measuring complexity of brain activity during maturation and provides age-dependent normal values as a basis for further study.

2.2 Nonlinear Dynamics in Schizophrenia

To gain insight into the nonlinear cortical dynamics in schizophrenia, we examined, using this methodology, the EEG of normal subjects and schizophrenics. Never-medicated, first-episode patients were compared with an unmedicated

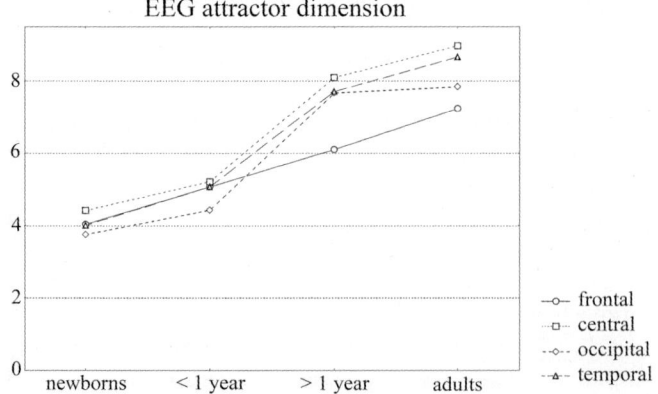

Fig. 1. Evolution of dimensional complexity with age [38]

group with chronic disease. EEG was examined under conditions of rest and cognitive stimulation. All schizophrenic patients studied had their diagnosis ascertained by means of DSM-III-R (APA 1987) criteria. Drug-naïve patients in the reported studies were examined during their first episode of disease and had never received neuroleptics. Chronic schizophrenic patients had, on the average, a disease history reaching back several years. All patients in this group had been treated with neuroleptics before, the substances having been withdrawn for five terminal half-life times prior to the EEG studies. The rest condition consisted of fixation of a blue circle on a computer monitor. In a shape perception task, one of three simple geometrical shapes was displayed and geometry of the shape was then varied continuously. Subjects were instructed to watch the shape and press both buttons simultaneously at the moment of symmetry. In a mental calculation task, six one- or two-digit numbers and binary operators "+" or "−" were displayed, and subjects were asked to mentally perform the calculation and respond afterward to two displayed results, indicating the correct one.

The results of these studies can be summarized as follows:

1. Nonlinearity of cortical dynamics could be ascertained in all studies, using the surrogate data method. However, there was no evidence for low-dimensional deterministic chaos. Generally, only weak correlations between correlation dimension and spectral power were found. No correlations were found between the Lyapunov exponent and power values. However, mutual dimension and coherence shared a significant amount of variance (about 30%, depending on study).
2. Cognitive activation led, in all examined groups, to a highly significant rise in dimensional complexity. The mental arithmetic task showed frontal activation and an activity maximum at T3. During moving shape perception, occipital activation and a right parietal activity maximum were seen. Both tasks activated central, parietal and temporal areas (Fig. 2). Mutual dimension suggested activation of a bilateral temporal-right frontal network in calculation. The Lyapunov exponent was not significantly changed by mental tasks.

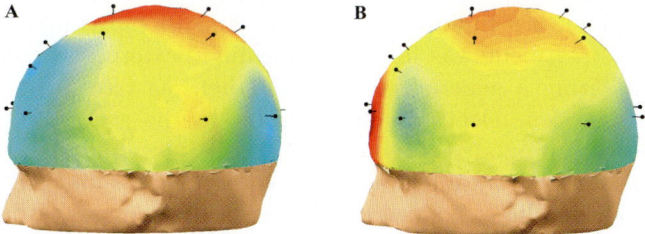

Fig. 2. Dimensional complexity during two tasks: (**A**) mental arithmetic; (**B**) processing of geometric shapes. Data from [39], plots from [40]

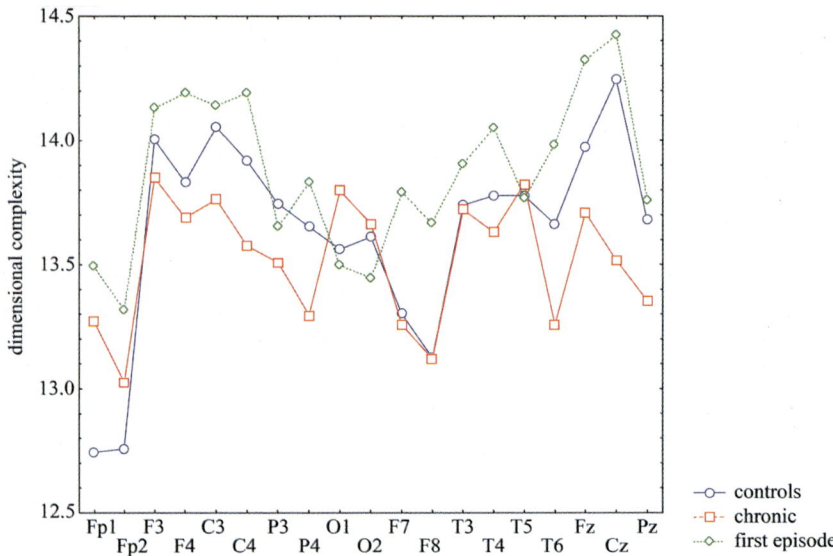

Fig. 3. Dimensional complexity in first-episode and chronic schizophrenia compared to normal controls [39]

3. At rest, schizophrenic patients show highly significantly different patterns of nonlinear complexity. In drug-naïve patients, an increase was found over the frontal brain area and a reduction occipitally. By contrast, chronic unmedicated patients showed an overall reduced complexity compared to controls (Fig. 3). Mutual dimension was increased in schizophrenic patients relative to controls, more so in the drug-naïve patients.

4. This pattern extended to cognitive activation tasks, where drug-naïve patients exhibited increased correlation dimension over the frontal brain area, compared to controls, during both tasks. Similar to the results at rest, chronic unmedicated patients showed decreased dimensional complexity over the centro-parietal area, again in both tasks.

5. When a group of chronic schizophrenic patients were followed longitudinally during antipsychotic treatment with atypical neuroleptics (zotepine or clozapine), a significant increase in dimensional complexity over the course of treatment was found. A dynamic pattern of change was seen, first exhibiting a generalized increase, then, as treatment continued, a decrease at some electrode sites relative to the early treatment phase. A stable and pronounced increase of correlation dimension was localized at the posterior frontal and anterior temporal areas on both sides. In addition, mutual dimensions were also significantly increased by treatment. The first Lyapunov exponent was not changed by therapy.

6. When atypical and typical neuroleptics (haloperidol or fluphenazine) were compared cross-sectionally in the treatment of chronic schizophrenia, we

found that classical neuroleptics differed from atypical substances, in that they produced a global, as opposed to localized, increase in dimensional complexity. Connectivity was also increased more by the typical antipsychotics.

The demonstration of significant nonlinearity in our data confirmed previous results. While we found no evidence for the presence of deterministic chaos, the clear evidence for the presence of nonlinearity together with the weak correlation of nonlinear and spectral measures indicates that significant information is accessible using EEG analysis with methods from nonlinear systems science that is not available from classical parameters. The relatively high association of mutual dimension and coherence suggests, however, that distance interactions of cortical regions may be described, as least in part, as linear.

Our results regarding cognitive activation in normal controls seem to indicate activation of cortical network that can be interpreted in terms of underlying neurophysiology, as ascertained by animal studies and examination of humans using other methods. For example, the parietal activation maximum seen may correspond to activity of the dorsal, or "where?" pathway of visual cognition, as suggested by Ungerleider and Mishkin [64]. Frontal activation in mental calculation is also predicted by a number of studies.

Based on these results, we conjectured that increases in correlation dimension may correspond to an increase in the number of independently active neural assemblies in a given cortical area. Such an increase may accompany cognitive activation of a specific area. However, as our results in drug-naïve schizophrenic patients show, such an increase may also be found associated with a pathological condition. The reduced correlation dimension found in chronic schizophrenic patients, on the other hand, could indicate a functional reorganization to preserve cortical processing integrity or be related to unstable nonlinear dynamics of the cortical network. Results under cognitive activation are consistent with this interpretation.

We posited that short-range synaptic connectivity may act as an "order parameter" in the sense used in synergetics [27], profoundly affecting global behavior of the underlying cortical network. Since neuroleptics interfere with dopaminergic neurotransmission, which has been conceptualized to affect the signal-to-noise ratio in the prefrontal cortex [42], the effects under neuroleptic treatment are interpreted to reflect the behavior of the cortical network during pharmacological manipulation of this order parameter. In particular, the dynamic nature of the alteration observed longitudinally suggests, in our view, such an interpretation. Finally, the localized nature of the increase in complexity produced by atypical neuroleptics may be related to claims of higher therapeutic efficacy of this class of drugs in treating cognitive dysfunction in schizophrenia.

3 Direct Demonstration of a Phase Transition in Human Brain

Brain function depends on the formation of, and rapid transition between, large-scale distributed neural assemblies, striking examples of self-organization supported by nonlinear phase transitions, in which a system state far from equilibrium becomes increasingly unstable (susceptible to small non-specific disturbances), eventually switching to a more stable state [27, 56]. Indeed, nonlinear properties of isolated neuronal membranes, cells and small assemblies have received much interest [15, 53], and in the human, correlative evidence comes from the observation of changes in electrophysiological compound measures that accompany altered stability at the behavioral level [30, 68]. The goal of the present study [44] was a direct demonstration of a nonlinear transition in the intact human brain. For this, we employed an interventional strategy of creating small local disturbances in cortical areas mapped out with functional neuroimaging, enabling us to investigate the hallmark characteristic of nonlinear systems, "sensitive dependence on initial conditions".

As a model system, we employed Kelso's motor coordination experiment [32], in which stability and instability are well characterized on the behavioral level, to investigate the corresponding properties on the neural level. Twelve subjects performed continuous metronome-paced movements of the index fingers of both hands in one of two patterns: either in mirror (fingers moving in phase) or in a parallel pattern (out of phase). Previous work on this well-researched paradigm [32] had demonstrated that when the frequency of the movement is increased, the mirror pattern is stable, whereas the parallel pattern destabilizes. If the frequency is high enough, it will switch into the mirror (stable) pattern. By varying one control parameter (frequency), the stability of the two movement patterns could thus be differentially manipulated.

During task performance, relative cerebral blood flow data (rCBF) were obtained using PET. Since increasing movement frequency increases the instability in the parallel pattern, but leaves the mirror pattern unaffected, the interaction of frequency and pattern was explored as the primary statistical target measure. Based on previous work showing that synaptic firing rate increases as the system nears the transition point and that changes in rCBF parallel changes in synaptic firing [54], we hypothesized that differential increases of blood flow would be found.

Significantly increased blood flow with increasing instability was found in premotor area (PMA) and supplementary motor area (SMA), cingulate, Broca's area, the left supramarginal gyrus and cerebellum (see Fig. 4). The difference between patterns became more pronounced as frequency (and thereby instability) increased. No such interaction was present in the primary sensorimotor cortex (S_1), which, in contrast, showed monotonically increasing blood flow with frequency regardless of pattern. The brain area most strongly associated with the target interaction effect was the dorsal PMA bilaterally.

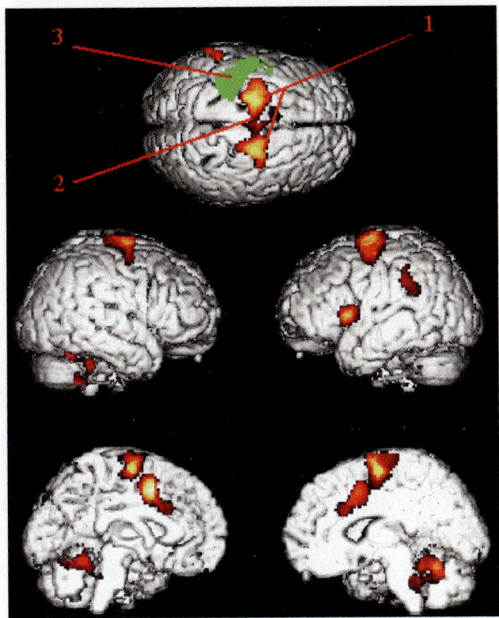

Fig. 4. Brain areas showing a significant interaction of movement pattern and frequency. Sites chosen for the TMS experiment are labeled (1) PMA, (2) SMA and (3) M_1/S_1 [44]

This was in accordance with previous work comparing in-phase and anti-phase movements at single movement frequencies [31, 55]. Convergent evidence from a multitude of studies shows that the premotor area is pivotal in motor feedback control. It has been implicated in the timing, selection, preparation and temporal control of movements [25, 66, 72]. Theoretical formulations have ascribed to the PMA a central role in the maintenance of 'motor set'. Lesions in this area lead to a disintegration of motor control [20].

The effect seen in the medial wall, comprising the posterior (behind the anterior commissure) SMA and a locus in the cingulate gyrus bilaterally, was again in accordance with results of previous nonparametric neuroimaging studies [55, 59, 63]. SMA lesions have been shown to impair bimanual coordination [9, 19]. The region of the cingulate identified here has been implicated in similar functions as the SMA to which it is closely adjacent, possibly with a larger involvement in motor execution.

Neuroimaging can only establish evidence of a correlation between behavioral instability and a cortical activation pattern. PET-guided transcranial magnetic stimulation offers the possibility of studying the underlying causal mechanisms by creating transient neural disturbances in the cortical regions of interest. We employed this approach to investigate whether the observed blood flow changes reflect instability on a neural level. In ten of the studied subjects we delivered short excitatory TMS pulses over the SMA and the PMA bilaterally during mirror and parallel 2 Hz movement. We then compared the movement phase before and after the stimulus. To control for movement effects of the procedure, a peripheral stimulation of the brachial plexus was used. To demonstrate regional specificity, a cerebral control region was also stimulated. We chose the primary somatosensory hand area, S_1, which exhibited a strong frequency-related blood flow increase, but no interaction with the movement pattern. We hypothesized that the excitatory stimulus in the PMA and SMA should be able to disrupt the unstable (parallel) but not the stable (mirror) pattern, whereas peripheral and S_1 activation would not demonstrate this effect.

Analysis of the movement before the stimulus confirmed that instability, ANOVA comparing pre- and post-stimulus movement phase showed that stimulation during parallel movement led to a significantly greater shift in relative phase (to the mirror pattern) ($F(1,9) = 32.1$, $p < 0.0005$) than during mirror movement, where hardly any change occurred. Stimulation of the SMA and the PMA had significantly greater effect than stimulation at the control sites (plexus and S_1) ($F(1,9) = 24.2$, $p < 0.001$). Post hoc testing confirmed significantly greater phase shifts when the PMA and SMA were stimulated during parallel movement than in all other stimulus conditions ($p < 0.05$, LSD test). As predicted, the TMS disturbance disrupted behaviorally unstable, but not stable movement patterns when applied over the PMA and SMA, but not the control areas. Double-pulse stimulation was sufficient in our paradigm to induce a disturbance in the unstable movement pattern. Since this mode of stimulation induces a comparatively minor disruption compared with the repetitive TMS paradigms more commonly employed [24], this result can be interpreted as a demonstration of a sensitive dependence on initial condition, a hallmark of nonlinear behavior, on the neural level.

The outlined theory would also predict that increasing behavioral instability should be associated with increased susceptibility for disturbance in the neural domain. To test this, we next performed a parametric TMS study in nine of our subjects. Movements at 1, 1.7 and 2 Hz in a parallel pattern were subjected to TMS stimulation at intensities of 70, 80, 90 and 100% at the right PMA in a fully factorial design. Repeated-measures ANOVA showed a significant ($F(6,54) = 6.25$, $p < 0.00005$) frequency by stimulation intensity interaction (Fig. 5). This demonstrated that, as predicted, the magnitude of the phase shift was dependent on the degree of instability of the pattern and on the strength of the pulse. This clear interaction of behavioral instability and TMS intensity effect indicates that as the system becomes more unstable, it also becomes more susceptible to nonspecific disturbance.

This combined neuroimaging-TMS study demonstrated neuronal dynamics that conformed to the predictions of nonlinear systems theory, including the property of "sensitive dependence on initial condition" on the neural level. From the point of view of classical neuroscience, the results are best described as a motor control feedback system placed under different demands by mirror

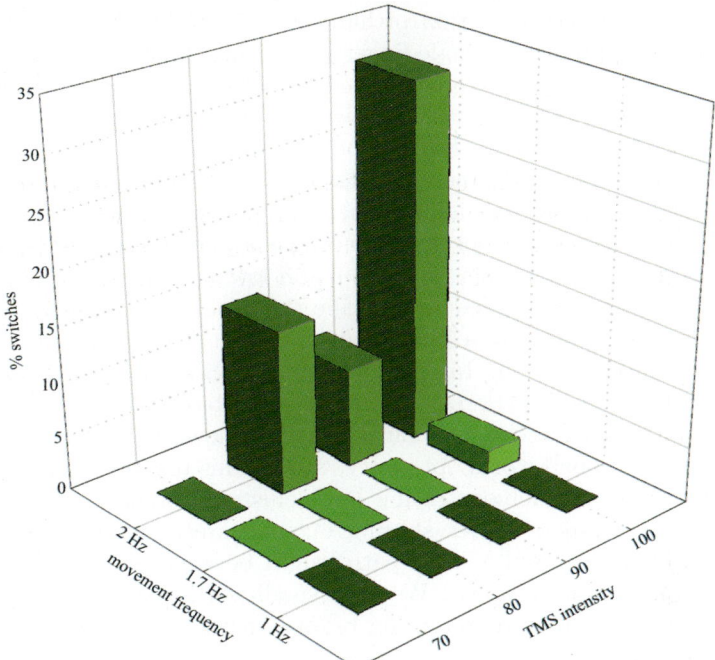

Fig. 5. Number of induced switches after TMS stimulation of varying intensity of the right PMA during parallel movement of varying frequency [44]

and parallel movements, leading to an increased susceptibility to disturbance during the latter by TMS. The present analysis shows these two interpretations to be entirely consistent, underscoring the potential of nonlinear dynamics as a unifying account of large-scale behavior of complex systems.

4 Adaptive Fractal Small-World Human Brain Functional Networks

The work summarized up to this point showed that ongoing neural activity in human brain is consistent with the presence of nonlinear dynamics and that one hallmark phenomenon, phase transitions, can be demonstrated directly using an interventional multimodal neuroimaging approach. In a recent and ongoing effort, we have begun to explicitly characterize the global topology supporting neural interactions in human brain and derive dynamical consequences which suggest that, at least in healthy humans during rest and the performance of simple motor task, brain function is positioned at the "edge of criticality", where self-organizational and adaptive processes are facilitated [6].

We used a wavelet analysis of magnetoencephalographic (MEG) signals [49], recorded from a set of 275 points overlying the scalp surface, to provide a time-frequency decomposition of human brain activity, again during rest and during performance of single-finger synchronized and syncopated-tapping tasks. Twenty-two healthy right-handed subjects participated in this study. The wavelet decomposition (using the maximum overlap discrete wavelet transform (MODWT), with the Daubechies 4 wavelet [1, 10]) was followed by a correlation analysis in the wavelet domain [71] to reveal which MEG signals represented similar physiological activity, or were functionally connected, in each of six distinct wavelet scales or frequency intervals (corresponding to the well-known EEG bands γ, β, α, θ, δ and low δ).

Graph theory was applied to describe the topological properties of adjacency matrices derived by a binary thresholding of the continuous wavelet correlation matrices [7, 16]. In particular, we pursued the possibility that human brain functional networks, like many other biological networks ranging from social interactions over cellular interactions down to metabolic networks, might exhibit a "small-world" property [2, 69]. This topology is characteristic of complex networks demonstrating both clustered or cliquish interconnectivity within groups of nodes sharing many nearest neighbors in common (like regular lattices) and a short path length between any two nodes in the network (like random graphs). We used small-world metrics to characterize these undirected graphs representing brain functional networks (Fig. 6) (for review, see [5]).

The average degree $\langle k \rangle$ of each graph was found by summing the edges between nodes throughout the whole brain network and dividing by the total number of nodes N, i.e., $\langle k \rangle = \frac{1}{N} \sum_{i=1}^{N} k_i$ where k_i is the number of other nodes that node v_i is connected to. The average path length L between any two nodes in the system was determined using Dijkstra's algorithm which gives the shortest path between any two nodes [12]. The clustering coefficient was first introduced by Watts and Strogatz as a metric to determine the interrelatedness of a region's (node's) neighbors [57, 69]. It can be constructed by letting $m(v_i)$ to be the number of opposite edges of a node v_i and $t(v_i)$ to be the number of potential opposite edges of v_i, defined as $t(v_i) = k_i(k_i - 1)/2$ while the degree k_i of the ith node is greater than 2 [57]. The clustering coefficient of node v_i is then defined as $c(v_i) = \frac{m(v_i)}{t(v_i)}$, while the clustering coefficient C of the graph is the average of the clustering coefficients of all nodes v_i, $C = \frac{1}{N} \sum_{i=1}^{N} c(v_i)$.

Random graphs with a Gaussian degree distribution will have clustering coefficients given by $C_{\text{rand}} = \frac{<k>}{N}$ [2]. The path lengths of a random graph are given by $L_{\text{rand}} = \frac{\ln N}{\ln(<k>)}$ [2]. A small-world network will be characterized by an average clustering greater than a random network and an average path length approximately equivalent to a comparable random graph, i.e., it will have a σ value greater than 1 where σ is defined as

$$\sigma = \frac{\frac{C}{C_{\text{rand}}}}{\frac{L}{L_{\text{rand}}}} \tag{2}$$

as defined in [29].

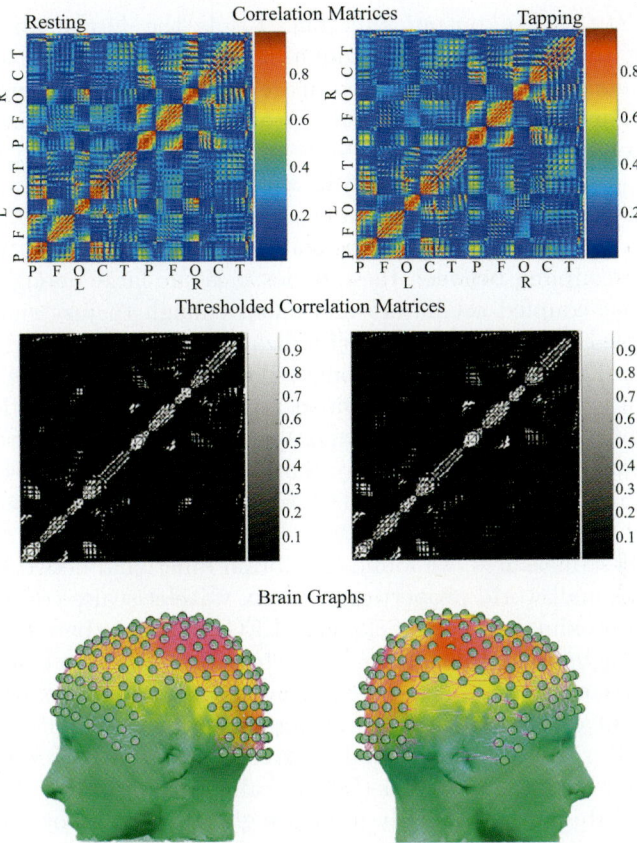

Fig. 6. Graph theoretic methodology for MEG-derived human brain functional networks for the resting state (*left column*) and the motor state (*right column*); *First row* wavelet correlation matrices (this figure shows only one matrix derived from one wavelet scale); *Second row* thresholded correlation matrices where white shows pairs of areas which should be connected; and *Third row* graphical depiction of highly correlated pairs being connected, with the darker red color indicating high node degree

We can consider the dynamical consequences of an arbitrary network topology by first defining the diagonal matrix D whose elements are equal to the degree of the respective nodes: $D(i, i) = k_i$, where k_i is the degree of the ith node. We can define the Laplacian matrix of the graph as $L = D - A$, where A is the adjacency matrix, i.e., the thresholded wavelet correlation matrix for a given wavelet scale where 1s represent a connection, 0s represent no connection and the rows and columns represent nodes.

We suppose each individual node can be coupled to each other node through some equation of motion: $\dot{x}_i = F(x_i) + \phi \sum_{j=1}^{N} L_{ij} H(x_j)$, $i = 1, \ldots, N$, where $\dot{x}_i$ is the activity of the ith node, ϕ is a uniform coupling strength, $H(x)$ is an output function, L_{ij} is the Laplacian matrix, and $\dot{x} = F(x)$ is the motion of an individual node independent of the system [4]. Although neural assemblies in the brain are likely to be coupled non-uniformly, results from uniformly coupled systems provide generalized dynamical tendencies of the system. A synchronized state of the system will then be characterized by $\{x_i(t) = x^*(t), \forall\ i\}$. Since the rows of L have zero sum, the smallest eigenvalue λ_1 is zero, and if the network is connected, $\lambda_2 > 0$. In fact, the magnitude of λ_2 is a measure of the connectivity of the graph. Furthermore, because the coupling between these nodes must be large enough to synchronize the least coupled oscillators and small enough to just synchronize the highly coupled oscillators, the synchronization threshold will depend on λ_2 and λ_N [45]. Thus, in addition to connectivity, we can determine the synchronizability of a specific network, defined as $S = \frac{\lambda_2}{\lambda_N}$, where λ_N is the largest eigenvalue of the Laplacian matrix and λ_2 is the second-smallest eigenvalue of the Laplacian. Fully synchronized systems of various types of oscillators have $0.01 < S < 0.2$; systems with $S \sim 0.01$ are close to the transition from global order to disorder [4].

Using these measures, we found that brain functional networks were characterized by small-world properties at all six wavelet scales considered, corresponding approximately to the classical EEG frequency bands. Global topological parameters (path length, L, and clustering, C) were conserved across scales, most consistently in the frequency range 2–37 Hz (δ–β), implying a scale-invariant or fractal small-world organization. The highest frequency γ network had greater synchronizability, greater clustering of connections and shorter path length than networks in the scaling regime of (lower) frequencies. Behavioral state did not strongly influence global topology or synchronizability; however, motor task performance was associated with emergence of long-range connections in both β and γ networks. Long-range connectivity, e.g., between frontal and parietal cortex, at high frequencies during a motor task may facilitate sensorimotor binding [36, 37]. Human brain functional networks demonstrate a fractal small-world architecture that supports critical dynamics and task-related spatial reconfiguration while preserving global topological parameters.

The critical link back to dynamics was afforded by the coupling analysis (Fig. 7). This dynamical analysis showed that networks were located close to the threshold of order/disorder transition in all frequency bands. The synchronizability of the networks in all scales and states was close to the threshold of 0.01, which marks the lower limit of the transition zone from globally ordered to disordered behavior in systems of coupled oscillators [4]. The γ network in both states and the β network in the motor task both have synchronizability very close to 0.01, implying that higher frequency networks in particular have critical dynamics "on the edge of chaos" which would then favor their rapid, adaptive reconfiguration in the face of changing environmental demands. This quantitative result led us to speculate that deviation of global topological and dynamical network parameters from the narrow range identified in this study

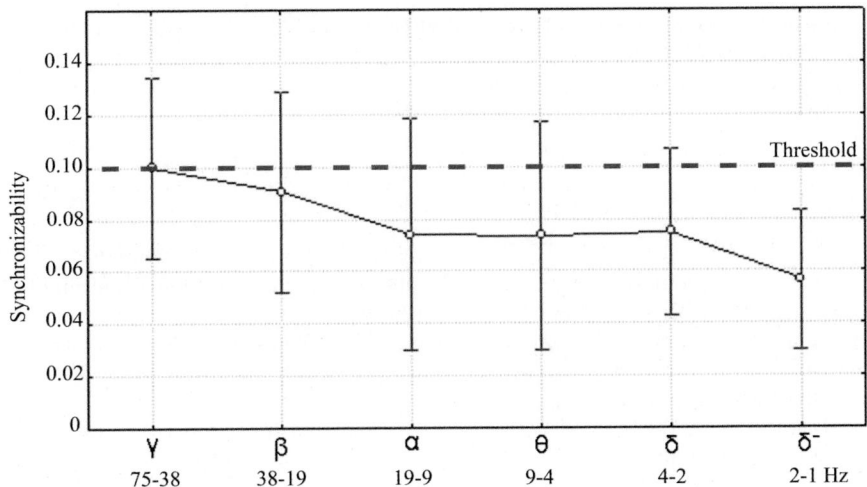

Fig. 7. Synchronizability at each of the six wavelet scales studied (relevant frequency band given in the x-axis). *Dotted red line* shows the synchronization threshold at $S = 0.10$

might be found in neuropsychiatric pathology – for example, epilepsy might be associated with synchronizability > 0.01 or neurodegenerative disorders or schizophrenia might be associated with increased path length and reduced clustering. These predictions are currently being tested.

Acknowledgments

The initial set of EEG studies was performed at the Centre for Psychiatry, University of Giessen Medical School, with help from Stefan Krieger, Stephanie Lis and Bernd Gallhofer. The multimodal study of nonlinear transitions in the human brain was performed with Karen Berman and Göran Hajak, NIMH, in collaboration with Leonardo Cohen and Ulf Zieman, NINDS. The studies of small-world networks in human brain were led by Danielle S. Bassett and performed in collaboration with Edward T. Bullmore, Sophie Achard and Thomas Duke, University of Cambridge, and with support from Shane Kippenham, Tom Holroyd, Zaid Saad, Fred Carver and Brad Zoltick, NIMH. Funding support from the NIMH Intramural Research Program is gratefully acknowledged.

References

1. Achard S, Salvador R, Whitcher B, Suckling J, Bullmore E (2006) A resilient, low-frequency, small-world human brain functional network with highly connected association cortical hubs. J Neurosci 26:63–72

2. Albert R, Barabási AL (2002) Statistical mechanics of complex networks. Rev Mod Phys 74:47–98
3. Babloyantz A, Salazar JM, Nicolis C (1985) Evidence of chaotic dynamics of brain activity during the sleep cycle. Phys Lett A 111:152–156
4. Barahona M, Pecora LM (2002) Synchronization in small-world systems. Phys Rev Lett 89:054101
5. Bassett DS, Bullmore ET (2006) Small-world brain networks. Neuroscientist 12:512–523
6. Bassett DS, Meyer-Lindenberg A, Achard S, Duke T, Bullmore ET (2006) Adaptive reconfiguration of fractal small-world human brain functional networks. Proc Natl Acad Sci USA 103:19518–19523
7. Bollobás B (2001) Random graphs. Cambridge University Press, Cambridge
8. Breakspear M (2006) The nonlinear theory of schizophrenia. Aust NZ J Psychiat 40:20–35
9. Brinkman C (1981) Lesions in supplementary motor area interfere with a monkey's performance of a bimanual coordination task. Neurosci Lett 27:267–270
10. Bullmore ET, Fadili J, Maxim V, Sendur L, Whitcher B, Suckling J, Brammer MJ, Breakspear M (2004) Wavelets and functional magnetic resonance imaging of the human brain. Neuroimage 23:S234–S249
11. Buzug T, Pawelzik K, von Stamm J, Pfister G (1994) Mutual information and global strange attractors in Taylor-Couette flow. Physica D 72:343–350
12. Dijkstra EW (1959) A note on two problems in connexion with graphs. Numer Math 1:269–271
13. Dunki RM, Ambuhl B (1996) Scaling properties in temporal patterns of schizophrenia. Physica A 230:544–553
14. Edelman GM (1992) Bright air, brilliant fire: on the matter of the mind. Basic Books, New York
15. Elbert T, Ray WJ, Kowalik ZJ, Skinner JE, Graf KE, Birbaumer N (1994) Chaos and physiology: deterministic chaos in excitable cell assemblies. Physiol Rev 74:1–47
16. Erdös P, Rényi A (1959) On random graphs. Publ Math Debrecen 6:290–297
17. Fletcher P (1998) The missing link: a failure of fronto-hippocampal integration in schizophrenia. Nat Neurosci 1:266–267
18. Frank GW, Lookman T, Nerenberg MAH, Essex C (1990) Chaotic time series analyses of epileptic seizures. Physica D 46:427–433
19. Freund HJ (1987) Differential effects of cortical lesions in humans. Ciba Found Symp 132:269–281
20. Freund HJ, Hummelsheim H (1985) Lesions of premotor cortex in man. Brain 108:697–733
21. Friston KJ (1996) Theoretical neurobiology and schizophrenia. Brit Med Bull 52:644–655
22. Friston KJ (1998) The disconnection hypothesis. Schizophr Res 30:115–125
23. Frith C (1995) Functional imaging and cognitive abnormalities. Lancet 346: 615–620
24. Gerloff C, Corwell B, Chen R, Hallett M, Cohen LG (1997) Stimulation over the human supplementary motor area interferes with the organization of future elements in complex motor sequences. Brain 120:1587–1602
25. Gerloff C, Corwell B, Chen R, Hallett M, Cohen LG (1998) The role of the human motor cortex in the control of complex and simple finger movement sequences. Brain 121:1695–1709

26. Grassberger P, Procaccia I (1983) Measuring the strangeness of strange attractors. Physica D 9:189–208
27. Haken H (1983) Synergetics. Springer-Verlag, Berlin, Heidelberg, New York
28. Haken H (1996) Principles of Brain Functioning. Springer, Berlin
29. Humphries MD, Gurney K, Prescott TJ (2006) The brainstem reticular formation is a small-world, not scale-free, network. Proc Biol Sci 273:503–511
30. Jirsa VK, Fuchs A, Kelso JAS (1998) Connecting cortical and behavioral dynamics: bimanual coordination. Neural Comput 10:2019–2045
31. de Jong BM, Willemsen AT, Paans AM (1999) Brain activation related to the change between bimanual motor programs. Neuroimage 9:290–297
32. Kelso JAS (1984) Phase transitions and critical behavior in human bimanual coordination. Am J Physiol 15:1000–1004
33. Koukkou M, Lehmann D, Federspiel A, Merlo MC (1995) EEG reactivity and EEG activity in never-treated acute schizophrenics, measured with spectral parameters and dimensional complexity. J Neural Transm Gen Sect 99:89–102
34. Lee KH, Williams LM, Breakspear M, Gordon E (2003) Synchronous gamma activity: a review and contribution to an integrative neuroscience model of schizophrenia. Brain Res Rev 41:57–78
35. Lopes da Silva FH (1993) EEG analysis: theory and practice. In: Niedermeyer E, Lopes da Silva FH (eds) Electroencephalography. Williams & Wilkins, Baltimore
36. von der Malsburg C (1981) The Correlation Theory of Brain Function. MPI Biophysical Chemistry, Internal Report 81–82
37. von der Malsburg C (1999) The what and why of binding: the modelers perspective. Neuron 24:95–104
38. Meyer-Lindenberg A (1996) The evolution of complexity in human brain development: an EEG study. Electroen Clin Neuro 99:405–411
39. Meyer-Lindenberg A (1999) Nichtlineare Dynamik im EEG gesunder Probanden und schizophrener Patienten (Nonlinear dynamics in the EEG of healthy subjects and schizophrenic patients). PhD Thesis, Justus-Liebig-University, Giessen
40. Meyer-Lindenberg A (2003) Nichtlineare Rauschunterdrckungsverfahren: algorithmische Aspekte und Anwendung auf Biosignale (Nonlinear noise supression methods: algorithms and application to biosignals). MSc, University of Hagen, Hagen
41. Meyer-Lindenberg A, Bauer U, Krieger S, Lis S, Vehmeyer K, Schuler G, Gallhofer B (1998) The topography of non-linear cortical dynamics at rest, in mental calculation and moving shape perception. Brain Topogr 10:291–299
42. Meyer-Lindenberg A, Kohn PD, Kolachana B, Kippenhan S, McInerney-Leo A, Nussbaum R, Weinberger DR, Berman KF (2005) Midbrain dopamine and prefrontal function in humans: interaction and modulation by COMT genotype. Nat Neurosci 8:594–596
43. Meyer-Lindenberg A, Poline JB, Kohn PD, Holt JL, Egan MF, Weinberger DR, Berman KF (2001) Evidence for abnormal cortical functional connectivity during working memory in schizophrenia. Am J Psychiat 158:1809–1817
44. Meyer-Lindenberg A, Ziemann U, Hajak G, Cohen L, Berman KF (2002) Transitions between dynamical states of differing stability in the human brain. Proc Natl Acad Sci USA 99:10948–10953
45. Motter AE, Zhou CS, Kurths J (2005) Enhancing complex-network synchronization. Europhys Lett 69:334–340

46. Osborne AR, Provenzale A (1989) Finite correlation dimension for stochastic systems with power-law spectra. Physica D 35:357–381
47. Paulus MP, Geyer MA, Braff DL (1996) Use of methods from chaos theory to quantify a fundamental dysfunction in the behavioral organization of schizophrenic patients. Am J Psychiat 153:714–717
48. Peitgen HO, Jürgens H, Saupe D (1992) Chaos and Fractals. New Frontiers of Science. Springer, New York, Berlin, Heidelberg, London, Paris, Tokyo, Hong Kong, Barcelona, Budapest
49. Percival DB, Walden AT (2000) Wavelet methods for time series analysis. Cambridge University Press, Cambridge
50. Perrin F, Pernier J, Bertrand O, Giard MH, Echallier JF (1987) Mapping of scalp potentials by surface spline interpolation. Electroen Clin Neuro 66: 75–81
51. Pritchard WS, Duke DW (1992) Measuring chaos in the brain: a tutorial review of nonlinear dynamical EEG analysis. Int J Neurosci 67:31–80
52. Prichard D, Theiler J (1994) Generating surrogate data for time series with several simultaneously measured variables. Phys Rev A 73:951–954
53. Rabinovich MI, Abarbanel HD (1998) The role of chaos in neural systems. Neuroscience 87:5–14
54. Rose G, Siebler M (1995) Cooperative effects of neuronal ensembles. Exp Brain Res 106:106–110
55. Sadato N, Yonekura Y, Waki A, Yamada H, Ishii Y (1997) Role of the supplementary motor area and the right premotor cortex in the coordination of bimanual finger movements. J Neurosci 17:9667–9674
56. Schöner G, Kelso JA (1988) Dynamic pattern generation in behavioral and neural systems. Science 239:1513–1520
57. Shank T, Wagner D (2004) Approximating clustering-coefficient and transitivity. Technical Report 2004–9. Universität Karlsruhe Fakultät fur Informatik, Karlsruhe
58. Stam CJ, van Woerkom TC, Pritchard WS (1996) Use of non-linear EEG measures to characterize EEG changes during mental activity. Electroen Clin Neuro 99:214–224
59. Stephan KM, Binkofski F, Halsband U, Dohle C, Wunderlich G, Schnitzler A, Tass P, Herzog H, Sturm V, Zilles K Seitz RJ, Freund HJ (1999) The role of ventral medial wall motor areas in bimanual co-ordination. Brain 122:351–368
60. Takens F (1981) Detecting strange attractors in turbulence. In: Rand DA, Young LS (eds) Dynamical systems and turbulence. Springer, Berlin
61. Theiler J (1994) On the evidence for low-dimensional chaos in an epileptic electroencephalogram. Phys Lett A 196:335–341
62. Theiler J, Eubank S, Longtin A, Galdrikian B, Farmer JD (1992) Testing for nonlinearity in time series: the method of surrogate data. Physica D 58: 77–94
63. Toyokura M, Muro I, Komiya T, Obara M (1999) Relation of bimanual coordination to activation in the sensorimotor cortex and supplementary motor area: analysis using functional magnetic resonance imaging. Brain Res Bull 48:211–217
64. Ungerleider LG, Mishkin M (1982) Two cortical visual systems. In: Ingle DG, Goodale MA, Mansfield RJW (eds) Analysis of visual behaviour. MIT Press, Cambridge

65. Versavel M, Leonard JP, Herrmann WM (1995) Standard operating procedure for the registration and computer-supported evaluation of pharmaco-EEG data. 'EEG in Phase I' of the Collegium Internationale Psychiatriae Scalarum (CIPS). Neuropsychobiology 32:166–170
66. Viviani P, Perani D, Grassi F, Bettinardi V, Fazio F (1998) Hemispheric asymmetries and bimanual asynchrony in left- and right-handers. Exp Brain Res 120:531–536
67. Wackermann J, Lehmann D, Dvorak I, Michel CM (1993) Global dimensional complexity of multi-channel EEG indicates change of human brain functional state after a single dose of a nootropic drug. Electroen Clin Neuro 86:193–198
68. Wallenstein GV, Nash AJ, Kelso JAS (1995) Slow cortical potentials and their frequency and phase characteristics preceding bimanual finger movements. Electroencephalogr Clin Neurophysiol 94:50–59
69. Watts DJ, Strogatz SH (1998) Collective dynamics of 'small-world' networks. Nature 393:440–442
70. Weinberger DR, Berman KF, Suddath R, Torrey EF (1992) Evidence of dysfunction of a prefrontal-limbic network in schizophrenia: a magnetic resonance imaging and regional cerebral blood flow study of discordant monozygotic twins. Am J Psychiat 149:890–897
71. Whitcher B, Guttorp P, Percival DB (2000) Wavelet analysis of covariance with application to atmospheric time series. J Geophys Res 105:941–962
72. Wise SM (1985) The primate premotor cortex: past, present, and preparatory. Annu Rev Neurosci 8:1–19
73. Wright JJ, Liley DTL (1996) Dynamics of the brain at global and microscopic scales: Neural networks and the EEG. Behav Brain Sci 19:285–320

Large-Scale Network Dynamics in Neurocognitive Function

Anthony R. McIntosh

Rotman Research Institute of Baycrest, University of Toronto, Toronto, ON, Canada, M6A 2E1

1 Overview

The study of human mental function is, without a doubt, at the edge of a new frontier thanks largely to neuroimaging (e.g., functional MRI, magneto- and electroencephalography). Access to human neurobiology can potentially provide the critical link between psychological theories of various cognitive functions and the concomitant physiology. Indeed, many psychological theories find their ultimate verification where the constructs appear to have a direct neural instantiation, yet few believe in an isomorphic relation between brain and mind. As such, there is a gap between how psychological constructs are represented, and how the operations of the nervous system fit with such representations. The fundamental challenge is captured by a quote from William James ([40] p. 28), "A science of the relations of the mind and brain must show how the elementary ingredients of the former correspond to the elementary functions of the later." The modern neuroscientist, and particularly a cognitive neuroscientist endowed with the new insight into human neurophysiology, must determine what features of the human brain are central in translating the biological representations to mental phenomena.

The purpose of this chapter is to present a framework for the exploration of critical linking features between brain and mind. Cognitive functions are the outcome of the complex dynamical interaction within distributed brain systems. These features arise from the anatomy and physiology of the brain. The anatomy enables a system that has a maximal capacity for both information segregation and integration. The physiological property of response plasticity, where optimal neural responses can change depending upon stimulus significance or internal network states, modifies the information as it is passed to different levels of the system. The distribution of information in the brain – here meant to indicate that information which is carried by signals among neurons (e.g., firing rate, temporal synchrony) – allows several parts of the brain to contribute to a broad range of mental function. The combination of anatomical architecture and physiology forms the basis for a *neural context*, wherein the potential contribution of one neural element to an operation is sculpted by its interactions with other elements. Such contextual influences emphasize the dynamic nature of neural function, where the relevance of that element changes as the mental function unfolds. Examples of contextual effects will be presented, and serve to develop a second linking feature: the *behavioral catalyst*, where certain neural elements are critical for a particular mental function when they enhance the transition between mental states.

2 Anatomical Considerations

Neurons are connected to one another both locally and at a distance. Most other systems in the body show some capacity for cell to cell communication, but the nervous system appears to be specialized for rapid transfer of signals. This means that a single change to the system is conveyed to several parts of the brain simultaneously and that some of this will feed back onto the initial site. There are obvious extremes to just how "connected" a system can be and the nervous system occupies some intermediate position. Local cell networks are highly interconnected, but not completely so, and this means that adjacent cells can have common and unique connections. As a more explicit illustration, consider the three networks in Fig. 1. The rightmost network has nodes that are completely connected and can be considered a redundant network. The leftmost network has disconnected elements, resulting in subnetworks that will operate independently. The central network most closely resembles the nervous system, where certain nodes are densely connected while others are more sparse. This configuration has been referred to as *degenerate* by Tononi and colleagues [71, 72], and I have used a term borrowed from graph theory – *semiconnected* – to designate this particular property of local cell networks [50].

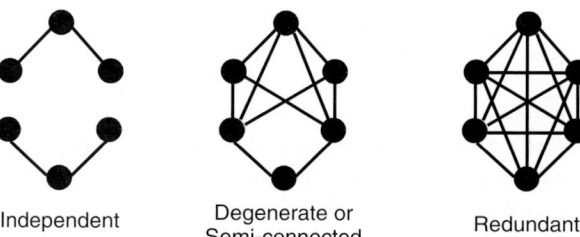

| Independent | Degenerate or Semi-connected | Redundant |

Fig. 1. Diagram of wiring diagrams for three networks. The Independent network (far left) has two sets of independent nodes which can exchange information within subnetworks but cannot between. The Redundant network (far right) has each node connected to every other node, enabling information exchange among all nodes, but with little capacity for the information to change (i.e., all nodes behave the same). The 'degenerate or semi-connected' system shows a balance between extremes and thus has the potential for optimal segregation and integration of information

Although the local networks in different cortical areas show cytoarchitectonic variation, the cellular components and internal connectivity of cortical circuits are generally similar throughout the cortex. What distinguishes the function of any local cortical network is its topological uniqueness, i.e., its particular pattern of interconnectivity with other networks. The unique set of local networks with which a given local cortical network is directly connected has been called its "connection set" [9] or "connectional fingerprint" [59]. The implication, stated simply, is that regions can only process information they receive. Cells in the medulla tend not to respond to visual stimuli because they are not connected to visual structures. Anatomy determines whether a given ensemble is capable of contributing to a process.

The connections between local neural ensembles are sparser than the intra-ensemble connectivity. Estimates of the connections in the primate cortical visual system suggest that somewhere between 30–40% of all possible connections between cortical areas exist [24]. Large-scale simulation studies show that this sparseness is a computation advantage for the nervous system in that it allows for a high degree of flexibility in responses at the system level even when the responses of individual units are fixed [70]. Additional analyses of the anatomical connections of large-scale networks in the mammalian cerebral cortex have demonstrated a number of distinct topological features that lead to systems with maximal capacity for both information segregation and integration [63, 64].

Neural network theories of brain function have, to varying degrees, emphasized these two basic organizational principles: segregation and integration. At each level in a functional system, there is a segregation of information processing such that units within that level will tend to respond more to a particular aspect of a stimulus (e.g., movement, color). At the same time the information is simultaneously exchanged with other units that are anatomically connected, allowing first for units to affect one another and second for the information processed within separate units to be integrated.

3 Functional Consideration

Neural plasticity is an established phenomenon. With maturation and following damage, there are massive changes in the functional responses of neural elements as the system adapts to the external and internal environment. These changes can take place over a few days or year. There is a type of plasticity that is more short-lived. Cells can show a rapid shift in response to afferent stimulation that is dependent on the behavioral situation in which they fire. This *transient response plasticity* occurs over a much shorter time scale compared to recovery from damage. Physiological investigation has consistently shown transient plasticity in the earliest parts of the nervous system, from single cells in isolate spinal cord preparations to primary sensory and motor structures. The change can occur within a few stimulus presentations [22, 23] and is likely a ubiquitous property of the central nervous system [75].

The generality of this feature in the brain is often neglected in considerations of the mind-brain link. Transient response plasticity is a rudimentary feature of an adaptable system. The degree of this plasticity is constrained by anatomy where, as noted above, a neural element has a restricted range of information it can respond to. A consequence of the anatomical structure is that adjacent neurons have similar response properties (e.g., orientation columns in primary visual cortex) whereas neurons slightly removed may possess overlapping, but not identical, response characteristics. These broad tuning curves are characteristic for most sensory system cells and those in motor cortices. The broad tuning curves result from anatomy, where cells share some similar and some unique connections. Interestingly, anatomy also ensures that response plasticity also has a graded distribution. This has important implications for the representational aspects of the brain. Rather than having each neuron code sharply for a single feature, the distributed response takes advantage of a division of labor across many neurons enabling representations to come from the aggregate of neuronal ensembles.

Electrophysiological studies of motor and sensory cortices have provided some examples of aggregate operations achieved through *population coding* [2, 31, 60, 77]. Population coding is likely used for higher-order cognitive functions, but on a larger scale than for sensory or motor functions. When neural populations interact with one another, these rudimentary functions combine to form an aggregate that represents cognitive processes. Cognitive operations are not localized in an area or network of regions, but rather emerge from dynamic network interactions that depend on the processing demands for a particular operation.

4 Neural Context

As noted in the opening section of this chapter, the anatomical and functional properties of the brain combine to enable a contextual dependency in how specific neural elements impact on a mental function. This neural context is the local processing environment of a given neural element that is created by modulatory influences from other neural elements [12]. Neural context allows the response properties of one element in a network to be profoundly affected by the status of other neural elements in that network. As a result of neural context, the relevance of a given neural element for cognitive function typically depends on the status of other interacting elements [9, 49]. By this definition, the processing performed by a given brain area may be modulated by a potentially large number of other areas with which it is connected. Because brain areas are most often bidirectionally connected, the neural context of each connected area emerges spontaneously from its interactions.

Furthermore, to the extent that basic sensory and cognitive operations share similar brain constituents, they experience similar neural contextual influences.

Neural context refers only to the context that arises within the brain as a result of interactions between neural elements. Another contextual influence that interacts with neural context comes from the environment: *situational context* [12]. Unlike neural context, situational context represents a host of interrelated environmental factors, including aspects of the sensory scenes and the response demands of both the external and internal milieus. A red light presented to a person in isolation usually means nothing, but a red light presented to that person while driving an automobile elicits a situationally specific response. Situational context is most often what researchers have in mind when they examine "contextual effects" on the brain [3, 4, 17, 30, 36].

In most normal circumstances, neural context is shaped by situational context. The environments in which animals and humans must survive have a high degree of structural complexity, an important consequence of which is a fundamental uncertainty in the organism's perceptuo-motor interactions with those environments. Complete information is never available to allow total certainty about the state of the environment and the optimal course of action in it. The limited information that the organism has about its environmental situation usually renders ambiguous its perceptual interpretation of environmental entities and the appropriate actions to be directed toward them. The ability to utilize and manipulate information about the organism's situational context can dramatically reduce uncertainty, thereby enhancing the organism's interactions with the environment and lending survival advantage to its species.

Since the complexity of the environment's structure spans multiple structural and temporal scales, situational context must affect all types of cognitive function, including sensation, perception, emotion, memory, planning, decision making, and action generation. It is therefore reasonable to infer that neural context should also be of primary importance in the implementation of those functions by the brain. In other words, just as situational context can have effects at multiple scales and across multiple behaviors, so too is neural context expected across all spatial and temporal scales in the brain and across all behaviors.

5 Empirical Examples

As a principle of brain function, neural context can be most easily demonstrated in relatively simple nervous systems, such as those of invertebrates. While these systems admittedly do not have the broad behavioral repertoire of primates, if contextual effects are indeed central to neural network operation, they should be present in simpler organisms. Neural context has been demonstrated in the Aplysia abdominal ganglion where the same neurons fire during performance of quite different behaviors [76]. What appears to differentiate these behaviors is not the activity of a particular neuron, or group of neurons, but rather the overall activity patterns of an entire network. Such observations have been made in other invertebrate species across highly dissimilar behaviors [61], suggesting that the observed behavioral variation resides in the large-scale dynamics of entire networks rather than dedicated circuits [46].

It has been hypothesized [10] that cortical context emerges from multiple recurrent interactions among cortical areas. An important prediction from this hypothesis is that the representation of categorical information in the cortex should be reflected by patterns of activity distributed across large cortical expanses rather than by the activity in a single specific area. Category-specificity has rapidly become a major focus in human neuroimaging research, exemplified by studies demonstrating face–category-specific responses in the fusiform gyrus [42]. However, a drawback of many such studies is that they employ very strict univariate statistical criteria that conceal all but the largest amplitudes in the activity patterns. Nonetheless, studies that have characterized the distributed response to faces have reported that greater category-specificity is revealed by the entire activity pattern in occipital and temporal cortices than by any specific area [35, 39]. Importantly, these studies have determined that the specificity of the distributed response is not dramatically altered if the regions typically associated with the category of interest are excluded. Further analysis of the distributed patterns emphasize that the most effective category specificity is captured by the specific pattern of voxel activity across the entire ventral temporoccipital region than by any single region [33, 58].

While these data seem to contradict the evidence that one can identify category-specific effects in the brain [62], this underscores an important issue in the study of representations in the brain. The brain operates at many spatial scales, synaptic, cellular, ensemble, population and what can be considered mesoscopic involving multiple populations [25, 52]. Within each of these there can be multiple time scales [7, 8]. Thus, signals that may appear discrete and carrying a very narrow piece of information (e.g., a category), become part of a larger collection of interacting units that together convey far more precise information, similar to what was presented early as population coding. It is, therefore, imperative to consider very carefully what signals exist across several spatiotemporal scales in the determination of the optimal mapping of perceptual, behavioral, or cognitive representations to brain dynamics.

6 Large-Scale Interactions and Neural Context

As contextual effects depend on the interactions between neural elements, the optimal manner to assess neural context is by estimating functional or effective connectivity. Functional connectivity emphasizes pairwise interactions, often in terms of correlations or covariances. Effective connectivity incorporates additional information, such as anatomical connections, and considers an interaction of several neural elements simultaneously to explicitly quantify the effect one element has on another. Both functional or effective connectivity were introduced in the context of electrophysiological recordings from multiple cells [1] and have been used in reference to neuroimaging data [27, 29, 38].

One well-established functional distinction in the brain is between object and spatial visual pathways. The foundation for this dual organization can be traced at least as far back as Kleist in the 1930s [44]. One of its strongest expressions to date is in the dorsal and ventral cortical processing streams, described by Ungerleider and Mishkin [73], which correspond to spatial and object processing pathways, respectively.

A similar duality was identified in humans with the aid of positron emission tomography (PET) [34]. In this experiment, a match-to-sample task for faces was used to explore object vision. For spatial vision, a match-to-sample task for the location of a dot within a square was used. The results from the right hemisphere analysis are presented in Figure 2 (left hemisphere interactions did not differ between tasks). Effects along the ventral pathway from cortical area 19v extending into the frontal lobe were stronger in the face-matching model whereas interactions along the dorsal pathway from area 19d to the frontal lobe were relatively stronger in the location-matching model. Among

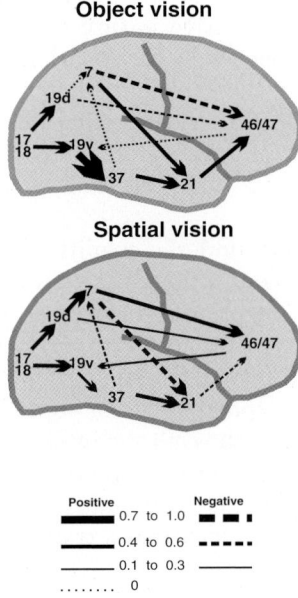

Fig. 2. Effective connectivity between cortical areas in the right hemisphere for object and spatial vision operations. The numbers on the cortical surface refer to Brodmann areas (d = dorsal, v = ventral). The arrows represent the anatomical connections between areas and the magnitude of the direct effect from one area to another is proportional to the arrow width for each path (adapted from [53])

posterior areas, the differences in influences were mainly in magnitude. Occipitotemporal interactions between area 19v and area 37 were stronger in the face-matching model while the impact of area 17/18 to 19d and the occipitoparietal influences from area 19d to area 7 was stronger in the location-matching model.

The model allowed for interactions between the dorsal and ventral pathways with connections from area 37 to area 7 and from area 7 to area 21. In the right hemisphere, the interactions among these areas showed task-dependent differences in magnitude and sign. The temporoparietal influence of area 37 on area 7 was relatively stronger in the location-matching model. The parietotemporal influence of area 7 on area 21 showed a difference in sign between the two functional models. These results show that while the strongest positive interactions in each model may have been preferentially located within one or the other pathway, the pathways did not function independently, but exerted contextual modulatory influences on one another.

Another important result of this study is that, although the prefrontal cortex (PFC) did not show a difference in mean activity between tasks, processes involving PFC shifted depending on the task. The influence of the dorsal and ventral pathways on the frontal cortex was similar in magnitude for the two tasks, but the origin of the positive and negative influences differed, implying that the qualitative nature of influence on the frontal lobe was different (positive influences in the location-matching model were from areas 7 and 19d, while in the face-matching model these were from area 21). In terms of neural context, these results demonstrate that it is not an area's activity per se that is the key to understanding its contribution to a task, but rather its pattern of interaction with other areas in large-scale networks.

Network interactions that underlie cognitive operations are observable as differences in the effective connections between elements of the network. As illustrated above, if visual attention is directed to the features of an object, effective connections among ventral posterior cortical areas tend to be stronger, whereas visual attention directed to the spatial location of objects leads to stronger interactions among dorsal posterior areas. Another way that cognitive operations may be observed is through the modulation of effective connections that occurs when one area provides an enabling condition to foster communications between other areas. Such enabling effects may represent a primary mechanism whereby situational context is translated into neural context.

7 Attention

In a study of anterior cingulate cortex (ACC) effective connectivity, Stephan et al. [65] examined whether hemispheric functional asymmetry was determined by a word stimulus (short words, with one letter colored red) itself or by the task, i.e., the situational context. In one instance, subjects judged whether the word contained the letter "A", ignoring the red letter, and in another instance, they made a visuospatial judgment indicating whether the red letter was to the right or left of center. A direct comparison of the activity (measured with fMRI) revealed strong hemispheric differences. The letter task produced higher activity in the left hemisphere, while the visuospatial task produced higher activity in the right hemisphere. The ACC was similarly active in both tasks relative to baseline, but showed distinctly different patterns of effective connectivity between tasks. Specifically, during the letter task, the ACC was coupled to the left prefrontal cortex (PFC); during the visuospatial task, the ACC was linked with the right posterior parietal cortex (PPC). These data are a compelling example of how situational context (in this case, task demands) can modulate the neural context within which a cortical area (i.e., the anterior cingulate) operates.

8 Working Memory

While working memory is often considered a unique psychological construct, another perspective emphasizes its close relation to attention [13, 21, 48]. Both working memory and sustained attention involve activity in overlapping regions of PPC, PFC, and ACC. In an fMRI study of the relationship between attention and working memory, Lenartowicz and McIntosh [47] used two variants of a two-back working memory task: a standard version with strong attentional demands and a cued version that more strongly promoted memory retrieval. Activation of ACC was found in both tasks, though it was more sustained in the standard condition. However, the regions functionally connected to the ACC, and the relation of the connectivity patterns to memory performance, differed completely between tasks. In the standard task, the observed pattern was related to a speed–accuracy tradeoff, with strong functional connection of ACC to PFC and PPC. In the cued task, the connectivity pattern was related only to better accuracy, and involved functional connections with middle and inferior PFC, and inferior temporal cortex. By virtue of these different patterns of functional connectivity, the contribution of ACC to attention- and memory-driven performance was similarly changed. In other words each task invoked a different neural context within which the ACC interacted, resulting in two very different behavioral profiles. Conversely, the difference in neural context reflected the difference in the functional role that ACC fulfilled.

9 Awareness and Neural Dynamics

In two studies of sensory associative learning, we obtained evidence that both learning crossmodal association and the awareness of such associations related to interactions among distributed cortical regions. In the first study, two tones were used having differential relations to the visual stimuli. One tone was a strong predictor of the presentation of a visual stimulus (Tone+), and the other tone a weak predictor (Tone−) [55]. In this PET rCBF study, brain activity was measured in response to isolated presentations of the Tone+ and Tone− as subjects learned. Much to our surprise, subjects in our sample divided perfectly in half into those who were aware of the stimulus associations and those who were not. The index of awareness came from debriefing questionnaires. Furthermore, only the Aware subjects learned the differentiation between the tones, while the Unaware subjects showed no behavioral evidence of learning.

In examining the underlying brain activity that supported learning and awareness, we observed that the strongest group difference in brain activity elicited by the tones was in left prefrontal cortex (LPFC) near Brodmann area 9. In Aware subjects, LPFC activity showed progressively greater activity to Tone− than to Tone+. Ventral and medial occipital cortices and right thalamus showed progressively greater activity to Tone+ than compared to Tone−. In Unaware subjects, no consistent changes were seen in LPFC or in any of the other regions. At first, these results seem to confirm the prominent role of PFC in monitoring functions [15, 66], and especially its putative role in awareness [45]. However, PFC activation has also been found in tasks where there was no overt awareness, such as in the previous sensory learning task, and in implicit novelty assessment [5]. It was thus possible that interactions of PFC with other brain regions, present in Aware but not in Unaware subjects, would better describe the neural system underlying awareness in this task.

When the interactions of LPFC was assessed between the two groups, we observed a remarkable difference in the strength and pattern of functional connections among several brain areas including right PFC, bilateral superior temporal cortices (auditory association), occipital cortex, and medial cerebellum. These areas were much more strongly correlated in Aware, than Unaware, subjects. To explore some of the network interactions within an anatomical reference, effective connectivity was assessed with structural equation modeling [53] from a subset of regions identified in the functional connectivity analysis.

There were significant changes in the effective connections for Aware subjects, including robust interactions involving LPFC (Figure 3). During the last Tone+ scan, feedback to occipital cortex was positive from temporal and prefrontal cortex, which may reflect implicit and explicit expectancy of the upcoming visual discrimination [54]. In the Tone− scan, this feedback switched to negative, which could reflect the knowledge that there would be no visual event following presentation of the Tone−. The functional network for Unaware differed from Aware subjects, but there were no significant changes in effective connections across experiment for the Unaware group. There were non-zero interactions in the functional network, but the involvement of LPFC

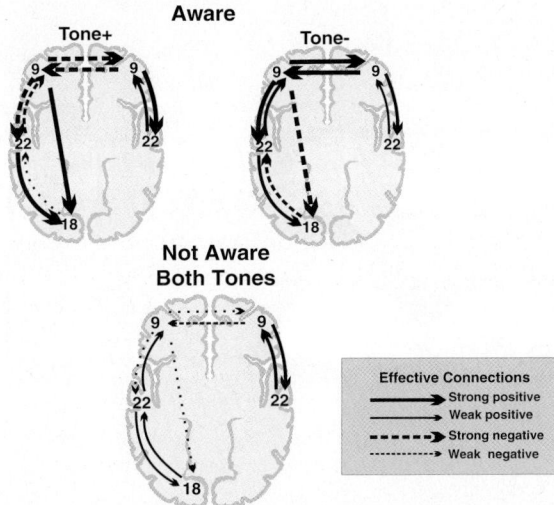

Fig. 3. Functional networks from late phases of training in a differential sensory conditioning task. Networks from two groups are shown. Aware subjects showed strong difference in effective connections involving left prefrontal area 9 and other regions that distinguished between the two tones. Conversely, the network for the Unaware subjects did not differ between tones and showed no strong left prefrontal involvement. Line thickness may be interpreted according to the legend at the bottom of the figure

was weak. This confirms that LPFC was not interacting systematically across subjects in the Unaware group.

9.1 Reversal Learning and the Medial Temporal Lobe

A more salient demonstration of neural context came from the follow-up study, looking again at differential sensory associative learning, but where the contingencies between the tones and visual stimuli were reversed part way through the study [56]. As before, subjects were classified as Aware or Unaware based on whether they noted that one of two tones predicted a visual event. Only Aware subjects acquired and reversed a differential response to the tones, but both groups showed learned facilitation.

When we related brain activity (index by blood flow measured with positron emission tomography) to behavior in each group, we observed that MTL activity related to facilitation in both groups. This was curious given the suggestion that the MTL is critical for learning with awareness, but not when learning proceeds without awareness [18, 19]. Given the principle of neural context, it was possible that this common regional involvement in the two groups was an expression of contextual dependency. We then examined the functional connectivity of the MTL and observed completely different

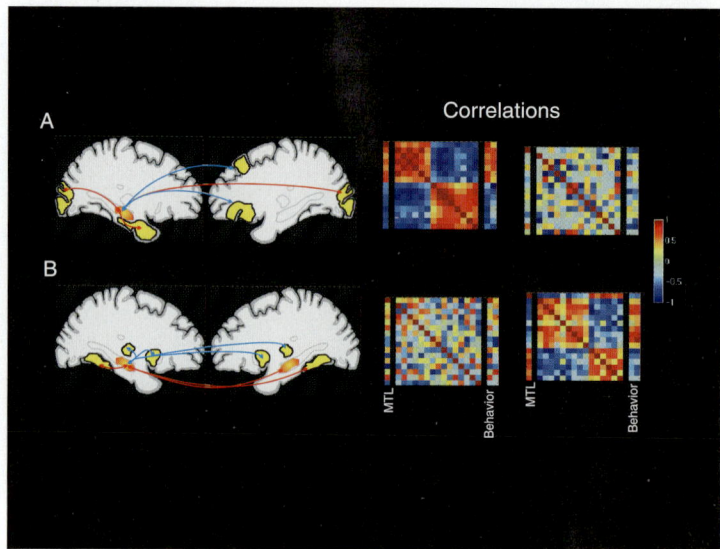

Fig. 4. Summary of the dominant functional connections of the left medial temporal lobe (MTL) in learning with (A) and without (B) awareness. Panel A shows dominant interactions, indicated by bidirectional arrows, between MTL and occipital, temporal and prefrontal cortices when learning proceeds with awareness. Panel B shows dominant interactions, between MTL with thalamus, basal ganglia, and contralateral MTL when learning proceeds without awareness. The key feature is the common involvement of the MTL, but because the regions to which it is functionally connected differ. Correlations (color coded for the range of +1 (deep red) to −1 (deep blue) as per legend) highlight the specificity of the functional connectivity to each group. The first column represents the correlation of LMTL activity with the other voxels. The middle columns represent the correlations among the voxels. The last two columns represent the correlation values for the voxels with behavior

interactions of the MTL between groups (Figure 4). In the Aware group, dominant MTL interactions were observed for prefrontal, occipital and temporal cortices, while in Unaware subjects, MTL interactions were more spatially restricted to inferotemporal, thalamus and basal ganglia. The functional connectivity of each region from the spatial pattern for each group with the MTL, with other regions in the pattern, as well with behavior are presented in Figure 4 as pseudo-colored images to illustrate two important points. First, the regions unique to each group showed strong overall functional connectivity, and second the regions related strongly with behavior only for the group from which the regional pattern was obtained. The MTL was thus part a large-scale interactive system related to learning in both groups, but only in one case was learning accompanied by awareness resulting from different constituents of the interacting systems. In other words, the differences in neural context serve as a possible explanation for the involvement of the MTL in both groups.

As stated earlier, there are many scales at which brain function can be captured and related to behavior. The MTL shows a distributed pattern of interactions that relates to learning with or without awareness, but this focuses only on how one region (albeit rather large), interacts at a mesoscopic scale. Neuroimaging data, with its broad spatial coverage, can give a comprehensive depiction of the dependency between interregional interactions and behavior. In mapping brain activity to behavior, we were able to identify two reliable spatial patterns in the Aware group that related (1) to general associative learning (Facilitation) and (2) to the differential association of the two tones (Discrimination). For the Unaware group, a single spatial pattern different from the Aware group, related to general associative learning, was identified (Facilitation), with a second pattern showing robust relations to behavior only in the early phase of the experiment (Facilitation 2). As with the assessment of MTL functional connectivity above, we explored the functional connectivity between the two behavioral patterns within each group

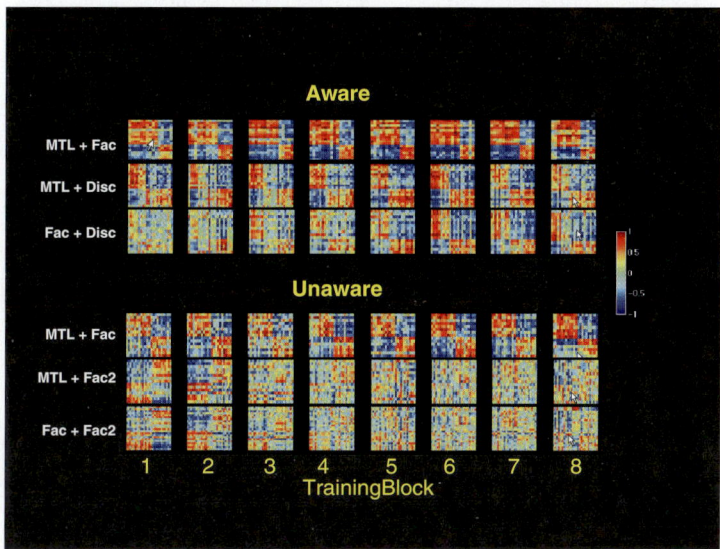

Fig. 5. Pseudocolor correlation matrices between voxels forming the functional network related to the MTL, or to behavior in subjects who learned with or without awareness (Aware and Unaware groups). The voxels in the "Fac" pattern related to learned facilitation in both groups. The "Disc" pattern was unique to Aware subjects and related to learning discrimination and the "Fac2" pattern was unique to Unaware subjects and contained additional voxels related to learned facilitation. Correlations are plotted for each of the 8 training blocks MTL activity, seed LV1 maxima and RT to T1 + V and T2 + V. Each block-specific correlation matrix is color coded for the range of +1 (deep red) to −1 (deep blue), as per the legend on the right of the figure. Each pixel within the matrices represents the correlation of activity between two voxels computed across subjects

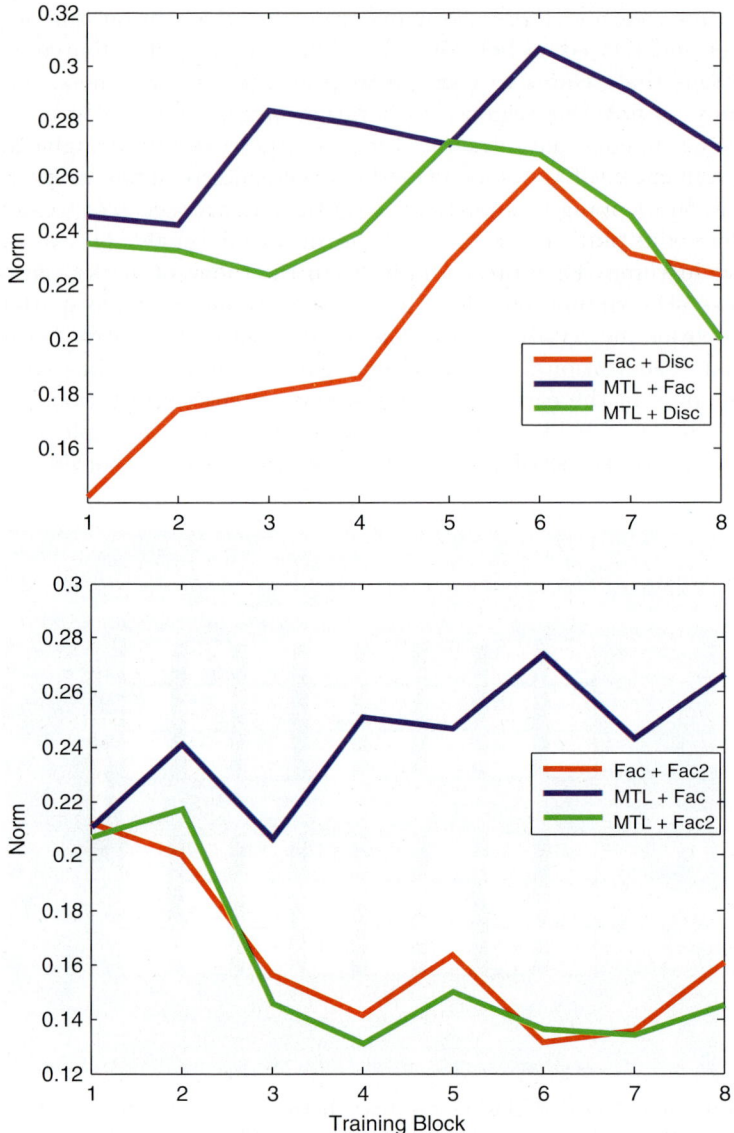

Fig. 6. Norms (first eigenvalue), scaled by the number of elements in the matrix, from the correlation matrices presented in Figure 5

and with the regional pattern for the MTL analysis. This analysis examined whether there was an even broader pattern of functional connectivity that could be related to learning for the two groups. These inter-pattern correlations, computed at each of the 8 time points during the experiment, are displayed as pseudo-colored images in Figure 5. For the Aware group, inter-pattern correlations for the MTL and Facilitation pattern and MTL and Discrimination pattern are strong across all 8 time points. However, the correlations between the two behavior patterns (i.e., Facilitation and Discrimination) became progressively stronger with time. For the Unaware group, only the MTL and Facilitation patterns maintained strong correlations across the experiment with the other two sets of correlations becoming progressively weaker across time. The impression from the correlation images can be quantified by computing the norm (first eigenvalue) of each correlation matrix. These norms, which are presented in Figure 6, show an increase for all three sets of correlations for the Aware group. For the Unaware group, the correlation between MTL and the Facilitation pattern is stable across time and the norms decline for the other two sets, indicating overall weaker correlations.

The demonstration of large-scale correlation patterns related to learning and awareness brings back an earlier point about the scales at which brain functions can occur. Had the analysis stopped at the exploration of the MTL, the likely interpretation would have emphasized the role of MTL functional connectivity in learning with awareness, much the same as the preceding study looking at prefrontal functional connectivity. This however, ignores the richness of the data where it is obvious that the pattern is far more distributed, engaging a wide expanse of the brain. Thus, just as with visual category perception, awareness and learning are embodied in the combined actions and interactions across the brain. It is not simply whether an area is active or not, but how its activity contributes to the broader neural pattern that needs to be the emphasis in considering the mind–brain link.

10 Critical Functions and Behavioral Catalysts

One problem for the notion of a neural context is the observation that the expression of certain behaviors or cognitive states appears to rely critically on specific brain areas. The overt expression of declarative memory, for example, depends on the integrity of the MTL. Such dependencies have led many researchers to speculate that such regions form a part of the neurocognitive system whose constituents subserve that function, but this implies a rather static view of brain function.

An alternative perspective, and one consistent with the idea of neural context, is to consider neural dynamics as a critical feature to understanding functional dependencies between things such as the MTL and memory. Most often studied at the cellular level, such dynamics are thought to be vital in enabling neurons to code for rapid temporal shifts in the environment and to make rapid adjustments of effectors at time scales much smaller than any single cell can achieve [57]. It is quite likely that such dynamics are at play across many levels of organization in the brain, with a similar general outcome: that it allows for rapid integration of information and responding. These same dynamics are also likely to underlie contextual effects between interacting populations that will manifest as changes in similar large-scale behaviors, such as attentional states, perceptions, memory types, and quite likely consciousness [11]. We have speculated that shifting between behavioral states may require the integrity of certain key regions, which when damaged would result in a deficit in that state. Such a region may not necessarily participate in the processing within that particular state, but rather enable the transition – it is a *behavioral catalyst* [51].

The likely singular feature of such catalysts is their anatomical relation to regions that are processing the primary information in the state in question. In the awareness study, the MTL was engaged in learning with or without awareness, interacting with regions that seemed to be related to learning in either attentive state. The MTL is anatomically connected with regions that were part of both patterns, providing the potential for the MTL to catalyze the transition between two different networks and thereby the movement from learning without awareness to learning with awareness. The critical point at which the MTL is needed is when learning moves to the conscious state. Before this, the MTL can be engaged, by virtue of its anatomical links, but not be critical for behavioral expression.

Considering regions that are critical for the expression of a function as potential catalysts emphasizes the dynamic nature of brain function. The temporal expansion of any behavior or cognitive function can be viewed as a series of transitions that require specific regions to be intact [32, 43]. In some cases, this dependence may reflect a network node that transmits information between regions (e.g., the lateral geniculate nucleus in the visual system). In other cases, areas enable the change in dominant interactions from one set of regions to another. These are the catalysts.

The temporal dimension of catalyst operations is illustrated in a very simplistic schematic (Figure 7). A 10-node network contains 3 subnetworks (A, B, C) and a catalyst node (X). Anatomically, X bridges A and C, but shows no connection with B (although the precise anatomy is less important for the present illustration). The graph in Figure 7b presents hypothetical activity profiles for the three subnetworks and the catalyst node X. Subnetwork B is active from the outset, and its impact is entirely on Subnetwork A. As Subnetwork A continues to be active, it begins to activate node X, which, after a threshold is reached, would in turn activate Subnetwork C, enabling it to process signals from Subnetwork B. This may come through modulatory effects from X that reduces thresholds on the recipient node of Subnetwork C. At the same time, feedback from X to Subnetwork A would reduce the efficacy of signals from Subnetwork B, resulting in a decline of its activity. As Subnetwork B continues to send signals, Subnetwork C maintains activity and a new function is instantiated.

The physiological plausibility of such state changes may be suspect, but I would again emphasize that effects may look different across spatial and temporal scales. A local increase in synaptic excitation may actually manifest as a general decrease of ensemble activity [74]. In terms of effective connec-

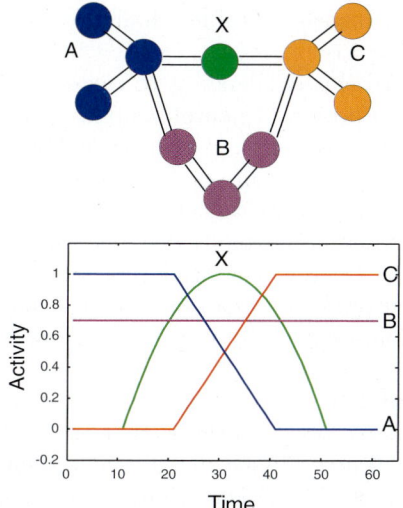

Fig. 7. Hypothesized action of a catalyst node (X) on the activity and interactivity of three subnetworks. The top of the figure depicts the connections between the three subnetworks A (blue), B (magenta), C (orange), and the catalyst node (green). The activity (hypothetical units) across time of the three subnetworks and the catalyst node is plotted in graph (hypothetical units). At a critical point in time, the catalyst node facilitates the transition of interactions from A and B to C and B

tivity, at the mesoscopic level, the same region can have a facilitory effect or suppressive effect depending on situational or neural context. Indeed, we have shown that feedback effects can move from suppression to facilitation as a function of learning [54], while others have shown that mutually antagonist processing stream can facilitate one another if the situational context requires it [14]. All of this is meant to emphasize that, while there may be general principles that govern brain function, their expression may be different across spatial and temporal scales.

11 Conclusions

I have made the argument that the anatomical structure of the brain enhances segregation and integration of information, and this paired with transient plasticity and population coding, makes for a whole system that is specialized for a broad range of mental functions. There are other features that I have not presented here that are necessary to consider in developing the mind–brain link, such as predictive modeling [20, 28, 37], and nonlinear dynamics [6, 26, 41, 43]. With these general features, the same neural mechanisms that underlie perceiving a face, or hearing a siren or symphony, also give rise to attention, memory and consciousness. Such a general mechanism produces a formidable, but not insurmountable, challenge for cognitive neuroscience theory.

A direct response to this challenge is to revise the concepts that link the brain to cognition. The notion of neural context and catalysts is an attempt to put forward such new concepts. While psychological theories tend to differentiate cognitive processes from operations such as sensation and action, it is important to recognize that sensation, perception, attention and memory are usually intimately intertwined. The inconvenient overlap in activation patterns for putatively different cognitive functions [16] is an indication that there are common neurophysiological mechanisms that support these diverse cognitive functions. Thus, the division between psychological constructs is not absolute in the brain.

Another important new concept that embodies general brain principles is evident in Tononi's [67, 68] hypothesis that the highest form of cognition, consciousness, can be directly linked to a neural system's capacity to integrate information. This capacity is a direct result of the physiology and anatomy of the nervous system. Formulated this way, consciousness, and therefore most of cognition, reflects a potentiality of a nervous system, rather than a special property that only exists in the primate brain. Indeed, the potential can be measured quantitatively [69], enabling a more direct assessment of the dynamics of a nervous system and the richness of the behavioral functions it generates. The interesting implication of considering cognitive functions from this perspective is that it dissociates the basic biological capacity from the phenomenology of our personal experience. It seems somewhat dissatisfying to conceive of the mind–brain link in this way, but in doing so, it opens a new venue for investigation of information integration, contextual effects and potential catalysts without having to label such endeavors as looking only at sensation, or attention, or memory. The positive consequence is that one can then measure the brain using the same quantitative tools and neural principles to characterize the transition between different cognitive functions. This necessarily yields a picture of human mental function that is more integrated across cognitive and behavioral domains, and at the same time, grounded in emerging principles of brain operation.

References

1. Aertsen A, Bonhoeffer T, Kruger J (1987) Coherent activity in neuronal populations: analysis and interpretation. In: Caianiello ER (ed) Physics of cognitive processes. World Scientific Publishing, Singapore, pp. 1–34
2. Averbeck BB, Latham PE, Pouget A (2006) Neural correlations, population coding and computation. Nat Rev Neurosci 7:358–366
3. Bar M (2004) Visual objects in context. Nat Rev Neurosci 5:617–629
4. Beck DM, Kastner S (2005) Stimulus context modulates competition in human extrastriate cortex. Nat Neurosci 8:1110–1116

5. Berns GS, Cohen JD, Mintun MA (1997) Brain regions responsive to novelty in the absence of awareness. Science 276:1272–1275

6. Breakspear M (2004) "Dynamic" connectivity in neural systems: theoretical and empirical considerations. Neuroinformatics 2:205–226

7. Breakspear M, Stam CJ (2005) Dynamics of a neural system with a multiscale architecture. Philos Trans R Soc Lond B Biol Sci 360:1051–1074

8. Breakspear M, Bullmore ET, Aquino K, Das P, Williams LM (2006) The multiscale character of evoked cortical activity. Neuroimage 30:1230–1242

9. Bressler S (2003) Context rules. Commentary on Phillips WA & Silverstein SM, Convergence of biological and psychological perspectives on cognitive coordination in schizophrenia. Behav Brain Sci 26:85

10. Bressler SL (2004) Inferential constraint sets in the organization of visual expectation. Neuroinformatics 2:227–238

11. Bressler SL, Kelso JAS (2001) Cortical coordination dynamics and cognition. Trends Cogn Sci 5:26–36

12. Bressler S, McIntosh AR (2007) The role of neural context in large-scale neurocognitive network operations. In: Jirsa VK, McIntosh AR (eds) Handbook of Brain Connectivity. Springer, Berlin

13. Bressler SL, Tognoli E (2006) Operational principles of neurocognitive networks. Int J Psychophysiol 60:139–148

14. Buchel C, Coull JT, Friston KJ (1999) The predictive value of changes in effective connectivity for human learning. Science 283:1538–1541

15. Burgess PW, Shallice T (1996) Response suppression, initiation and strategy use following frontal lobe lesions. Neuropsychologia 34:263–272

16. Cabeza R, Nyberg L (2000) Imaging cognition II: an empirical review of 275 PET and fMRI studies. JCNs 12:1–47

17. Chun MM (2000) Contextual cueing of visual attention. Trends Cogn Sci 4:170–178

18. Clark RE, Squire LR (1998) Classical conditioning and brain systems: the role of awareness. Science 280:77–81

19. Clark CM, Squire LR (2000) Awareness and the conditioned eyeblink response. In: Woodruff-Pak DS, Steinmetz JE (eds) Eyeblink classical conditioning Vol I: applications in humans. Kluwer Academic Publishers, Norwell, MA, pp. 229–253

20. Dayan P, Hinton GE, Neal RM, Zemel RS (1995) The Helmholtz machine. Neural Comput 7:889–904

21. Deco G, Rolls ET (2005) Attention, short-term memory, and action selection: a unifying theory. Prog Neurobiol 76:236–256

22. Dorris MC, Pare M, Munoz DP (2000) Immediate neural plasticity shapes motor performance. J Neurosci 20:RC52

23. Edeline JM, Pham P, Weinberger NM (1993) Rapid development of learning-induced receptive field plasticity in the auditory cortex. Behav Neurosci 107:539–551

24. Felleman DJ, Van Essen DC (1991) Distributed hierarchical processing in the primate cerebral cortex. Cereb Cortex 1:1–47

25. Freeman WJ (2000) Mesoscopic neurodynamics: from neuron to brain. J Physiol Paris 94:303–322

26. Freeman WJ, Holmes MD (2005) Metastability, instability, and state transition in neocortex. Neural Netw 18:497–504

27. Friston K (1994) Functional and effective connectivity: a synthesis. HBM 2:56–78
28. Friston KJ (1997) Transients, metastability, and neuronal dynamics. Neuroimage 5:164–171
29. Friston KJ, Frith C, Fracowiak R (1993) Time-dependent changes in effective connectivity measured with PET. HBM 1:69–79
30. Garraux G, McKinney C, Wu T, Kansaku K, Nolte G, Hallett M (2005) Shared brain areas but not functional connections controlling movement timing and order. J Neurosci 25:5290–5297
31. Georgopoulos AP, Schwartz AB, Kettner RE (1986) Neuronal population coding of movement direction. Science 233:1416–1419
32. Haken H (1996) Principles of brain functioning: a synergetic approach to brain activity, behavior and cognition. Springer, Berlin
33. Hanson SJ, Matsuka T, Haxby JV (2004) Combinatorial codes in ventral temporal lobe for object recognition: Haxby (2001) revisited: is there a "face" area? Neuroimage 23:156–166
34. Haxby JV, Grady CL, Horwitz B (1991) Two visual processing pathways in human extrastriate cortex mapped with positron emission tomography. In: Lassen NA, Ingvar DH, Raichle ME, Friberg L (eds) Brain work and mental activity (Alfred Benzon Symposium 31). Munksgaard, Copenhagen, pp. 324–333
35. Haxby JV, Gobbini MI, Furey ML, Ishai A, Schouten JL, Pietrini P (2001) Distributed and overlapping representations of faces and objects in ventral temporal cortex. Science 293:2425–2430
36. Hepp-Reymond M, Kirkpatrick-Tanner M, Gabernet L, Qi HX, Weber B (1999) Context-dependent force coding in motor and premotor cortical areas. Exp Brain Res 128:123–133
37. Hinton GE, Dayan P (1996) Varieties of Helmholtz machine. Neural Netw 9:1385–1403
38. Horwitz B (2003) The elusive concept of brain connectivity. Neuroimage 19:466–470
39. Ishai A, Ungerleider LG, Martin A, Schouten JL, Haxby JV (1999) Distributed representation of objects in the human ventral visual pathway. Proc Natl Acad Sci USA 96:9379–9384
40. James W (1890) The principles of psychology. Dover Publications Inc, Boston, MA
41. Jirsa VK, Kelso JA (2000) Spatiotemporal pattern formation in neural systems with heterogeneous connection topologies. Phys Rev E Stat Phys Plasmas Fluids Relat Interdiscip Topics 62:8462–8465
42. Kanwisher N, McDermott J, Chun M (1997) The fusiform face area: a module in human extrastriate cortex specialized for face perception. J Neurosci 17:4302–4311
43. Kelso JAS (1995) Dynamic patterns: the self-organization of brain and behavior. MIT Press, Cambridge, MA
44. Kleist K (1935) Ueber Form und Orstsblindheit bei Verletzungen des Hinterhautlappens. Deutsch Z Nervenheilk 138:206–214
45. Knight RT, Grabowecky MF, Scabini D (1995) Role of human prefrontal cortex in attention control. Adv Neurol 66:21–34
46. Kristan WB, Jr., Shaw BK (1997) Population coding and behavioral choice. Curr Opin Neurobiol 7:826–831

47. Lenartowicz A, McIntosh AR (2005) The role of anterior cingulate cortex in working memory is shaped by functional connectivity. JCNs 17:1026–1042

48. McElree B (2001) Working memory and focal attention. J Exp Psychol Learn Mem Cogn 27:817–835

49. McIntosh AR (1999) Mapping cognition to the brain through neural interactions. Memory 7:523–548

50. McIntosh AR (2000) From location to integration: how neural interactions form the basis for human cognition. In: Tulving E (ed) Memory, consciousness, and the brain: the tallinn conference. Psychology Press, Philadelphia, PA

51. McIntosh AR (2004) Contexts and catalysts: a resolution of the localization and integration of function in the brain. Neuroinformatics 2:175–182

52. McIntosh AR (2007) Mesoscale brain dynamics in memory coding and representation. In: Roediger HL, Dudai Y, Fitzpatrick S (eds) Science of memory: concepts. University Press, Oxford

53. McIntosh AR, Gonzalez-Lima F (1994) Structural equation modeling and its application to network analysis in functional brain imaging. HBM 2:2–22

54. McIntosh AR, Cabeza RE, Lobaugh NJ (1998) Analysis of neural interactions explains the activation of occipital cortex by an auditory stimulus. J Neurophysiol 80:2790–2796

55. McIntosh AR, Rajah MN, Lobaugh NJ (1999) Interactions of prefrontal cortex related to awareness in sensory learning. Science 284:1531–1533

56. McIntosh AR, Rajah MN, Lobaugh NJ (2003) Functional connectivity of the medial temporal lobe relates to learning and awareness. J Neurosci 23:6520–6528

57. Milton JG, Mackey MC (2000) Neural ensemble coding and statistical periodicity: speculations on the operation of the mind's eye. J Physiol Paris 94:489–503

58. O'Toole AJ, Jiang F, Abdi H, Haxby JV (2005) Partially distributed representations of objects and faces in ventral temporal cortex. J Cogn Neurosci 17:580–590

59. Passingham RE, Stephan KE, Kotter R (2002) The anatomical basis of functional localization in the cortex. Nat Rev Neurosci 3:606–616

60. Pasupathy A, Connor CE (2002) Population coding of shape in area V4. Nat Neurosci 5:1332–1338

61. Popescu IR, Frost WN (2002) Highly dissimilar behaviors mediated by a multifunctional network in the marine mollusk Tritonia diomedea. J Neurosci 22:1985–1993

62. Spiridon M, Kanwisher N (2002) How distributed is visual category information in human occipito-temporal cortex? An fMRI study. Neuron 35:1157–1165

63. Sporns O, Kotter R (2004) Motifs in brain networks. PLoS Biol 2:e369

64. Sporns O, Zwi JD (2004) The small world of the cerebral cortex. Neuroinformatics 2:145–162

65. Stephan KE, Marshall JC, Friston KJ, Rowe JB, Ritzl A, Zilles K, Fink GR (2003) Lateralized cognitive processes and lateralized task control in the human brain. Science 301:384–386

66. Stuss DT, Benson DF (1987) The frontal lobes and control of cognition and memory. In: Perecman E (ed) The frontal lobes revisited. The IRBN Press, New York, pp. 141–158

67. Tononi G (2004) An information integration theory of consciousness. BMC Neurosci 5:42.

68. Tononi G (2005) Consciousness, information integration, and the brain. Prog Brain Res 150:109–126
69. Tononi G, Sporns O (2003) Measuring information integration. BMC Neurosci 4:31.
70. Tononi G, Sporns O, Edelman GM (1992) Reentry and the problem of integrating multiple cortical areas: simulation of dynamic integration in the visual system. Cereb Cortex 2: 310–335
71. Tononi G, Sporns O, Edelman GM (1994) A measure of brain complexity: relating functional segregation and integration in the nervous system. Proc Natl Acad Sci USA 91:5033–5037
72. Tononi G, Sporns O, Edelman GM (1999) Measures of degeneracy and redundancy in biological networks. Proc Natl Acad Sci USA 96:3257–3262
73. Ungerleider LG, Mishkin M (1982) Two cortical visual systems. In: Ingle DJ, Goodale MA, Mansfield RJW (eds) Analysis of visual behavior. MIT Press, Cambridge, MA pp. 549–586
74. van Vreeswijk C, Sompolinsky H (1996) Chaos in neuronal networks with balanced excitatory and inhibitory activity. Science 274:1724–1726.
75. Wolpaw JR (1997) The complex structure of a simple memory. Trends Neurosci 20:588–594
76. Wu JY, Cohen LB, Falk CX (1994) Neuronal activity during different behaviors in Aplysia: a distributed organization? Science 263:820–823
77. Young MP, Yamane S (1992) Sparse population coding of faces in the inferotemporal cortex. Science 256:1327–1331

Neural Indices of Behavioral Instability in Coordination Dynamics

Olivier Oullier[1] and Kelly J. Jantzen[2]

[1] Laboratoire de Neurobiologie Humaine (UMR 6149), Aix-Marseille Université, Marseille, France
[2] Department of Psychology, Western Washington University, Bellingham, WA 98225, USA

> *So, the whole reason the French people can't really dance, is because they haven't got the beat in their blood. And why don't they live and feel the beat? It's because their language has no tonic accent [...] This is something about the energy you can find in music. I mean specifically African music. As I understand, it's dynamic and bouncy because it's driven by the beat. And it's syncopated of course. The down-beat is actually the upbeat*
>
> Martin Solveig (Sur la Terre, 2002)

1 Introduction

In studies of coordination dynamics, behavioral coordination has proven a rich entry point for uncovering principles and mechanisms of human action [46, 104]. Within this conceptual and theoretical framework, coordination is defined in terms of collective (or coordination) variables that specify the spatiotemporal ordering between component parts. In the vicinity of critical points, emergent behavior is governed by the low-dimensional dynamics of these collective variables [33]. Seminal studies of motor coordination conducted in the late 1970s used nonlinear dynamics as a framework to understand bimanual coordination [44, 45, 52, 56]. The influential results of this work demonstrated the self-organized nature of coordinated rhythmic behavior by showing that the global pattern generated by the combined movement of individual fingers is captured at the collective level by the value of an *order parameter* that, in this and many cases, turns out to be the relative phase between the coordinated elements. The low-dimensional dynamics of this self-organized system is revealed via manipulating a nonspecific parameter referred to as *control parameter* that guides the system through its various states without directly specifying those states. A quantitative change of the control parameter gives rise to a qualitative change of the order parameter *via* a *nonequilibrium phase transition* [55]. Such transitions, together with other key features, including *critical slowing down* and *multi-stability*, are classic hallmarks of self-organizing systems [46]. Intuitively, transitions may be thought of as a means for the system to adopt a more comfortable regime when constraints become too high, just like when walking becomes uncomfortable and one naturally transitions into running without adopting an intermediate pattern [20].

In the present chapter, we illustrate the key ideas and features of coordination dynamics of brain and behavior using a simple experimental setting: coordinating one's movement with external information. We avoid a lengthy review of the sensorimotor coordination literature (recent detailed treatments being available elsewhere [36, 43, 78, 99]). Instead, we will highlight three studies that bring into focus the specific neural basis of coordination. The point is not simply to image the brain and offer a neural metaphor of sensorimotor coordination but to propose (elements of) mechanisms of sensorimotor coordination that have been addressed in a mostly speculative way within the behavioral literature (cf. several chapters in [43]). The experiments we report here represent an approach to describe and understand the neural foundations of fundamental coordination laws described elsewhere in this volume (cf. [27, 80]). In section two, we outline the empirical and theoretical foundations on which coordination dynamics is formed. In section three, we briefly review the imaging literature relevant to coordination dynamics and our discussion. Subsequent sections introduce recent experimental studies that we feel are important for understanding cerebral contributions to coordination dynamics.

2 Behavioral Stability

Coordination dynamics treats the problem of sensorimotor coordination between oneself and their environment as a pattern-forming process [48]. In the paradigmatic case, a temporal coupling is required between a finger flexion/extension movement and a periodic stimulus. Although any number of spatiotemporal relationships may be possible, two dominant patterns emerge under the instructional constraint to maintain a one-to-one relationship between movements and metronome.

Synchronization is defined by the temporal coincidence between peak finger flexion and an environmental stimulus such as a beep or a light flash. In the ideal case, the relative phase difference between flexion of the finger and each metronome pulse is 0°. Intentional synchronization requires only a small number of cycles to establish and can be maintained in a relatively easy way when performed between 1 and 3 Hz [24]. This is also the case for nonintentional synchronization, a feature that humans often exhibit spontaneously toward external events such as a song played on the radio, a conversation, a moving object, or movements of other persons [6, 65, 70, 71, 73, 98].

Syncopation requires each movement to be performed directly between consecutive beats, i.e., with a 180° relative phase between finger flexions and metronome pulses. Within a bi-stable regime defined for movement rates below approximately 2 Hz, both syncopation and synchronization are accessible. However, whereas synchronization can be carried on quite accurately up to 4 Hz, syncopation cannot be maintained over 2 Hz [25, 26]. Increasing movement rate beyond an individually defined critical frequency moves the system into a monostable regime where spontaneous switches from syncopation to synchronization are observed [48]. It is noteworthy that, for movement frequencies under 0.75 Hz, both synchronization and syncopation are no longer rhythmic but more a series of discrete, reactive movements [23].

The key dynamic features of the behavioral coordination paradigm are depicted in the top panels of Fig. 1A–C. Participants start out coordinating in a syncopated fashion with a metronome presented at approximately 1.0 Hz (Fig. 1A). The temporal pattern is maintained with relatively little variability across movement cycles (Fig. 1B and C). As the rate of the metronome progressively increases (in this case using a step size of 0.25 Hz), syncopation becomes progressively less stable (more variable). At the critical frequency (in this case 2 Hz), the syncopated pattern becomes unstable and spontaneous transitions to synchronization ensue [23, 24, 25, 26, 48].

Initial theoretical considerations addressing syncopation to synchronization transitions transposed the HKB model for bimanual coordination dynamics [33] to nonsymmetrical oscillators [48], symmetry being a key feature in coordination dynamics [1, 16, 90]. This model revealed the tendency of the coupled system to functionally explore the patterns that can be adopted in the vicinity of the phase transition. In this region, the system can adopt potentially either coordination pattern. This feature is called *bi-stability* and is accompanied at the coordination level by *critical fluctuations* (Fig. 1, yellow overlay). These fluctuations (of the order parameter) are expressed through a temporary increase of variability of the relative phase between the pulse and the flexion. Critical fluctuations happen because of the temporary loss of stability of the pattern induced by the increase (or decrease) of a control parameter [54]. They constitute a key feature of dynamical systems and reveal the proximity of the phase transition from one pattern to another [31, 32]. However, if rate is decreased, there is no "back-transition" from syncopation to synchronization the way it would happen from running to walking for example [20]. Overall, syncopation is intrinsically less stable than synchronization – the variability of the metronome-flexion relative phase being higher – even in a frequency range where both patterns can be maintained accurately (between 0.75 and 2 Hz).

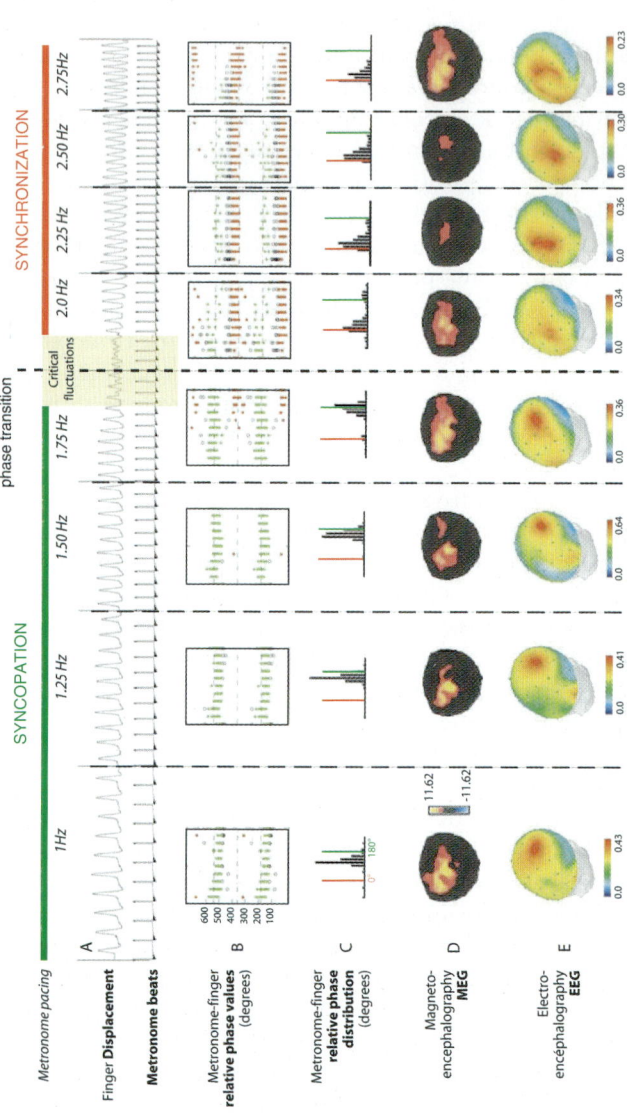

Fig. 1. Shared behavioral and neural dynamics of nonlinear phase transitions (adapted from [78]). In the paradigmatic sensorimotor coordination experiment, participants start out syncopating (top green bar) by timing finger flexion/extension movements (A, top trace) with a periodic auditory metronome (A, bottom trace). Metronome frequency is increased every ten cycles from 1 to 2.75 Hz in 0.25 Hz increments. (B) The relative phase between the peak finger flexion and metronome onset provides a measure of the coordination pattern which tends toward 180° (syncopation, green circles) and 0° (synchronization, red circles). (C) The distribution of relative phase values across a frequency plateau. The perfect synchronization and syncopation are indicated by the red and green vertical lines, respectively. Panels A–C illustrate a phase transition from syncopation to synchronization between 1.75 and 2.0 Hz. (D) Magnetoencephalography (MEG) maps of β-power (20–30 Hz) for a single participant at each frequency plateau. (E) Electroencephalography (EEG) maps resulting from the averaging of all participants at each frequency plateau. Large-scale changes in the spatiotemporal pattern of neural activity observed in both MEG and EEG are coincident with the spontaneous reorganization of the behavioral pattern (A, B, & C are adapted from [30]; D is adapted from [63], and E is adapted from [62])

3 Neural Indices of Behavioral (In)stability

Studies in the dynamical systems framework employed sensorimotor co-ordination tasks to uncover the link between the dynamics of behavior and the dynamics of the brain, connecting these levels by virtue of their shared dynamics [29, 47, 91]. The high temporal resolution of electroen-cephalography (EEG) and magnetoencephalography (MEG) was exploited to quantify the relationship between behavioral and spatiotemporal patterns of neural activity. These data offer a conceptual link between the large-scale neural dynamics emerging from billions of neurons (and their countless interconnections) and the behavioral dynamics revealed in coordination ex-periments. Common features of the dynamics expressed at both levels of de-scription, including phase transitions, were taken as evidence that similar principles of (self) organization govern pattern formation in brain and behav-ior. Of particular initial interest was the identification of qualitative changes in the pattern of neural activity that occurred simultaneously with transi-tions between coordination patterns (see Fig. 1D and E). The goal of such experiments was not to present uniquivocal relationships between coordina-tion processes occurring at those two distinct levels of analysis, nor to propose a causal chain, but to highlight the existence of shared coordination dynamics expressed at the kinematic and neural levels [49].

The foregoing brain imaging studies provided initial evidence that sen-sorimotor coordination and the underlying fast time scale cortical processes share a similar dynamics. However, the poor spatial resolution of the EEG and MEG precludes detailed information concerning the specific neural structures involved in coordination. Subsequent work has employed functional magnetic resonance imaging (fMRI) in an attempt to identify the broad cortical and sub-cortical networks underlying sensorimotor coordination and the relationship between activity across such networks and the associated behavioral dynamics. The majority of such studies investigate differences in cortical and subcortical activation associated with synchronized and syncopated patterns of coordina-tion when performed at a single low movement rate, typically 1.25 Hz. The selected rate of coordination reveals intrinsic differences in the stability of coordination [23, 26, 48] while avoiding transitions between patterns.

A general coordination network common to both synchronization and syn-copation (Fig. 2A and B) includes controlateral sensorimotor cortex (M1/S1), bilateral superior temporal gyrus (STG), supplementary motor area (SMA), thalamus, putamen, as well as medial and ipsilateral cerebellum [38, 64]. The intrinsically less stable syncopated pattern consistently demonstrates signifi-cantly greater activity in dorsolateral premotor cortex, supplementary motor area, anterior prefrontal and temporal cortices, and controlateral cerebellum [38, 39, 64, 75]. The differences in cerebral activity underlying synchroniza-tion and syncopation (Fig. 2) first revealed by Mayville and colleagues [64] has since then been replicated – most often as a control – in several studies involv-ing more than a hundred participants [37, 38, 39, 40, 41, 64, 74, 75, 76, 77].

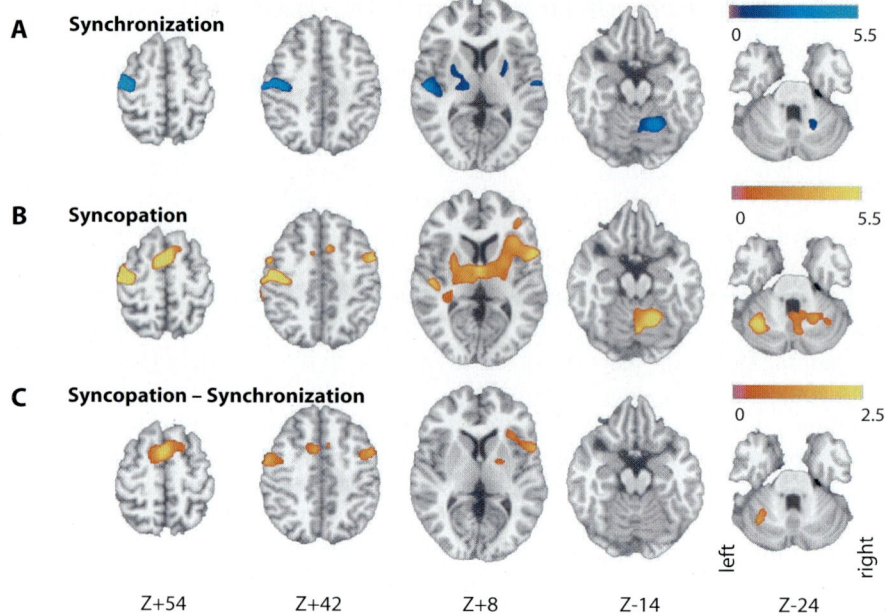

Fig. 2. Hemodynamic correlates of synchronization, syncopation and their comparison. Statistical parametric T-maps ($N = 14$, corrected $p < 0.05$) for synchronizing (A) and syncopating (B) the right hand with an auditory metronome presented at 1.25 Hz. (C) Areas demonstrating significantly greater activity for syncopation compared to synchronization. The Z-axis locations shown give the inferior–superior distance from the AC–PC line in Talairach space

The critical question that emerges from these initial studies of sensorimotor coordination is, what is the individual and combined role of distributed brain regions in forming, maintaining, and switching between coordination patterns? The relationship between network activity and pattern dynamics as revealed by fMRI has been reviewed recently by Jantzen and Kelso [36]. In this review, it is argued that multiple cortical and subcortical regions function in concert with the stability of patterns of coordination. The complimentary goal of the following experiments is to unpack, as it were, the cortical and subcortical network in an attempt to understand what specific (if any) role different components of the network play in mediating coordination and ultimately how brain areas function together in the service of goal-directed coordinated action. The specific three issues on which we focus are as follows:

1. Do neural differences between patterns of coordination reflect the dynamics of a single pattern-forming process? Or do they indicate that different coordination patterns are supported by fundamentally different cognitive strategies?

2. Is there a generalized distributed neural system that supports behavioral pattern formation and change across a range of different coordination contexts?
3. What is the role played by cognitive, motor, and perceptual processes in mediating behavioral stability?

4 On the Potentially Discrete Nature of Syncopation

Although the foregoing experiments and the associated neural results have been considered within the dynamical systems framework [36], several alternative information-processing hypotheses have been offered to account for the same results. We recognize that information processing and dynamical systems approaches to understanding coordination are not necessarily mutually exclusive. However, different functional roles may be attributed to brain regions identified in imaging studies depending on the conceptual framework adopted [106]. As such, it is important to explore the behavioral and neural predictions associated with different conceptual approaches as a means of testing the underlying framework.

With respect to basic forms of behavioral coordination, it has been suggested that synchronization is carried out relatively automatically with little planning or monitoring as if it were organized as a continuous rhythmic sequence that is only adjusted by sensory feedback [64]. In contrast, syncopation may involve the planning and execution of each movement as an individual perception-action cycle. For instance, syncopation may require one to react to a stimulus, anticipate the following one, and precisely time halfway between based on a representation of the interstimulus interval. The resulting increase in cognitive demand may account for the greater attentional load for the off-the-beat pattern [103] (see also [101] for a review), and the additional cognitive processes may account for neural differences between patterns [35].

In support of the above hypothesis, the specific neural areas repeatedly identified during syncopation compared to synchronization are functionally linked to task demands on motor planning, preparation, selection, and timing [38, 39, 64, 75]. Moreover, there are parallels to work investigating differences between continuous and discrete movements, at the behavioral [19, 23, 34] and cerebral levels [22, 93, 95, 96]. Work at both levels has led to the conclusion that rhythmic movements are not a sequence of discrete ones [87, 107]; however, the question remains as to whether syncopation is discrete or rhythmic in nature.

In order to address the question of the potential discreteness of syncopation compared to synchronization, the neural underpinning of three specific but nevertheless potentially generalizable situations has been investigated [76, 77]. We wondered whether, at the neural level, syncopation is supported by cortical networks of similar to those of a single rhythmic sequence (synchronization) or of a set of individual discrete movements [76]. Blood oxygen-dependent signals (BOLD) were acquired during performance of three experimental conditions: synchronization (*time your peak flexion on each metronome event*), syncopation (*time your peak flexion between two metronome events*), and reaction (*perform a flexion movement in response to each metronome event*). Responses were recorded as changes in pressure in air pillows that participants compressed between their right thumb and index finger while listening to an auditory stimulus presented at a constant pace of 1.25 Hz. As previously mentioned, coordinating at 1.25 Hz avoids transitions from syncopation to synchronization [48] and, importantly, from anticipatory to reactive patterns [23]. Of primary interest was the comparison between patterns of activation associated with syncopation and those associated with both the synchronization and reaction conditions.

The basic pattern of neural activity associated with synchronization and syncopation (when compared to resting) was similar to that reported earlier in this chapter (see Fig. 2) [64]. The reaction condition recruited a broad set of neural areas that overlap with both synchronization and syncopation partially reflecting the common need to perform finger flexion independent of coordination constraints. If reaction and syncopation conditions have similar processing requirements, they may also share similar patterns of neural activity and demonstrate similar differences when compared to synchronization. When statistically compared to synchronization, both reaction and syncopation demonstrated increased BOLD activity across a number of spatially restricted brain areas including cingulate gyrus, thalamus, middle frontal gyrus, and SMA (Fig. 3). A subset of these brain regions are associated with motor preparation and planning and, as such, may argue in favor of the cycle-by-cycle strategy underlying syncopation hypothesized in previous studies.

Interestingly, the reaction condition also exhibits an additional network compared not only to synchronization but also to syncopation. When the discrete (reactive) condition was compared to synchronization and syncopation, significant increases in activity were found in bilateral middle frontal gyrus and inferior parietal cortex as well as the bilateral basal ganglia, thalamus, and ipsilateral cerebellum (Fig. 4). Brain imaging studies have reported co-activation of the prefrontal and parietal cortices in task involving working memory and response selection [8]. Moreover, the parietal-premotor/prefrontal network identified is reminiscent of the pattern of activity associated with visually guided movements via the dorsal visual stream [83]. Given the multi-sensory nature of the parietal cortex [68], it is not surprising that a similar network mediates movements guided based on auditory stimulation. Therefore, it appears that the activity in these areas reflects the stimulus-driven nature of the reaction movement rather than the more cyclic movement underlying synchronization. The fact that activity across this reaction network is not observed during syncopation indicates that the off-the-beat movements may be performed based on an internal representation of the coordination pattern and not directed solely by external stimulus input [76].

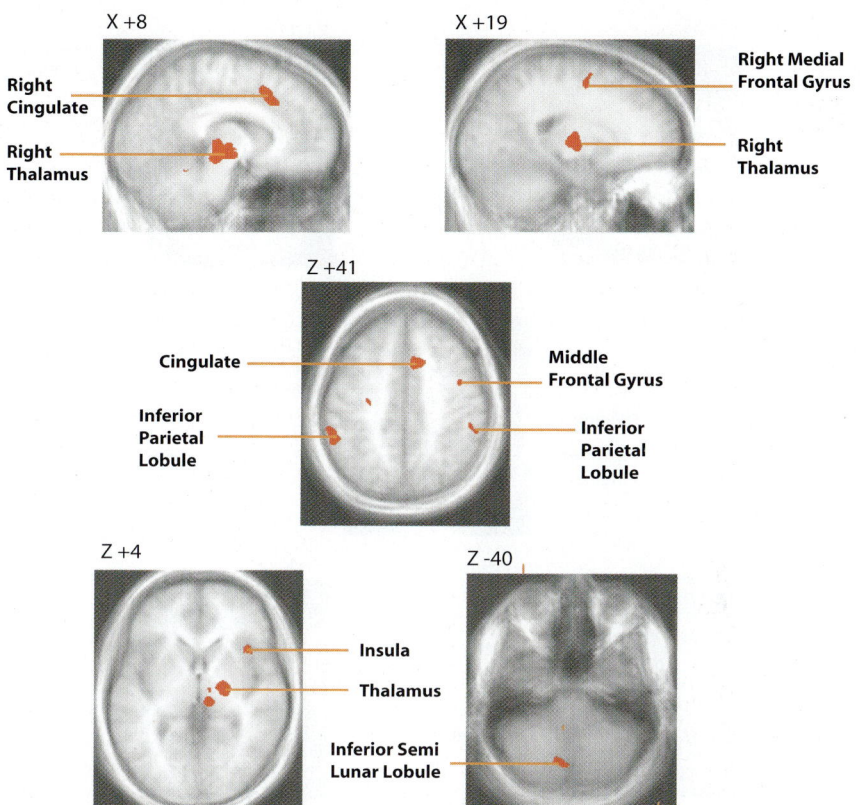

Fig. 3. Reaction and syncopation compared to synchronization. Areas showing significantly greater activity ($N = 11$) for a unimanual reaction and a syncopation task compared to synchronization. Those areas represent the overlap between the "reaction versus synchronization" and "syncopation versus synchronization" comparisons ($p < 0.005$ corrected to 0.01). The X- and Z-axes locations shown give the right–left and the inferior–superior distance from the AC–PC line in Talairach space, respectively

This simple study contrasting synchronization, syncopation, and reaction found a specific network of activity that underlies discrete tasks (reaction). This network of reaction-specific activity in the brain closely resembles the one previously found by Schaal and colleagues [87] (Figs. 2c and d p. 1139) distinguishing discrete from rhythmic tasks. Importantly, reaction-specific neural activations do not overlap with those observed during syncopation. Together such imaging findings indicate that in spite of sharing common cortical signatures syncopation differs from a purely discrete/reactive task. In support of this conclusion, recent imaging work suggests that differences in the pattern of neural activity between coordination patterns can be explained in terms of

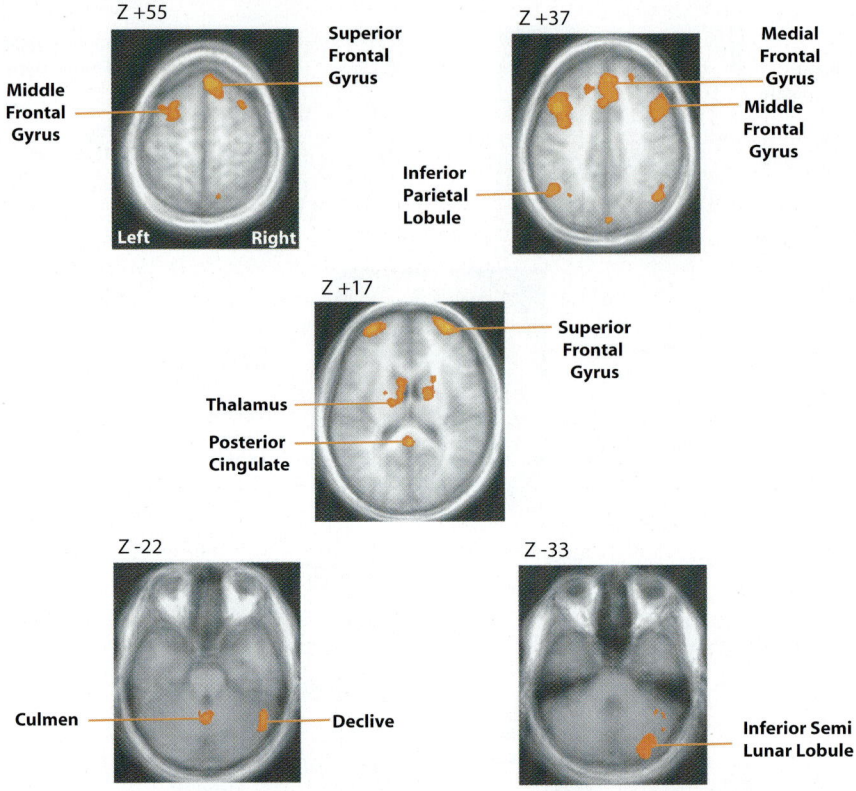

Fig. 4. Reaction compared to synchronization and syncopation. Areas showing significantly greater activity over 11 participants for a reaction task compared to synchronization and syncopation to an auditory metronome pacing at a constant 1.25 Hz rate (p <0.005 corrected to 0.01). The Z-axis locations shown give the inferior–superior distance from the AC–PC line in Talairach space

the stability of the collective pattern formed as opposed to reflecting demand on discrete information processing components [36, 67].

5 Generalized Neural Circuits Underlying Behavioral Pattern Formation

Phenomenological evidence for the claim that behavioral coordination is a pattern-forming process governed by self-organizing principles has been provided across a broad range of experimental settings and contexts [43, 46, 104]. These settings include, among others, bimanual coordination [44, 45, 56, 57, 85, 91], hand–foot coordination [42, 53, 86], postural coordination [2, 4, 5, 79], interpersonal coordination [18, 58, 72, 73, 81, 89, 88], human [20, 21] and animal locomotion [90], man–animal coordination [60], or sensorimotor coordination between a person and her unimodal [3, 5, 23, 48] or multimodal environment [59].

Although overwhelming, such behavioral evidence does not necessarily imply that coordination is mediated at the neural level by a single underlying circuitry. However, a growing number of imaging studies suggest that a common brain network may be recruited across coordination contexts. Brain imaging studies have consistently demonstrated greater and more extended neural activity in association with performance of less compared to more stable patterns of coordination. This is true for the unimanual sensorimotor case [64] as well as multilimb coordination when the two hands must coordinate with each other [94, 108, 109] or with other limbs [13]. In the latter, greater activity is generally found for the less stable antiphase pattern compared to the more stable in-phase one.

The hypothesis of a generalized stability-dependent cortical network is supported by recent studies investigating how brain dynamics evolve when confronted with changes in coordination stability. Two groups [15, 67] reported results of parametric investigations into the neural dynamics governing bimanual coordination during manipulations of coordination pattern and movement rate. This combination of conditions provides experimental control over coordination stability since stability decreases with increasing rate during the antiphase and not the in-phase pattern. A third complementary study investigated stability-dependent cortical activity in a unimanual sensorimotor coordination experiment [41]. The key finding for the present discussion is that a similar set of brain regions systematically increased their level of activity as stability of the pattern decreased regardless of whether coordination was uni- or bimanual. Amodal responsiveness to pattern stability was particularly evident in supplementary motor area and lateral premotor cortex [41, 67], suggestive of a generalized role for premotor regions in maintaining pattern stability across coordinative contexts.

The relationship between increased BOLD amplitude and decreasing pattern stability demonstrated by previous studies [14, 41, 64, 67] suggests that activity in premotor cortex may reflect growing demand for sensorimotor integration and organization related directly to stability of the relative phase between components. Therefore, Jantzen and colleagues conclude that correspondence between neural activity and coordinative stability may be interpreted as a neural signature of the relative phase dynamics and taken as evidence that complex coordinated action may be organized at the collective level through integration across a broadly distributed but highly interconnected network [36].

Although the foregoing results are provocative, direct evidence supporting the existence of common neural signatures requires the within study comparison of similar coordination patterns performed using different sensory and motor components. Here we discuss preliminary data [74] from such a study comparing BOLD activity patterns observed during the production of both uni- and bimanual coordination paradigms. Subjects coordinated with an auditory metronome delivered at a constant pacing frequency (1.25 Hz). Four pairs of unstable stable patterns were compared: unimanual syncopation versus unimanual synchronization (right hand and left hand separately), bimanual syncopation versus bimanual synchronization (i.e., both hands in-phase with each other but 180° out of phase with the metronome versus both hands in-phase with each other and the metronome), and bimanual antiphase versus bimanual in-phase.

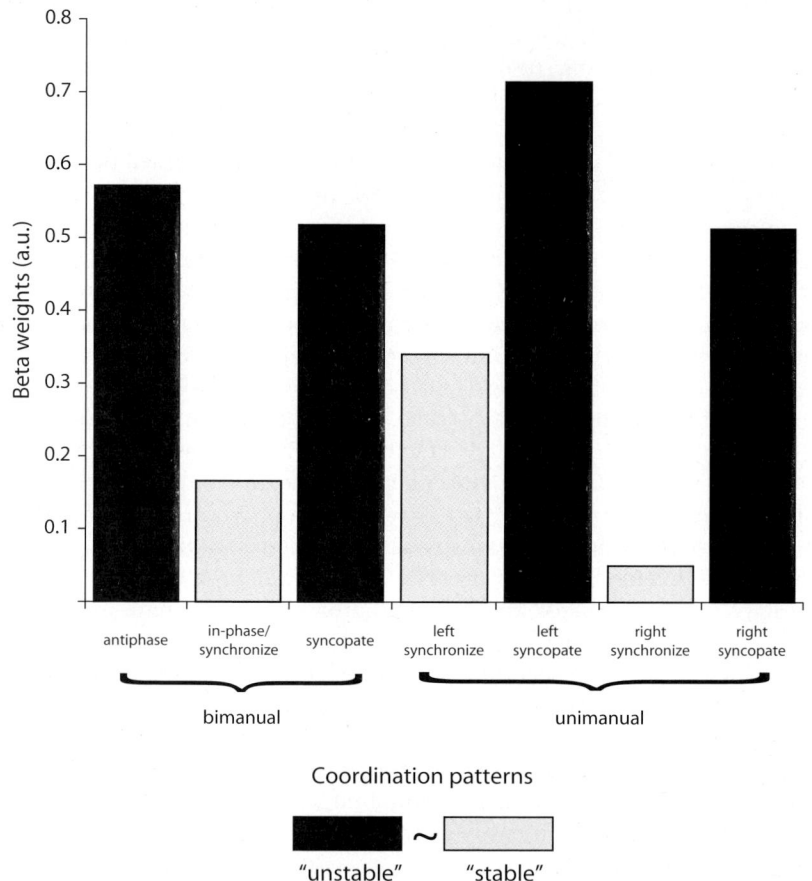

Fig. 5. SMA activity in each condition. From *left* to *right*, beta weights in the SMA (11 subjects) for bimanual antiphase, bimanual in-phase (=bimanual synchronization), bimanual syncopation, left synchronization, left syncopation, right synchronization, right syncopation. The less stable coordination patterns (antiphase and syncopation) are in *black* and their stable counterparts (in-phase and synchronization) are in *gray*

The critical finding from this study was the direct identification of cortical regions demonstrating increased BOLD amplitude during performance of less stable patterns of coordination regardless of the components being coordinated. Significantly greater activity was observed in SMA, premotor cortex, insula, and basal ganglia for less stable patterns of coordination regardless of whether the pattern was unimanual, bimanual, or interlimb. A case in point is the activity within the supplementary motor area, a region of particular interest because of its proposed specialized role for bimanual coordination (Fig. 5). In this study, we see that the syncopated conditions (unimanual right, left and bimanual) demonstrate much greater activity than the accompanying synchronized ones within the SMA. Moreover, the level of activity in the SMA is similar across complimentary unimanual and bimanual conditions. These data argue against the hypothesis that the SMA plays a specialized role in bimanual coordination by mediating the coupling between the hands (see [99, 100] for discussions). Instead, our results indicate that SMA activity is modulated in a predictable way across all conditions and reflects intrinsic pattern stability.

Such results are in agreement with recent studies demonstrating a direct link between pattern stability and neural activity across a frontal motor network that includes SMA and bilateral premotor cortex [15, 37, 41, 67]. Moreover, temporary disruption of the same regions during coordination perturbs unstable movement to a greater degree than stable movements, indicating the putative importance of premotor regions in maintaining unstable patterns [67, 97]. Importantly, the magnitude of the disruption increases with decreasing stability eventually culminating in provoked phase transitions from antiphase to in-phase [67]. The present finding predicts that premotor perturbations would result in similar disruptions in pattern stability across a range of coordination contexts.

6 Neural Indices of Behavioral Instability Persist in the Absence of Movement

In this section, we describe work designed to address the ongoing debate in the behavioral literature concerning the relative contribution of neuromuscular versus perceptual processes in determining the stability of uni- and bimanual coordination [43, 50, 61, 66, 72, 99]. Advocates of the former view have argued that coordination phenomena are governed by physical (neuromuscular-skeletal) limitations such as hand posture constraints [9], the type of muscles recruited [10, 11], and the tendency toward activation of homologous muscles [82]. At the other extreme is the belief that the stability of coordination is completely arbitrary with respect to the physical properties of the individual components, depending instead only on the perceptual relationship between them [66]. To a certain extent, such debate has proceeded largely independent of evidence concerning the underlying neural processes. Given the functional specialization of neural areas distinguishing between coordination modes [39, 64] in conjunction with recent behavioral studies favoring a context-dependent role of constraints on coordination stability [50, 72], it seems likely that such dichotomies may be enlightened by direct neurophysiological investigation.

We have already discussed imaging work using fMRI [38, 39, 64], magnetoencephalography (MEG) [29, 47, 51] and transcranial magnetic stimulation (TMS) [67, 97] showing that coordination patterns of differing behavioral stability are supported by different patterns of neural activity [36, 46, 48, 64, 75]. If stability differences and underlying differences in neural activity are due only to neuromuscular factors [110] that are ultimately linked to cerebral processes through the production of motor output and the resulting feedback, these neural differences should be extinguished (or at least greatly diminished) when coordination patterns are imagined but not performed. However, if both perceptual and motor processes play a role in determining the stability of coordination [50, 72, 85], some aspects of the coordination-dependent differences observed between the two neural activation patterns should persist, even in the absence of overt movement.

We employed a motor imagery paradigm in order to investigate the role of the generation of efferent motor signal and the associated sensory feedback in governing rhythmic coordination [75]. Neural correlates of physically executed and imagined rhythmic coordination were compared to better assess the relative contribution of hypothesized neuromuscular-skeletal mechanisms in modulating behavioral stability. The executed tasks were to coordinate index finger-to-thumb opposition movements of the right hand with an auditory metronome (pacing at the constant frequency of 1.25 Hz) in either a synchronized or syncopated fashion. Imagined tasks – without actual physical movement – required participants to imagine they were either synchronizing or syncopating to the metronome beats. Thus, the sensory stimulus and coordination constraints remained the same in both physical and imagination tasks, whereas the sensorimotor requirements did not.

This paradigm revealed two new findings. First, similar brain networks support the performance and imagination of rhythmic coordination tasks. Like actual performance, imagination of the synchronized and syncopated coordination modes resulted in activity within SMA, premotor cortex, inferior parietal lobe, superior temporal gyrus, inferior frontal gyrus, and basal ganglia. This finding is in concert with a wealth of previous neuroimaging studies demonstrating that imagination requires activation of similar brain regions involved during execution of the same motor act [17, 69, 84].

Second, well-established neural differences between synchronization and syncopation (Fig. 2, see also [27, 38, 64]) were also observed when one simply imagined the two coordination patterns (Fig. 6). As expected, executed syncopation resulted in significantly greater activity in SMA, lateral premotor cortex, cingulate, thalamus, inferior parietal lobe, and bilateral cerebellum when compared to executed synchronization (Fig. 6, red overlay). Importantly,

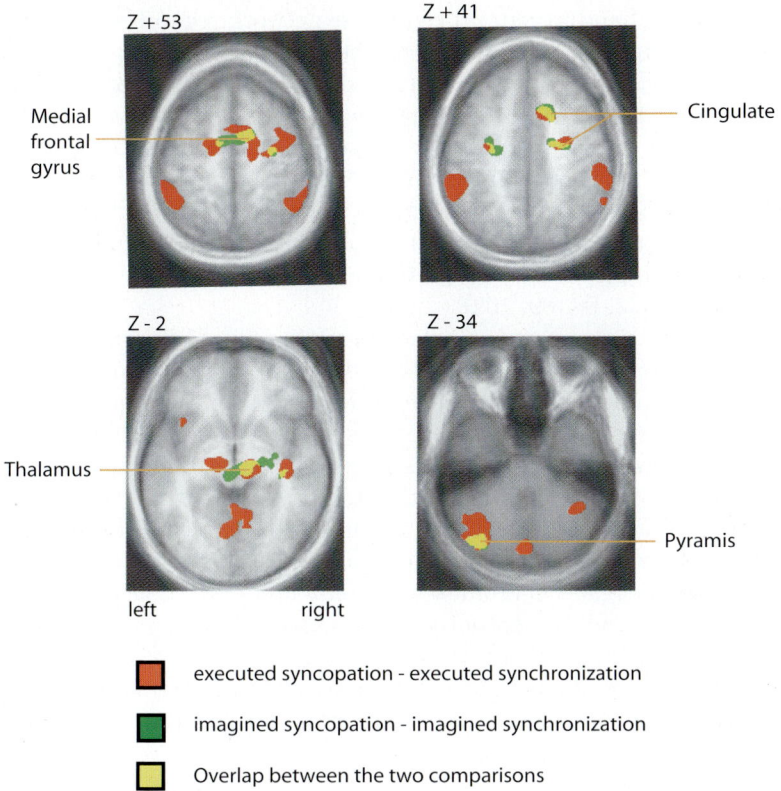

Fig. 6. Neural indices of stability for the performance and imagination of patterns of coordination (adapted from [75]). Significant areas for 14 participants are overlaid on top of selected axial slices of an average anatomical scan. Red overlays: brain areas that are significantly different between the execution of syncopation and synchronization (executed syncopation compared to executed synchronization). Multiple comparisons were corrected by adopting a per voxel criterion threshold of $p < 0.005$ and a volume threshold for an overall corrected error rate of $p < 0.05$. Green overlays: brain areas that are significantly different between the imagination of syncopation and synchronization (imagined syncopation compared to imagined synchronization). Yellow overlays: areas that are common to both comparisons. The Z-axis location shown gives the inferior–superior distance from the AC–PC line in Talairach space

imagined syncopation resulted in similar increases when compared to imagined synchronization (Fig. 6, green overlay). Remarkable overlap between executed and imagined comparisons (Fig. 6, yellow overlay) was observed in SMA and cingulate, dorsal premotor cortex, superior temporal gyrus, thalamus, and lateral portions of the cerebellum.

The present findings demonstrate that neural activity related to coordination constraints is observed across a premotor (SMA, cingulate, lateral premotor cortex) and ipsilateral cerebellar network regardless of whether the pattern is executed or only imagined. The identified network may thus be related to the organization of the spatiotemporal pattern as opposed to its execution per se. Moreover, this suggests that motor system constraints such as hand posture or muscle recruitment are not singularly and exclusively responsible for determining coordinative (in)stability. The imagination study corroborates those of interpersonal interactions where coordination between individuals exhibits typical hallmarks of self-organizing systems including *differential stability* between modes of coordination, even in the absence of reciprocal motor constraints [72, 88, 89, 102].

At the same time, there is no denying the body of literature showing that neuromuscular-skeletal constraints such as the nature of coupling between oscillators [5, 7] and/or the difference between their eigenfrequencies [28, 105] can serve to shape the dynamics of the system. In addition, Carson and colleagues [12] showed that the positions of external axes of rotations of the limbs, and the muscles involved, influence the stability of interlimb coordination. Under specific skeletal configurations, the in-phase pattern of coordination was found to be less stable than the antiphase one. This body of work suggests that biomechanical and neuromuscular constraints work together with central neural cognitive and perceptual processes to modulate intrinsic patterns of coordinative stability. When considered as a whole, the literature argues for an embedded (or embodied) view of coordination dynamics that cannot be considered fully in isolation of environmental and physical constraints.

7 Conclusions

In this chapter, we have considered three complimentary experiments whose goal was to refine our understanding of the relationship between central neural activity and coordination dynamics. The strategy was to isolate specific features of coordination in an attempt to identify the underlying neural processes. Our efforts focus primarily on premotor cortical areas and understanding what role they play during coordination. In the first experiment, we compared coordination with reaction and found that sensorimotor coordination between stimulus and metronome was neurally distinct from the stimulus response chain putatively underlying reaction. This comparison confirmed that syncopation is not simply a timed reaction to a series of individual stimuli. Instead, when considered in the light of mounting evidence, it appears that syncopation and synchronization are organized at the neural level based on pattern and its stability. In the second experiment, we showed that premotor cortical regions are sensitive to pattern stability across a variety of coordination contexts. In particular, supplementary motor area appears to be critical for organizing and maintaining pattern stability in general as opposed to the more specific role of mediating coupling between homologous limbs. Finally, we demonstrate that central neural activity related to behavioral dynamics persists in the absence of actual movement. This last finding demonstrates that coordination dynamics are partially rooted in neural process and not determined solely by neuromuscular-skeletal constraints.

Together the results of the foregoing experiments lay a strong foundation for understanding the neural mechanisms of complex coordinated action within a coordination dynamics framework. This and previous considerations [36] have begun to uncover the link between large-scale neural function and behavioral dynamics, thereby providing greater insight into how such behaviors are organized and maintained at the level of the brain. We end this chapter with the provocative claim that coordination dynamics and the tools it provides, offers the ideal entry point for investigating and understanding the neural mechanisms of complex rhythmic behavior. The evidence we have described represents initial steps toward supporting this claim, and we hope this chapter encourages additional work along similar lines.

Acknowledgments

The preparation of this manuscript was supported by the *National Institute of Neurological Disorders and Stroke Grant* R01-NS48229-01A1 (KJJ) and the *Ministère Délégué à l'Enseignement Supérieur et la Recherche* (OO). We thank Julien Lagarde and Carl Douglas for their help. Much of our own work presented in this paper was co-authored with Fred L. Steinberg and J.A. Scott Kelso who supervised and supported it with *National Institute for Mental Health* grants 42900 and 01386 (JASK). We are grateful to J.A. Scott Kelso for mastering the art of coordination at so many levels, as illustrated by our complementary pair that would not exist without him.

References

1. Amazeen PG, Amazeen EL, Turvey MT (2004) Symmetry and the devil. J Mot Behav 36:371–372
2. Balasubramaniam R, Wing AM (2002) The dynamics of standing balance. Trends in Cogn Sci 6:531–536
3. Balasubramaniam R, Wing AM, Daffertshofer A (2004) Keeping with the beat: Movement trajectories contribute to movement timing. Exp Brain Res 159:129–134
4. Bardy BG, Marin L, Stoffregen TA, Bootsma RJ (1999) Postural coordination modes considered as emergent phenomena. J Exp Psychol Hum Percept Perform 25:1284–1301
5. Bardy BG, Oullier O, Bootsma RJ, Stoffregen TA (2002) Dynamics of human postural transitions. J Exp Psychol Hum Percept Perform 28:499–514
6. Barsalou LW, Niedenthal PM, Barbey AK, Ruppert JA (2003) Social embodiment. Psychol Learn Motiv Adv Re Theor, 43:43–92
7. Buchanan JJ, Kelso JAS, de Guzman GC (1997) Self-organization of trajectory formation. 1. Experimental evidence. Biol Cybern 76:257–273

8. Bunge SA, Hazeltine E, Scanlon MD, Rosen AC, Gabrieli JDE (2002) Dissociable contributions of prefrontal and parietal cortices to response selection. Neuroimage 17:1562–1571

9. Carson RG, Chua R, Byblow WD, Poon P, Smethurst CJ (1999) Changes in posture alter the attentional demands of voluntary movement. Proc R Soc Lond B-Biol Sci 266:853–857

10. Carson RG, Riek S (1998) The influence of joint position on the dynamics of perception-action coupling. Exp Brain Res 121:103–114

11. Carson RG, Riek S (2001) Changes in muscle recruitment patterns during skill acquisition. Exp Brain Res 138:71–87

12. Carson RG, Riek S, Smethurst CJ, Parraga JF, Byblow WD (2000) Neuromuscular-skeletal constraints upon the dynamics of unimanual and bimanual coordination. Exp Brain Res 131:196–214

13. Debaere F, Swinnen SP, Beatse E, Sunaert S, Van Hecke P, Duysens J (2001) Brain areas involved in interlimb coordination: A distributed network. Neuroimage 14:947–958

14. Debaere F, Wenderoth N, Sunaert S, Van Hecke P, Swinnen SP (2003) Internal vs external generation of movements: Differential neural pathways involved in bimanual coordination performed in the presence or absence of augmented visual feedback. Neuroimage 19:764–776

15. Debaere F, Wenderoth N, Sunaert S, Van Hecke P, Swinnen SP (2004) Cerebellar and premotor function in bimanual coordination: Parametric neural responses to spatiotemporal complexity and cycling frequency. Neuroimage 21:1416–1427

16. de Guzman GC, (2004) It is a matter of symmetry. Personal communication to the authors.

17. Deiber MP, Ibanez V, Honda M, Sadato N, Raman R, Hallett M (1998) Cerebral processes related to visuomotor imagery and generation of simple finger movements studied with positron emission tomography. Neuroimage 7:73–85

18. de Rugy A, Salesse R, Oullier O, Temprado JJ (2006) A neuro-mechanical model for interpersonal coordination. Biol Cybern 94:427–443

19. de Rugy A, Sternad D (2003) Interaction between discrete and rhythmic movements: Reaction time and phase of discrete movement initiation during oscillatory movements. Brain Res 994:160–174

20. Diedrich FJ, Warren WH (1995) Why change gaits? Dynamics of the walk run transition. J Exp Psychol Hum Percept Perform 21:183–202

21. Diedrich FJ, Warren WH (1998) The dynamics of gait transitions: Effects of grade and load. J Mot Behav 30:60–78

22. Diedrichsen J, Hazeltine E, Ivry R, Kennerley S, Spencer B (2002) Comparing continuous and discrete movements with fMRI. Ann NY Acad Sci 978:509–510

23. Engstrøm DA, Kelso JAS, Holroyd T (1996) Reaction-anticipation transitions in human perception-action patterns. Hum Mov Sci 15:809–832

24. Fraisse P (1966) L'anticipation de stimulus rythmiques:Vitesse d'tablissement et pecision de la synchronisation. L'Année Psychologique 66:15–36

25. Fraisse P (1982) Rhythm and tempo. In: Deutsch D (ed) Psychology of music. Academic Press, New York, pp. 149–180

26. Fraisse P, Ehrlich S (1955) Note sur la possibilité de syncoper en fonction du tempo d'une cadence. L'Année Psychologique 55:61–65

27. Fuchs A, Jirsa VK (2008, this volume) J.A. Scott Kelso's contributions to our understanding of coordination. In: Fuchs A, Jirsa VK (eds) Coordination: Neural, behavioral and social dynamics. Springer, Berlin, pp. 329–348.

28. Fuchs A, Jirsa VK, Haken H, Kelso JAS (1996) Extending the HKB model of coordinated movement to oscillators with different eigenfrequencies. Biol Cybern 74:21–30

29. Fuchs A, Kelso JAS, Haken H (1992) Phase transitions in the human brain: Spatial mode dynamics. Int J Bifurcat Chaos 2:917–939

30. Fuchs A, Mayville JM, Cheyne D, Weinberg H, Deecke L, Kelso JAS (2000) Spatiotemporal analysis of neuromagnetic events underlying the emergence of coordinative instabilities. Neuroimage 12:71–84

31. Haken H (1983) Advanced synergetics. Springer-Verlag, Heidelberg

32. Haken H (1977) Synergetics: An Introduction. Springer-Verlag, Berlin

33. Haken H, Kelso JAS, Bunz H (1985) A theoretical-model of phase-transitions in human hand movements. Biol Cybern 51:347–356

34. Huys R, Jirsa VK, Studenka B, Rheaume N, Zelaznik HN (2008, this volume) Human trajectory formation: Taxonomy of motor primitives based on phase flow topology. In: Fuchs A, Jirsa VK (eds) Coordination: Neural, behavioral and social dynamics. Springer, Berlin, pp. 77–92.

35. Ivry RB, Spencer RMC (2004) The neural representation of time. Curr Opin Neurobiol 14:225–232

36. Jantzen KJ, Kelso JAS (2007) Neural coordination dynamics of human sensorimotor behavior: A Review. In: Jirsa VK, McIntosh AR (eds) Handbook of brain connectivity. Springer-Verlag, Berlin, pp. 421–461

37. Jantzen KJ, Oullier O, Marshall M, Steinberg FL, Kelso JA (2007) A parametric fMRI investigation of context effects in sensorimotor timing and coordination. Neuropsychologia 45:673–684

38. Jantzen KJ, Steinberg FL, Kelso JAS (2002) Practice-dependent modulation of neural activity during human sensorimotor coordination: A functional Magnetic Resonance Imaging study. Neurosci Lett 332:205–209

39. Jantzen KJ, Steinberg FL, Kelso JAS (2004) Brain networks underlying human timing behavior are influenced by prior context. Proc Nat Acad Sci USA 101:6815–6820

40. Jantzen KJ, Steinberg FL, Kelso JAS (2005) Functional MRI reveals the existence of modality and coordination-dependent timing networks. Neuroimage 25:1031–1042

41. Jantzen KJ, Steinberg FL, Kelso JAS (sumbitted) Neural dynamics during sensorimotor coordination are related to pattern stability. Manuscript submitted for publication

42. Jeka JJ, Kelso JAS (1995) Manipulating Symmetry in the Coordination Dynamics of Human Movement. J Exp Psychol Hum Percept Perform 21:360–374

43. Jirsa VK, Kelso JAS (2004) Coordinations dynamics: Issues and Trends. Springer-Verlag, Berlin

44. Kelso JAS (1981) On the oscillatory basis of movement. Bull Psychon Soc 18:63

45. Kelso JAS (1984) Phase-transitions and critical behavior in human bimanual coordination. Am J Physiol 246:1000–1004

46. Kelso JAS (1995) Dynamic patterns: The self-organization of brain and behavior. MIT Press, Cambridge

47. Kelso JAS, Bressler SL, Buchanan S, de Guzman GC, Ding M, Fuchs A, Holroyd T (1992) A phase-transition in human brain and behavior. Phys Lett A 169:134–144
48. Kelso JAS, DelColle J, Schöner G (1990) Action-perception as a pattern formation process. In: Jeannerod M (ed) Attention and Performance XIII. Erlbaum, Hillsdale, pp. 139–169
49. Kelso JAS, Engstrøm DA (2006) The complementary nature. MIT Press, Cambridge
50. Kelso JAS, Fink PW, DeLaplain CR, Carson RG (2001) Haptic information stabilizes and destabilizes coordination dynamics. Proc R Soc Lond B-Biol Sci 268:1207–1213
51. Kelso JAS, Fuchs A, Lancaster R, Holroyd T, Cheyne D, Weinberg H (1998) Dynamic cortical activity in the human brain reveals motor equivalence. Nature 392:814–818
52. Kelso JAS, Holt KG, Kugler PN, Turvey MT (1980) On the concept of coordinative structures as dissipative structures: II. Empirical lines of convergence. In: Stelmach GE, Requin J (eds) Tutorials in motor behavior. North Holland, Amsterdam, pp. 49–70
53. Kelso JAS, Jeka JJ (1992) Symmetry-breaking dynamics of human multilimb coordination. J Exp Psychol Hum Percept Perform 18:645–668
54. Kelso JAS, Scholz JP, Schöner G (1986) Nonequilibrium phase-transitions in coordinated biological motion – Critical fluctuations. Phys Lett A 118:279–284
55. Kelso JAS, Schöner G, Scholz JP, Haken H (1987) Phase-locked modes, phase transitions and component oscillators in coordinated biological motion. Physica Scripta 35:79–87
56. Kelso JAS, Southard DL, Goodman D (1979) On the nature of interlimb coordination. Science 203:1029–1031
57. Kelso JAS, Zanone PG (2002) Coordination dynamics of learning and transfer across different effector systems. J Exp Psychol Hum Percept Perform 28: 776–797
58. Lagarde J, de Guzman GC, Oullier O, Kelso JAS (2006) Interpersonal interactions during boxing: Data and model. J Sport Exerc Psychol 28:S108–S108
59. Lagarde J, Kelso JAS (2006) Binding of movement, sound and touch:multimodal coordination dynamics. Exp Brain Res 173:673–688
60. Lagarde J, Kelso JAS, Peham C, Licka T (2005) Coordination dynamics of the horse-rider system. J Motor Behav 37:418–424
61. Lee TD, Almeida QJ, Chua R (2002) Spatial constraints in bimanual coordination: Influences of effector orientation. Exp Brain Res 146:205–212
62. Mayville JM, Bressler SL, Fuchs A, Kelso JAS (1999) Spatiotemporal reorganization of electrical activity in the human brain associated with a timing transition in rhythmic auditory-motor coordination. Exp Brain Res 127: 371–381
63. Mayville JM, Fuchs A, Ding MZ, Cheyne D, Deecke L, Kelso JAS (2001) Event-related changes in neuromagnetic activity associated with syncopation and synchronization timing tasks. Hum Brain Mapp 14:65–80
64. Mayville JM, Jantzen KJ, Fuchs A, Steinberg FL, Kelso JAS (2002) Cortical and subcortical networks underlying syncopated and synchronized coordination revealed using fMRI. Hum Brain Mapp 17:214–229
65. McGarva AR, Warner RM (2003) Attraction and social coordination: Mutual entrainment of vocal activity rhythms. J Psycholinguist Res 32:335–354

66. Mechsner F, Kerzel D, Knoblich G, Prinz W (2001) Perceptual basis of bi-manual coordination. Nature 414:69–73

67. Meyer-Lindenberg A, Ziemann U, Hajak G, Cohen L, Berman KF (2002) Transitions between dynamical states of differing stability in the human brain. Proc Nat Acad Sci USA 99:10948–10953

68. Molholm S, Sehatpour P, Mehta AD, Shpaner M, Gomez-Ramirez M, Ortigue S, Dyke JP, Schwartz TH, Foxe JJ (2006) Audio-visual multisensory integration in superior parietal lobule revealed by human intracranial recordings. J Neurophysiol 96:721–729

69. Nair DG, Purcott KL, Fuchs A, Steinberg F, Kelso JAS (2003) Cortical and cerebellar activity of the human brain during imagined and executed unimanual and bimanual action sequences: A functional MRI study. Cognit Brain Res 15:250–260

70. Néda Z, Ravasz E, Brechet Y, Vicsek T, Barabasi AL (2000) The sound of many hands clapping – Tumultuous applause can transform itself into waves of synchronized clapping. Nature 403:849–850

71. Oullier O, Bardy BG, Stoffregen TA, Bootsma RJ (2002) Postural coordination in looking and tracking tasks. Hum Mov Sci 21:147–167

72. Oullier O, de Guzman GC, Jantzen KJ, Kelso JAS (2003) On context dependence of behavioral variability in inter-personal coordination. Int J Comput Sci Sport 2:126–128

73. Oullier O, de Guzman GC, Jantzen KJ, Lagarde J, Kelso JAS (2008) Social coordination dynamics: Measuring human bonding. Soc Neurosci, manuscript in press.

74. Oullier O, Jantzen KJ, Steinberg FL, Kelso JAS (2003) fMRI reveals neural mechanisms common to sensorimotor and bi-manual coordination. Society for Neuroscience CD-ROM 554.6

75. Oullier O, Jantzen KJ, Steinberg FL, Kelso JAS (2005) Neural substrates of real and imagined sensorimotor coordination. Cereb Cortex 15:975–985

76. Oullier O, Jantzen KJ, Steinberg F, Kelso JAS (2006) Neural correlates of rhythmic and reactive sensorimotor coordination. In: Hoppeler H, Reilly T, Tsolakidis E, Gfeller L, Klossner S (eds) European College of Sport Sciences. Sportverlag Strauss, Cologne, pp. 34–35

77. Oullier O, Jantzen KJ, Steinberg FL, Kelso JAS (submitted) Neural dynamics of continuous and disrete coordination. Manuscript submitted for publication

78. Oullier O, Lagarde J, Jantzen KJ, Kelso JAS (2006) Coordination dynamics: (in)stability and metastability in the behavioural and neural systems. Journal de la Société de Biologie 200:145–167

79. Oullier O, Marin L, Stoffregen TA, Bootsma RJ, Bardy BG (2006) Variability in postural coordination dynamics. In: Davids K, Bennett S, Newell KM (eds) Movement system variability. Human Kinetics, Champaign, pp. 25–47

80. Park H, Turvey M (2008, this volume) Imperfect symmetry and the elementary coordination law. In: Fuchs A, Jirsa VK (eds) Coordination: Neural, behavioral and social dynamics. Springer, Berlin, pp. 3–25

81. Richardson MJ, Marsh KL, Schmidt RC (2005) Effects of visual and verbal interaction on unintentional interpersonal coordination. J Exp Psychol Hum Percept Perform 31:62–79

82. Riek S, Carson RG, Byblow WD (1992) Spatial and muscular dependencies in bimanual coordination. J Hum Mov Stud 23:251–265

83. Rizzolatti G, Matelli M (2003) Two different streams form the dorsal visual system: Anatomy and functions. Exp Brain Res 153:146–157
84. Roth M, Decety J, Raybaudi M, Massarelli R, Delon-Martin C, Segebarth C, Morand S, Gemignani A, Decorps M, Jeannerod M (1996) Possible involvement of primary motor cortex in mentally simulated movement: A functional magnetic resonance imaging study. Neuroreport 7:1280–1284
85. Salesse R, Oullier O, Temprado JJ (2005) Plane of motion mediates the coalition of constraints in rhythmic bimanual coordination. J Mot Behav 37:454–464
86. Salesse R, Temprado JJ (2005) The effect of visuo-motor transformations on hand-foot coordination:evidence in favor of the incongruency hypothesis. Acta Psychol 119:143–157
87. Schaal S, Sternad D, Osu R, Kawato M (2004) Rhythmic arm movement is not discrete. Nat Neurosci 7:1137–1144
88. Schmidt RC, Carello C, Turvey MT (1990) Phase-transitions and critical fluctuations in the visual coordination of rhythmic movements between people. J Exp Psychol Hum Percept Perform 16:227–247
89. Schmidt RC, Richardson MJ (2008, this volume) Dynamics of interpersonal coordination. In: Fuchs A, Jirsa VK (eds) Coordination: Neural, behavioral and social dynamics. Springer, Berlin, pp. 283–310
90. Schöner G, Jiang WY, Kelso JAS (1990) A synergetic theory of quadrupedal gaits and gait transitions. J Theor Biol 142:359–391
91. Schöner G, Kelso JAS (1988) Dynamic pattern generation in behavioral and neural systems. Science 239:1513–1520
92. Solveig M (2002) Sur la terre. Mixture, Paris
93. Spencer RMC, Zelaznik HN, Ivry RB, Diedrichsen J (2002) Does the cerebellum preferentially control discrete and not continuous movements? Ann NY Acad Sci 978:542–544
94. Stephan KM, Binkofski F, Halsband U, Dohle C, Wunderlich G, Schnitzler A, Tass P, Posse S, Herzog H, Sturm V, Zilles K, Seitz RJ, Freund HJ (1999) The role of ventral medial wall motor areas in bimanual co-ordination – A combined lesion and activation study. Brain 122:351–368
95. Sternad D (2008, this volume) Towards a unified theory of rhythmic and discrete movements – Behavioral, modeling and imaging results. In: Fuchs A, Jirsa VK (eds) Coordination: Neural, behavioral and social dynamics. Springer, Berlin, pp. 105–133
96. Sternad D, Wei K, Diedrichsen J, Ivry RB (2007) Intermanual interactions during initiation and production of rhythmic and discrete movements in individuals lacking a corpus callosum. Exp Brain Res 176:559–574
97. Steyvers M, Etoh S, Sauner D, Levin O, Siebner HR, Swinnen SP, Rothwell JC (2003) High-frequency transcranial magnetic stimulation of the supplementary motor area reduces bimanual coupling during anti-phase but not in-phase movements. Exp Brain Res 151:309–317
98. Stoffregen TA (1985) Flow structure versus retinal location in the optical control of stance. J Exp Psychol Hum Percept Perform 11:554–565
99. Swinnen SP (2002) Intermanual coordination: From behavioural principles to neural-network interactions. Nat Rev Neurosci 3:348–359
100. Swinnen SP, Wenderoth N (2004) Two hands, one brain: Cognitive neuroscience of bimanual skill. Trends Cognitive Sci 8:18–25

101. Temprado JJ (2004) A dynamical approach to the interplay of attention and bimanual coordination. In: Jirsa VK, Kelso JAS (eds) Coordination dynamics: Issues and trends. Springer-Verlag, Berlin, pp. 21–39
102. Temprado JJ, Swinnen SP, Carson RG, Tourment A, Laurent M (2003) Interaction of directional, neuromuscular and egocentric constraints on the stability of preferred bimanual coordination patterns. Hum Mov Sci 22:339–363
103. Temprado JJ, Zanone PG, Monno A, Laurent M (1999) Attentional load associated with performing and stabilizing preferred bimanual patterns. J Exp Psychol Hum Percept Perform 25:1579–1594
104. Turvey MT (1990) Coordination. Am Psychol 45:938–953
105. Turvey MT, Rosenblum LD, Schmidt RC, Kugler PN (1986) Fluctuations and phase symmetry in coordinated rhythmic movements. J Exp Psychol Hum Percept Perform 12:564–583
106. Utall WR (2001) The new phrenology. MIT Press, Cambridge
107. van Mourik AM, Beek PJ (2004) Discrete and cyclical movements:unified dynamics or separate control? Acta Psychol 117:121–138
108. Wenderoth N, Debaere F, Sunaert S, Van Hecke P, Swinnen SP (2004) Parietopremotor areas mediate directional interference during bimanual movements. Cereb Cortex 14:1153–1163
109. Wenderoth N, Debaere F, Sunaert S, Swinnen SP (2005) Spatial interference during bimanual coordination: Differential brain networks associated with control of movement amplitude and direction. Hum Brain Mapp 26:286–300
110. Yue GH, Liu JZ, Siemionow V, Ranganathan VK, Ng TC, Sahgal V (2000) Brain activation during human finger extension and flexion movements. Brain Res 856:291–300

Brain–Computer Interfaces (BCI): Restoration of Movement and Thought from Neuroelectric and Metabolic Brain Activity

Surjo R. Soekadar[1], Klaus Haagen[2], and Niels Birbaumer[3,4]

[1] Neurostimulation Unit, Department of Psychiatry and Psychotherapy, University of Tübingen, Tübingen, Germany
[2] Department of Economy, University of Trento, Trento, Italy
[3] Institute of Medical Psychology and Behavioral Neurobiology, University of Tübingen, Tübingen, Germany
[4] National Institutes of Health (NIH), NINDS, Human Cortical Physiology, Bethesda, MD, USA

Abstract. This chapter provides an overview of the scientific and clinical progress in the development of non-invasive and invasive brain–computer interfaces (BCI). BCI uses electric, magnetic or metabolic brain activity for the activation and control of external devices and computers. Clinically, until now it has been successfully used as a communication system for totally paralyzed patients ("locked-in patients"), in restoration of movement after stroke or spinal cord injury and the treatment of epilepsy for example. Here we emphasize that BCI technology is a powerful tool to systematically induce neuroplastic changes and therefore has a significant potential to promote innovative approaches in neurorehabilitation. After a short introduction, the mechanisms underlying BCI control will be outlined and an overview of the available invasive and non-invasive BCI systems will be given. The differences and challenges in the use of BCI technology in healthy and patients with neurological disorders will be sketched. Newly developed approaches (i.e., using functional magnetic resonance imaging (fMRI) and near infrared spectroscopy (NIRS) to manipulate very localized and subcortical brain changes) and diverse applications of BCIs will be introduced. Besides a critical discussion of limitations and problems in BCI research and clinical application, ethical and quality of life issues will be addressed. The chapter ends with some remarks on future directions in the development of BCI systems introducing invasive and non-invasive neurostimulation techniques that can coequally initiate, enhance or stabilize neuroplastic changes induced by BCI use resulting in behavioral benefits.

1 Introduction

Since the discovery of the electroencephalogram (EEG) by Hans Berger published in 1929 [3] a broad scientific community aimed at deciphering the versatile code of thoughts that were meant to underlie the fluctuations of electrical potentials within the brain. Hans Berger was convinced that decryption is possible and that all thoughts would be accessible and "readable" once the code has been found. For many decades, however, only very limited information (such as state of awareness or pathological alterations) could be extracted from EEG. Inspired by Berger's vision, more than 40 years later Grey Walter developed the first automatic EEG frequency analyzer aiming at a reliable discrimination of language and thoughts. But only the recent advent of relatively powerful computers, innovative algorithms and bio-compatible materials combined in brain–computer interfaces (BCI) or brain–machine interfaces (BMI) have brought us a little bit closer to Berger's vision. Since a couple of years scientific interest in the development of BCI technology has experienced a tremendous growth, but over all BCI research is still in its infancy. Whereas 10 years ago only about eight scientific groups worked on BCI research, today there are more than hundred groups worldwide, many of them supported by federal funds.

There are several motives behind the enthusiasm – one of them is the hope that BCI research will allow disabled patients to communicate, move paralyzed limbs or better interact with their environment. It has been estimated that the prevalence and incidence of neurological diseases such as stroke will significantly increase in the Western societies due to demographic changes. Even today stroke is the most common cause for disability of the adult; 70–85% of stroke patients suffer from hemiparesis. After extensive conventional rehabilitation programs only 60% of these patients are able to manage their daily needs without assistance [48, 49]. All other patients (often with a full hemiplegia) rely on daily support and care. A restoration of motor control in these patients would be of great importance for their quality of life.

But at present time there are enormous unsolved problems that limit the practicability and success of BCI devices for daily use in a normal environment. The hitherto-attained achievements in clinical studies are instructive for further developments and the range of application. These achievements will be sketched and summarized here.

2 Mechanisms Underlying BCI Control

The origin of BCI systems can be seen in early works done on operant training of brain response in the context of neuro-feedback. In neuro-feedback the subject receives visual or auditory online feedback of his/her brain activity and tries to voluntarily modify a particular type of brainwave.

Once a person has learned to voluntarily control his/her brain response, this response can be more or less reliably used to activate external devices or computers.

An important aspect in the development of BCI technology lies in the fact that operant training of the brain is possible although there is no neuromuscular system involved. This at first glance contradicts the proposal of Skinner [77] who emphasized the distinction between operant (instrumental) conditioning underlying learning of voluntary control and classical (Pavlovian) conditioning [9]. In classical conditioning the modification of involuntary functions controlled by the autonomous nervous system (ANS) occurs. In 1969 Neal E. Miller and collaborators presented experimental evidence suggesting that instrumental ("voluntary") control of the ANS is possible [65]. Although these results could not be replicated during the next decades they lead to a vast amount of clinical trials aiming at developing treatments for many medical diseases such as high blood pressure, cardiac arrhythmias, vascular pathologies, renal failure and gastrointestinal disorders. While the most impressive clinical results using classical conditioning were achieved with electromyographic feedback in chronic neuromuscular pain [29], neuromuscular rehabilitation of various neurological conditions [8], particularly external sphincter control in enuresis end encopresis [43], and posture control in kyphosis and scoliosis [13, 25]; trials to control the ANS in order to treat essential hypertension [27, 64], increase heart rate control [20] and gastric motility [43] remained clinically unimpressive or negligible. Evidence became striking that only very limited positive effects of biofeedback on visceral pathology with clinically and statistically relevant changes occur. But there was one notable exception: neuro-feedback of brain activity [7, 10, 26].

This exception opens the door for voluntary regulation of measurable functions that are independent of musculoskeletal functions. This exception has, however, to be relativized as attempts to train completely locked-in patients suffering from amyotrophic lateral sclerosis (ALS) to voluntarily modulate their brain oscillations failed once they have reached the locked-in stadium. A possible explanation might be that the ability to reliably control a BCI depends on intact *contingencies* between volitional modulation of neuronal activity, motor output and sensory feedback. If the homeostatic function of the "reward" is not strong enough or lost, the ability of learning and voluntary control of a BCI is also likely to be extinguished. This occurs even if the cognitive system (attention, memory, verbal imagery and thinking) is intact. In this case *artificial contingencies* induced for example by neurostimulation might be a possibility to improve voluntary control of a BCI.

Although until now many researchers have assumed that invasive BCI methods that use single neuron activity after implantation of recording electrodes will provide far better control than non-invasive techniques, the results to date, however, do not support this assumption [90]. Speed and precision of BCI control by non-invasive technology is currently in the same range as the ones by invasive recordings. Basically, the most prominent problem is the high variability in the performance to control a BCI. It becomes clear that the development of reliable and practical BCI systems that can deliver high-speed communication and control is far from being possible (particularly in patients with neurological disorders where highly effective contingence between numerous neural systems has been disrupted).

3 Invasive and Non-invasive BCIs

While non-invasive BCI systems detect fluctuations of electrical potentials or magnetic fields from outside the skull, invasive BCI systems acquire their control signals directly in the brain tissue. Semi-invasive approaches use epidural electrode arrays. Until now four types of signals, three based on EEG activity and one based on MEG activity have been more thoroughly tested in non-invasive BCI research:

- Slow cortical potentials (SCP-BCI)
- Brain oscillations measured by EEG ranging from 4 to 40 Hz, primarily μ- or SMR (sensorimotor rhythm) and its harmonics (8–30 Hz from sensorimotor cortex) (EEG-BCI)
- Event-related brain potentials (ERPs), primarily the P300 (P300-BCI)
- Brain oscillations measured by MEG in the range of SMRs (MEG-BCI)

Besides these signals acquired in a non-invasive fashion by EEG or MEG three invasively acquired signals have been tested:

- Electrocorticogram (ECoG) from subdural implanted macroelectrodes or arrays
- Action potential spike trains from implanted microelectrodes
- Synaptic field potentials from implanted electrodes [67]

The beginning of studies regarding operant control of SCPs reaches back to the late 1970s when Birbaumer and co-workers (University of Tuebingen, Germany) demonstrated by numerous publications that strong and anatomically specific effects of self-induced changes in cortical activity on behavior and cognition occur. SCP control was learned through visual and auditory feedback and reward. Birbaumer et al. brought up solid neurophysiological evidence about the anatomical sources and physiologic functions of SCPs (see Fig. 1) [4, 11, 12, 14, 15]. An intriguing aspect of their work was the finding that SCPs from posterior parietal and occipital sources were resistant to operant learning whereas central and frontal SCPs could be voluntarily controlled after one to five training sessions [62]. The critical importance of the anterior brain systems to generate physiologic SCPs was confirmed by a number of following clinical studies. Patients with prefrontal dysfunctions such as attention-deficit disorder (ADD, [12]) or schizophrenia [74] had severe problems to gain control over SCPs. Those patients with extended prefrontal brain damage were entirely unable to learn SCP control [62]. Recent studies using fMRI showed that successful voluntary brain control correlated with activity changes in premotor as well as anterior parts of the basal ganglia [39, 40, 41]. Taken together, these findings are in line with the proposal that operant voluntary control of local cortical excitation thresholds depends on an intact prefrontal, motor or premotor cortical and subcortical system.

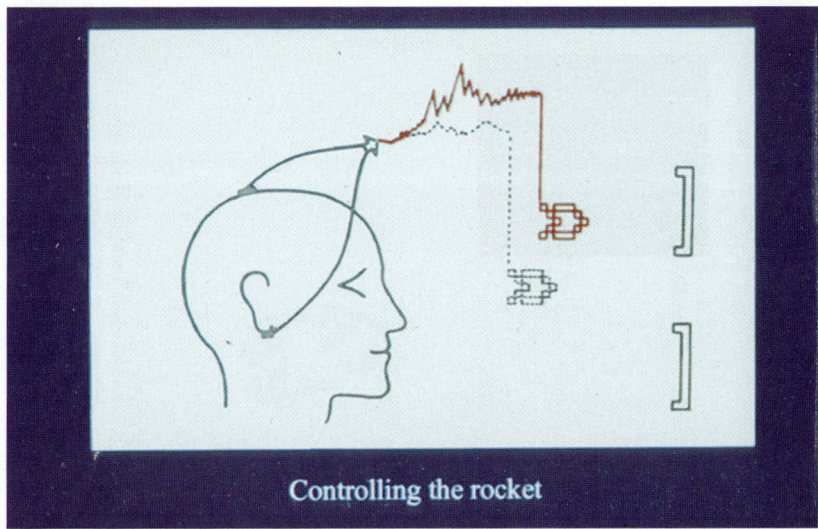

Fig. 1. The slow cortical potential BCI or thought translation device described by Birbaumer et al. [15]

In 1972 Sterman proposed that augmentation of sensorimotor rhythms (SMR) increases self-control of epileptic seizures [79, 82]. SMRs (defined by frequencies of 10–20 Hz recorded over the sensorimotor cortex) disappear during actual or imagined movements and seem to be closely associated with motor inhibition. It should be noted that – depending from the context – the SMR is also called the μ − trhythm following a suggestion of Gastaut [34] and Gastaut et al. [35]. It is, however, not clear whether the neurophysiological bases of the two phenomena are comparable. SMRs have been extensively investigated by the Pfurtscheller group in Graz and the Wolpaw group in Albany. The focality and versatility to cognitive manipulation of SMR makes it an ideal candidate to drive a BCI device. Pfurtscheller showed that voluntary control over SMRs could be achieved by imagination of movements, even in paralyzed patients: imagery of hand movements annihilated SMRs over the neuronal representation of the hand ([68], see Fig. 2). He called this phenomenon event-related desynchronization (ERD) and event-related synchronization (ERS). The disappearance of SMR can be used as a control signal for external devices, for example prostheses. The successful regulation of SMRs in completely paralyzed patients showed that the mechanisms to generate SMRs are not directly mediated by changes of the peripheral nervous system. However, the critical role of central motor activity in voluntary action and thought remains.

Another extensively tested BCI controller is the P300 event-related brain potentials (ERP)-BCI developed by Donchin ([22, 28], Fig. 3). The P300-BCI in contrast to many other BCI systems needs no training at all. The P300 is an extremely robust event-related potential (ERP) appearing

Senso-motor-rhythm (SMR) training

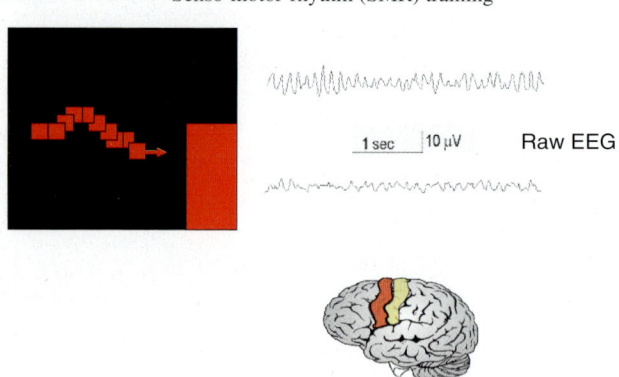

Raw EEG

Fig. 2. The SMR (sensorimotor-rhythm)-BCI described by Pfurtscheller et al. [68]

approximately 300 ms to 1 s following an unexpected sensory stimulus. The subject is instructed to concentrate on a letter in a letter matrix he/she wants to spell. Whenever the desired letter is among the lightened string of the matrix a P300 ERP appears in the EEG selecting the desired letter [76]. This system allows much faster selection of letters than any other BCI system. The P300 can be best recorded over anterior parietal areas. Today, the P300 is clinically widely used to evaluate attention deficits, cognitive disorders and lately in lie detectors.

Using 270-channel magnetoencephalography (MEG) Birbaumer and Cohen [7] developed a BCI that classifies magnetic field changes within the frequencies

P300 –Brain-Computer-Interface (BCI)

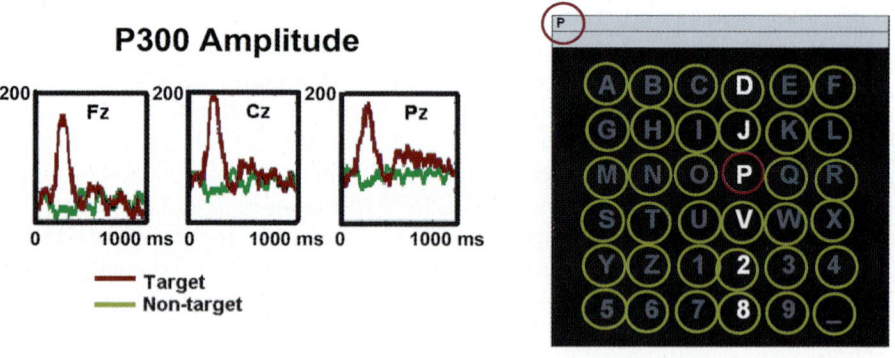

Fig. 3. The P300-BCI described by Farwell and Donchin [28]

of SMR or beta–gamma range with an excellent spatial and temporal resolution. It uses the most responsive MEG channels during actual or imagined movements. Signals are analyzed by the BCI2000 software developed by BCI group at the Wadsworth Laboratory in Albany, New York (Fig. 4). In contrast to EEG-based BCI systems the MEG-BCI allows recording of a much broader frequency range and allows the gradual distinction of anatomically specific cortical activation.

Most recently a BOLD signal-based fMRI-BCI has been introduced [18, 21, 86, 92]. In fMRI increases or decreases of paramagnetic load of blood flow can be measured, indicating local metabolic deficiencies of the vascular bed supplying active neurons. It has been shown that intracortical activity is highly correlated with local blood flow change and the BOLD signal [60]. Studies that simultaneously investigated voluntary SCPs and BOLD signals revealed that central SCPs are particularly associated with BOLD variations in the anterior basal ganglia and premotor cortex [40, 41]. In 2003 Weiskopf and Birbaumer et al. proposed that the development of fMRI-BCIs might be a powerful tool in the treatment of various disorders and diseases. Coherently DeCharms et al. demonstrated that voluntary self-regulation affects pain perception and behavior using a real-time fMRI-BCI that allows voluntary self-regulation of deeper brain structures, apart from cortical areas, such as amygdala, anterior cingulate, insula and parahippocampal gyrus. He showed that negative emotional valence ratings were intensified in people who voluntarily increased their insular activation. Modulation of activity in the anterior cingulate affected arousal whereas parahippocampal gyrus activation was associated with memory performance reduction.

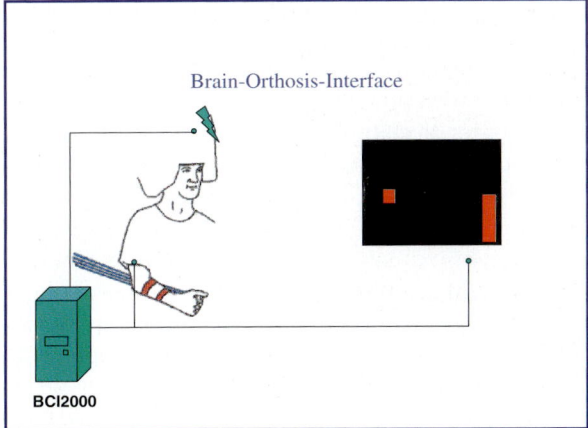

Fig. 4. Principle of the BCI that has been developed by Birbaumer and Cohen [7]: magnetic field potentials of the sensorimotor cortex are given as optical feedback by a moving cursor (on the *left, red square*). If the cursor hits the red bar on the right the hand orthosis will be opened, otherwise the hand orthosis will be closed

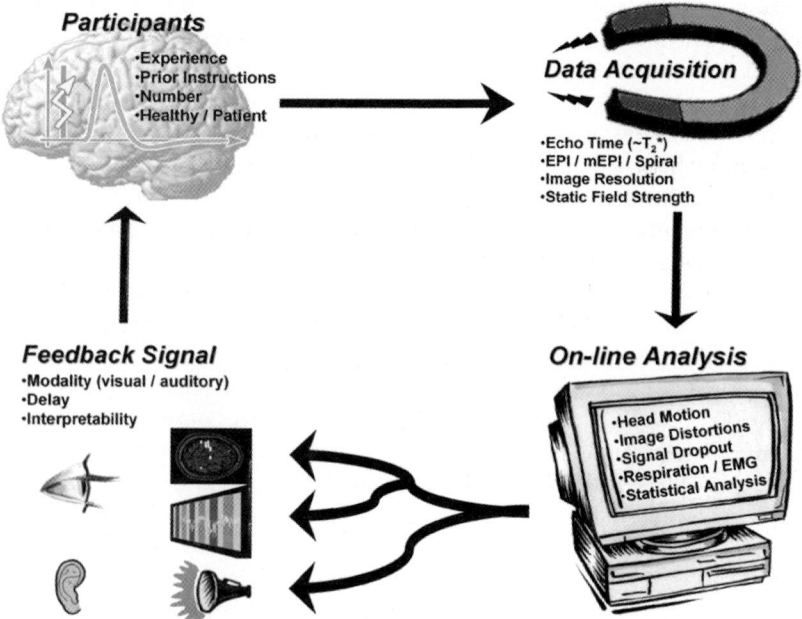

Fig. 5. The real-time fMRI-BCI [86]

Figure 5 shows the first fMRI-BCI system developed in Birbaumer's laboratory [86] using extremely fast echo-planar gradient sequences in a 1.5–3 T MR scanner [87, 88]. Subjects observe the visual feedback reflecting the movement-corrected BOLD changes of a circumscribed cortical or subcortical brain area.

In addition to the fMRI-BCI approach near infrared spectroscopy (NIRS) is also a non-invasive technique based on metabolic changes of the brain. NIRS uses multiple pairs or channels of light sources and light detectors operating at two or more discrete wavelengths near infrared range (700–1000 nm) to determine cerebral oxygenation and blood flow of localized regions of the brain. Typically the depth of brain tissue that can be measured is between 1 and 3 cm. The degree of increases in regional cerebral blood flow (rCBF) exceeds that of increases in regional cerebral oxygen metabolic rate (rCMRO$_2$) resulting in a decrease in deoxygenated hemoglobin in venous blood. Thus, increase in total hemoglobin and oxygenated hemoglobin with a decrease in deoxygenated hemoglobin is expected to be observed in activated areas during NIRS measurement. The optical parameter measured is attenuation of light intensity due to absorption by the intermediate tissue. The concentration changes of oxygenated hemoglobin and deoxygenated hemoglobin are computed from changes in the light intensity at different wavelengths, using the modified Beer–Lambert equation. Compared to other BCI approaches the advantage of the NIRS approach lies in its simplicity, flexibility and high signal-to-noise ratio.

Besides the described non-invasive BCI approaches several invasive and semi-invasive techniques have been developed during the last years. Whereas there is only little risk in applying non-invasive BCI systems, for implantation of electrodes or microelectrode arrays it is normally necessary to open the skull, respectively the dura mater. The reconstruction of movements from firing patterns of single cells of the motor cortex [67] or parietal neuronal pools [73] in animals were remarkably successful and induced notable enthusiasm. Monkeys learned to move cursors into moving goals on a computer screen in a predetermined sequence by successively activating motor, premotor and parietal motor neuron pools. In one particularly successful preparation 32 cells were sufficient to move an artificial arm and perform skilful reaching movements after extensive training [75]. The plasticity of the cortical circuits allows learned control of movements directly from the cellular activity even outside the primary or secondary homuncular representations of the motor cortex [84]. The animals studied in BCI research [67] were all intact animals that learned to move an artificial device or cursor for food reward in highly artificial laboratory situations. There are a couple of major problems that have not been solved so far. Some of them are related to the fact that the motor cortex contributes only about 40% of the whole cortical input to the corticospinal tract indicating that complex motor behavior is not exclusively represented in the motor cortex. Another problem is often referred to as the *decoding problem* meaning that the spike-firing pattern is ambiguous. Nevertheless, results of invasive BCI systems are impressive though generalization from the invasive animal BCI approach to people with neurological disorders is premature.

Taken together the following properties of BCI control should be emphasized:

(1) Voluntary control of circumscribed brain areas is possiblewithin several hours of training in healthy subjects
(2) The learned regulation is specific and does not coactivate unspecifically other brain sites
(3) The behavioral effects of self-induced local brain changes are functionally specific for the respective brain region
(4) The central motor systems need not be coactivated during brain self-regulation indicating that motor mediation of the brain is not a necessary prerequisite for its regulation and learned brain control should be possible in completely paralyzed people and animals unable to use motor projections for peripheral physiological regulation [24]
(5) Cognitive activity such as imagery may assist the acquisition of brain control but is not a necessary condition to manipulate brain regions locally

4 Application of BCIs in Restoration of Movement and Thought

4.1 Use of BCI Technology for Communication with Patients Suffering from Amyotrophic Lateral Sclerosis (ALS)

ALS was first described by Charles Bell in 1830 and is characterized by a combined degradation of upper and lower motor neurons. The progression of this disease is heterogenous. The mean range of survival lies at 3–4 years after diagnosis, only 4–10% of patients survive more than 10 years. The most common cause of death is respiratory failure. During the late stadium of this disease normal communication with patients becomes more and more difficult.

In the late 1990s Birbaumer et al. developed a "thought translation device" for these patients [38]. It allows even locked-in patients to keep up communication with their environment using a SCP-based brain–computer interface [15, 54, 55]. Lately also P300- and SMR-based BCI systems have been used. Until now 28 patients with ALS were trained with these BCI systems [5, 6]. Most patients achieved remarkable control of the BCI systems and were enabled to select letters and write words even when they reached the locked-in stadium. However, sufficient BCI training to reach acceptable BCI control was impossible in patients that have already reached the locked-in stadium ($n=7$). Only one patient was able to communicate for three sessions using a pH-based device [25, 89].

4.2 BCI Systems in the Restoration of Movement

Pfurtscheller's BCI group in Graz, Austria, was the first group that applied an SMR-based BCI to a quadriplegic patient with high spinal cord lesion. By voluntary modulation of SMRs the patient was enabled to control electrical stimulation of his hand and arm muscles. With this apparatus he could for example grasp [68].

In 2005 the magnetoencephalography (MEG-) BCI was developed by Niels Birbaumer, Germany, and Leonardo Cohen, USA. It is based on arbitrary modulation of sensorimotor rhythm (SMR) (Fig. 4). Birbaumer has shown that also patients with stroke can use this device [7]. A great advantage of the MEG-BCI is the high spatial and temporal resolution that additionally allows precise examination of cortical reorganizational processes during and after BCI training.

Up to now, Niels Birbaumer and Leonardo Cohen showed that four of five stroke patients are in principle able to control the device with a precision of 70–90% after 20 training sessions and that the precision and complexity of learned movements were comparable to those achieved with invasive recordings. Before the actual training, the patients have to imagine several distinct movements of the upper and lower extremity as well as the tongue. While doing this the ipsilesional area with the strongest oscillatory perilesional MEG response

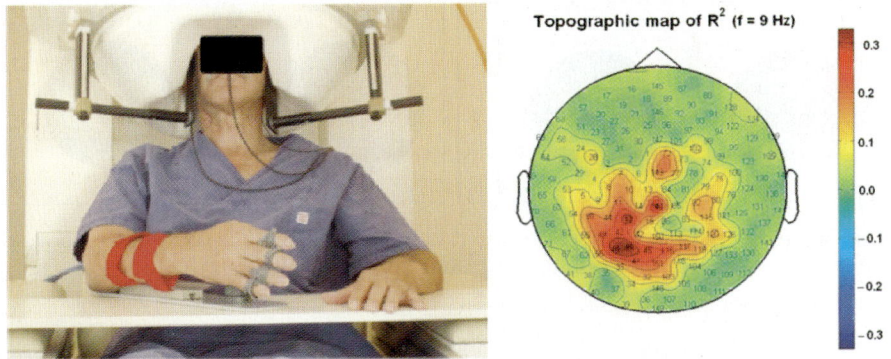

Fig. 6. *Left:* a stroke patient in a MEG moves his plegic arm by using the MEG-BCI. Right: sensorimotor activity is being recorded by 150 MEG sensors. The patient uses perilesional brain areas to control the BCI by arbitrary modulation of his brain oscillations (at 9 Hz)

will be selected. They range normally between 9 and 25 Hz. This area is represented by three MEG sensors (Fig. 6). Depending on their activity a cursor can be moved on a screen. For the training only these frequencies will be selected that allow optimal distinction between imagined hand opening and closing. Patients learn to control a hand-orthoses that can be moved within five grades.

During the following training sessions patients are asked to imagine opening and closing of their hand and to keep track of the moving cursor that indicates the corresponding brain activity's strength. Depending on the brain activity the cursor moves either up or down. If the cursor hits the red wall (Fig. 4) hand is opened in five grades according to the signal strength. Figure 7 shows two examples of learning curves of two hemiplegic stroke patients. Clear learning progress can be observed. The patient whose learning curve is shown on the right achieves correct hand movements in up to 90%. The curves show that there is a high variability of performance within the sessions.

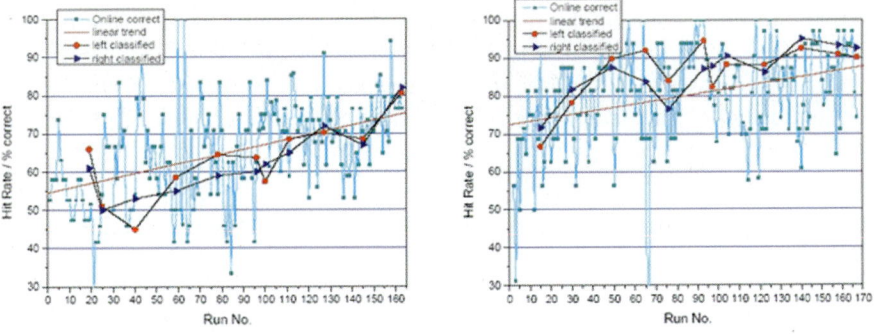

Fig. 7. Learning curves of two patients while training to control the BCI

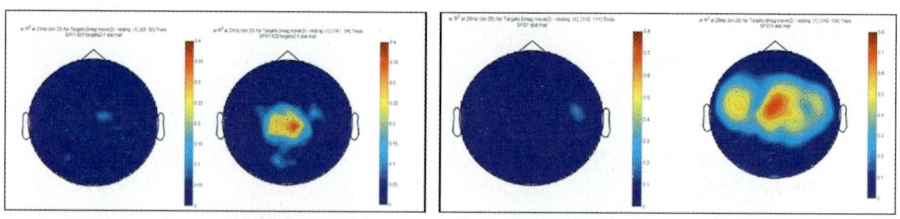

Fig. 8. *Left* = patient 1; *right* = patient 2; increase of SMR activity (also above the brain lesion), before the training (*left*) and after the training (*right*)

Figure 8 shows the "strength" of SMR activity before and after the training for opening the hand. It should be noted that there is an increase of SMR activity neighboring the focal brain lesion.

In the use of the MEG-BCI system it has been shown that BCI technology has the capability to initiate vast cortical reorganization. These findings have significant implications for the development of innovative approaches in neurorehabilitation as BCI technology might not only be used to restore lost functions by controlling artificial devices but also to systematically induce neuroplastic changes that consequentially lead to an improvement of brain functions itself (in motor control of paretic limbs for example).

Regarding *invasive* BCI approaches for movement restoration, several successful studies in healthy monkeys and rats to perform skilful movements with spike patters from single motor or parietal neurons to control BCI devices led to substantial optimism (see [67] for review). The perspective of fully implantable BCI devices allowing its use in a daily life environment is an appealing vision – but until now confronted with enormously difficult problems far from solution. In an encouraging experiment Hochberg et al. [42] implanted extremely densely packed microelectrode arrays of up to several hundred microelectrodes [23] in two quadriplegic human patients. Within a few training sessions, the patients learned to use neuronal activity from field potentials to move a computer cursor in several directions comparable to the tasks used for multidimensional cursor movements in the non-invasive SMR-BCI reported by Wolpaw and McFarland [91]. In contrast to the studies in healthy animals none of the invasive procedures allowed restoration of skilful movement in paralyzed animals or people in everyday-life situations. It is not clear why so far the human preparations (see also Kennedy et al. [51] for a single electrode preparation) have achieved only limited results in terms of application to activities of daily living [42].

4.3 BCI Technology in Attention Regulation and the Treatment of Epilepsy

Based on the finding by Sterman and colleagues [80, 81] that cats exhibited maximum SMRs during motor inhibition and spindle sleep stages indicating recruitment of inhibitory thalamocortical circuits that were incompatible with seizures, humans were systematically trained to voluntarily increase SMRs. After 20–100 sessions some of those trained patients were able to reduce frequency of seizures or even achieved full remission in drug-resistant epilepsy [78]. Unfortunately, Sterman's finding was never applied as a therapeutic tool in larger and well-controlled clinical studies.

In later studies patients with focal epileptic seizures were intensively trained to downregulate their cortical excitability by modulating their SCPs [53, 71, 72, 83]. Some patients were able to fully control their SCP changes. In average, patients could reduce the frequency of seizures by 50% compared to baseline or control condition combined with a significant gain in cognitive functioning and IQ. These clinically very successful studies confirmed the principle of Sterman's approach. It should be noted, however, that one patient with very high negative SCP amplitudes did not profit from the SCP-BCI. All other patients showed stable improvement at follow-up examinations, some remained seizure free.

As mentioned before the SCP-BCI was also applied in patients with attention-deficit disorder (ADD). In general these patients showed initially less SCP control than healthy controls. After training patients could control volitional increase of central-frontal negativity of SCPs associated with an improvement of ADD symptoms. These improvements were comparable to medication with amphetamines [32]. Training to modulate SMRs and frequencies in the beta-1-range led to similar behavioral results suggesting that control of different EEG activities converge to a common therapeutic pathway. The improvement might be associated with the general capacity to regulate attention by voluntary control of brain activity.

4.4 fMRI-BCI in Anxiety Disorder and Psychopaths

Whereas DeCharms et al. [21] investigated the behavioral effects of brain regulation on pain perception using real-time functional MRI (rtfMRI), in a series of pilot experiments of the Tuebingen BCI group used an rtfMRI-BCI approach to train healthy volunteers to modulate emotions of anxiety and fear. Previous imaging studies showed that fear and anxiety is accompanied by metabolic activation of the amygdala, anterior insula, cingulate and orbitofrontal cortex. It could be demonstrated that after training healthy subjects ($n=9$) could increase BOLD responses in the left anterior insula involved in the fear circuit (Fig. 9). Within 60–90 min of training there was a significant increase of BOLD responses in this area compared to a sham control and imagery control. The imagery control group is depicted in Fig. 9C. The experimental condition resulted in a more negative emotional valence rating of fear-evoking pictures of the IAPS series [58] during insula BOLD increase only. It is remarkable that the percentage of BOLD change in the target region of interest (ROI) – here the left insula – increased with training while the change in the right insula and the reference ROI was kept comparatively

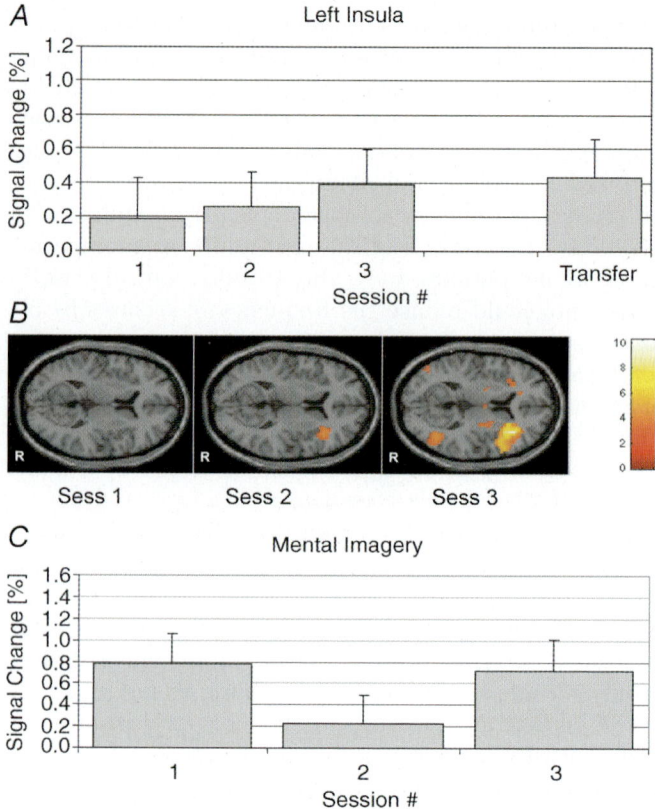

Fig. 9. Voluntary increase of brain activity in the insula: (**A**), increase in percentage BOLD during operant feedback training of the left insula averaged over nine healthy subjects and three sessions of 20 min each. (**B**) Horizontal section at the insular region; BOLD responses averaged over nine healthy subjects and three training sessions. (**C**) Control group of three subjects instructed to use emotional imagery without contingent feedback of the BOLD responses over three sessions

constant. It could be shown that the effect is not only area specific but also valence specific as only aversive slides change their valence while other types of emotional slides are not affected.

Examination of criminal psychopaths [16] showed a dramatic deficit in metabolic activity of the fear circuit: lateral orbital prefrontal cortex, amygdalae and anterior cingulate, anterior insula and superior parietal cortex were not activated during aversive classical (Pavlovian) conditioning. During the study the first three psychopaths investigated could learn to voluntarily increase activation of the anterior insula. The behavioral effects of the training are currently being investigated.

5 Ethical and Quality of Life Issues in BCI Research

Most ALS patients opt against artificial respiration and feeding and die of respiratory and other clinical complications. In many countries, doctors are allowed to assist the transition with sedating medication to ease respiration-related symptoms. If doctor-assisted suicide or euthanasia is legal, as it is in the Netherlands and Belgium, very few patients vote for continuation of life. The vast majority of family members and doctors believe that the quality of life in total paralysis is poor, that continuation of life constitutes a burden for the patient and that it is unethical to use emergency measures such as tracheostomy to continue life. The pressure on the patient to discontinue life is high. However, the evidence does not support hastened death decisions in ALS [1, 70]. Most instruments measuring depression and quality of life, such as the widely used Beck or Hamilton depression scales, are invalid for paralyzed people living in protected environments because most of the questions do not apply to the life of a paralyzed person ("I usually enjoy a good meal", "I like to see a beautiful sunset""). Special instruments have had to be developed for this population [56, 57]. In studies by Breitbart et al. [17] and at the University of Tuebingen by Kübler et al. [56, 57] only 9% of ALS patients showed long episodes of depression, most of them in the time period following the diagnosis and a period of weeks after tracheostomy. Figure 10 shows the results for depression (a) and for quality of life (b) rated by patients and family members and caretakers. It can be seen that ALS patients are not clinically depressed. In fact, they are in a much better mood than psychiatrically depressed patients without any life-threatening bodily disease. Likewise, patients rate their quality of life as much better than their caretakers and family members do, even when these patients are completely paralyzed and respirated. None of the patients of our sample (some of them in locked-in state (LIS)) requested

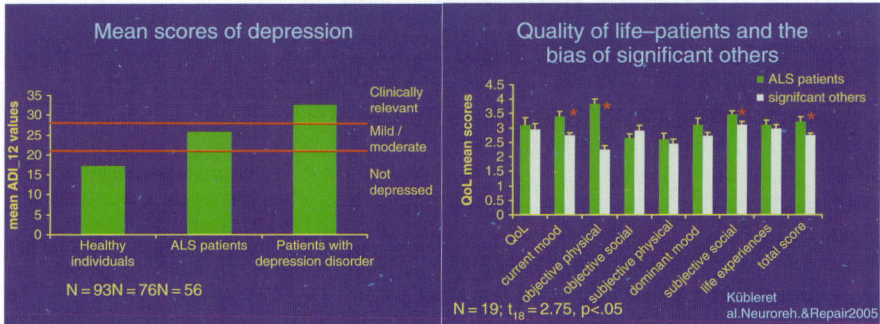

Fig. 10. Depression (**a**) and quality of life (**b**) in ALS. ALS patients are within the normal range for depression (**a**, *red lines*). (**b**) Quality of life is rated higher by patients (*green*) than significant others (*white*)

hastened death. It should be added that life expectancy with adequate communication through BCI or other means is much longer than the average of 5 years.

It could be argued that questionnaires and interviews reflect more social desirability and social pressure than the "real" behavioral–emotional state of the patient. The social pressure in ALS, however, directs the patient toward death and interruption of life support. The data therefore may *underestimate* the positive attitude in these groups. This hypothesis is strongly supported by a series of experiments with ALS patients at all stages of their disease using the International Affective Picture System (IAPS; [58]). Using a selection of IAPS pictures with social content, more positive emotions to positive pictures and less negative ratings to negative pictures were found in ALS than in matched healthy controls. Even more surprising are the brain responses to the IAPS slides (Fig. 11). fMRI measurement in 13 patients with ALS and controls demonstrated increased activation in the supramarginal gyrus and other areas responsible for empathic emotional responses to others, comparable to the "mirror neuron network" identified first by Rizolatti and colleagues [33] in patients. Furthermore, brain areas related to the processing of negative emotional information such as the anterior insulae and amygdala show less activation in ALS. These differences become stronger with progression of the disease 6 months later.

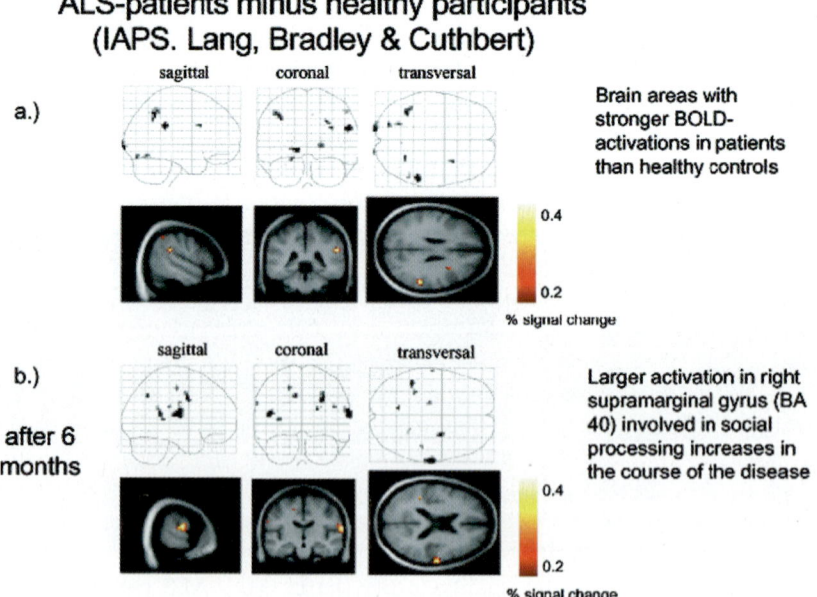

Fig. 11. Brain activation (with fMRI) in ALS patients to emotional slides in the course of disease progression (from Lulé et al., 2005 [61])

This evidence suggests that with progression of this fatal disease, emotional responses improve towards positively valenced social cues resulting in a more positive emotional state than in healthy controls. The positive responding and positive interaction of the social environment and caretakers to a fatally ill, paralyzed person may in part be responsible for the prosocial emotional behavior and for the modified brain representation of the ALS patients depicted in Fig. 11 as predicted by social learning theory [2].

Taken together, the results on emotional responding and quality of life in paralyzed ALS patients suggest a more cautious and ethically more responsive approach toward hastened death decisions and last-will orders of patients and their families. The data reported here also speak pervasively for the usefulness of BCIs in ALS and other neurological conditions leading to complete paralysis.

6 Future Direction in BCI Development

From the studies summarized in the cited work it is not clear which of the non-invasive BCIs is the most promising. In a recently completed study at the University of Tuebingen by Femke Nijboer, eight severely paralyzed patients with ALS were trained with SCP, P300 and SMR BCIs in a balanced crossover-within-subject design. Each training block of each BCI type lasted 20 sessions. The results were clear-cut: SCP-BCI improved control of brain activity within this short training period but was not good enough to select letters (70% minimum success rate). The fastest acquisition and fastest spelling rates were achieved with the P300 BCI. SMR was also successful but not as fast as the P300 system.

A highly relevant contribution for BCI research and application in general was recently published by Jackson et al. [47]. Based on a strict Hebbian concept, monkeys received intracranial electric stimulation of the motor hand area contingent upon a specific spike pattern related to a torque movement of the hand. The stimulated area, however, was distant from the recording area and was responsible for hand movements in the opposite direction. *Repetitive electric stimulation* instigated automatically at that distant region in close time proximity to a specific spike pattern (50 ms later) led to a conditioned reproduction of the spike pattern at the recording sites and a consequent change in the hand movement. This study demonstrates that the brain can be induced to reroute neural information responsible for a particular behavior if Hebbian connectivity between two arbitrary brain areas is built through *artificial but time-contingent stimulation*. Translated into the BCI situation, electric or magnetic stimulation of specific brain areas contingent upon a classified brain-signal pattern (EEG, MEG, BOLD, etc.) may circumvent interrupted neuronal pathways or structures with forced reorganization using this paired electrical stimulation technique. This example shows that the combination of BCI technology with neurostimulation is most instructive for future developments.

Neurostimulation has been used for five decades. However, its use has been increasing in recent years due to many technical innovations. Stimulation of the peripheral nerves or peripheral nerve stimulation was the first form of neurostimulation to be developed. Later it was replaced by an invasive technique which directly stimulated the spinal cord (SCS). Besides spinal cord stimulation and peripheral nerve stimulation, there are invasive and non-invasive forms of brain stimulation. The most prominent intracranial brain stimulation methods include deep brain stimulation (DBS) and motor cortex stimulation (MCS). DBS has been successfully used in Parkinson disease and lately also in mental disorders, such as depression [36] and obsessive–compulsive disorders. MCS showed to be effective in neuropathic pain [37]. The use of invasive neurostimulation entails the risk of complications, i.e., infections, bleedings and accidental brain injury. Therefore non-invasive forms of neurostimulation represent a promising alternative. The best-established techniques are transcranial magnetic stimulation (TMS) and transcranial direct current stimulation (tDCS). They lead to a lasting modulation of cortical excitability. By this it has been shown that anodal transcranial direct current stimulation (tDCS) affects cortical plasticity and learning in healthy persons as well as stroke patients. It has been indicated that enhancing activity in motor areas of the affected hemisphere by the use of tDCS [30, 46] leads to motor improvements of the paretic hand. A study by Hummel and Cohen [45] showed that non-invasive anodal tDCS applied to the affected hemisphere in stroke patients improved reaction time and pinch force compared to sham stimulation in a group of patients with subcortical stroke.

TMS presents concentrated magnetic fields at a fixed point on the scalp to induce electrical currents in the underlying cortex that can lead to neuronal discharge. It is used for over 20 years to investigate nervous propagation along the corticospinal tract, spinal roots and peripheral nerves (as relevant in diagnostic routines of multiple sclerosis, posttraumatic, neoplastic and compressive myelopathies, ALS, stroke, epilepsy and dystonia) and to evaluate excitatory/inhibitory intracortical circuits. By this it provides information on brain physiology and also pathophysiology. Starting from the observation that TMS elicits either long-lasting excitatory or inhibitory effects depending on the frequency of stimulation (low- or high-frequency repetitive TMS), attempts to use repetitive TMS (rTMS) as a therapeutic tool in treatment of neurologic and psychiatric disorders emerged. This includes depression [66], tinnitus [69], posttraumatic stress disorders [19], auditory hallucinations [44], chronic pain [59], epilepsy [85] and movement disorders [31]. Additionally, recent reports have suggested that rTMS in acute stages of stroke might improve clinical outcome [52, 63].

A major problem of all BCI use is the *variability of performance*. Users do much better on some days than on others, and performance can widely vary from trial to trial. It is therefore of highest relevance to stabilize variability to ensure that the behavior controlled by the BCI is always the desired one. The reduction of variability has implications not only for the practicability of

BCI use in everyday life but also for the effectiveness in neurorehabilitation to induce neuroplasticity.

In this context neurostimulation offers a very promising perspective: A study by Karim et al. [50] manipulated operant training of SCP control with anodal and cathodal polarization using tDCS. Anodal polarization before trials with required positivity substantially increased performance of SCP control by augmenting baseline negativity (depolarization) and thus increasing the amplitude of the self-produced positivity afterwards.

Hence, to precondition the state of the BCI user's brain by neurostimulation might be an effective way to decrease variability and therefore improve motor restoration or BCI reliability.

The diverse techniques of neurostimulation offer a variety of possibilities to enhance BCI systems. As mentioned before time-contingent repetitive electric stimulation can be used to reroute neural information and tDCS/TMS can be used to precondition the brain in the context of SCP or SMR control. Neurostimulation can coequally induce neuroplastic changes or stabilize those induced by BCI use resulting in behavioral benefit.

The limited results of invasive BCI techniques in patients with brain disorders are closely related to the neuroplastic nature of brains. This nature is equally a blessing and source of hope but also a curse. Seen from the positive side, besides the impressive achievements described above in clinical use of available BCI technology it has been proven that BCI technology is a powerful tool to systematically induce neuroplastic changes and therefore has a significant potential to promote innovative approaches in neurorehabilitation. However, future developments of BCIs in the restoration of movement and thought will be intimately connected to our understanding of neuroplasticity in the context of BCI use, the role of contingence and its selective modulation by neurostimulation.

Acknowledgments

This work was supported by the German Research Society (DFG) and the National Institutes of Health (NIH).

References

1. Albert S, Rabkin J, Del Bene M, Tider M, Mitsumoto H (2005) Wish to die in end-stage ALS. Neurology, 65:68–74
2. Bandura A (1969) Social learning of moral judgements. Journal of Personality and Social Psychology, 11(3):275–279
3. Berger H (1929) Ueber das Elektrenkephalogramm des Menschen. Archiv für Psychiatrie und Nervenkrankheiten, 87:527–570
4. Birbaumer N (1999) Slow cortical potentials: Plasticity, operant control, and behavioral effects. The Neuroscientist, 5(2):74–78

5. Birbaumer N (2006a) Brain–computer–interface research: Coming of age. Clinical Neurophysiology, 117:479–483
6. Birbaumer N (2006b) Breaking the silence: Brain–computer interfaces (BCI) for communication and motor control. Psychophysiology, 43:517–532
7. Birbaumer N, Cohen LG (2007) Brain–computer interfaces: Communication and restoration of movement in paralysis. Journal of Physiology, 579:621–636
8. Birbaumer N, Kimmel H (Eds.) (1979) Biofeedback and Self-Regulation. Hillsdale: Erlbaum
9. Birbaumer N, Schmidt RF (2005) Biologische Psychologie (6th ed). Berlin Heidelberg New York: Springer
10. Birbaumer N, Elbert T, Rockstroh B, Lutzenberger W (1986) Biofeedback of slow cortical potentials in attentional disorders. In W.C. McCallum, R. Zappoli, and F. Denoth (Eds.), Cerebral Psychophysiology: Studies in Event-Related Potentials (pp. 440–442). Amsterdam: Elsevier
11. Birbaumer N, Elbert T, Canavan A, Rockstroh B (1990) Slow potentials of the cerebral cortex and behavior. Physiological Reviews, 70:1–41
12. Birbaumer N, Roberts L, Lutzenberger W, Rockstroh B, Elbert T (1992) Area-specific self-regulation of slow cortical potentials on the sagittal midline and its effects on behavior. Electroencephalography and Clinical Neurophysiology, 84:353–361
13. Birbaumer N, Flor H, Cevey B, Dworkin B, Miller NE (1994) Behavioral treatment of scoliosis and kyphosis. Journal of Psychosomatic Research, 6:623–628
14. Birbaumer N, Flor H, Lutzenberger W, Elbert T (1995) Chaos and order in the human brain. In G. Karmos and M. Molnar (Eds.), Perspectives of Event-Related Potentials Research (EEG Suppl. 44) (pp. 450–459). Amsterdam: Elsevier
15. Birbaumer N, Ghanayim N, Hinterberger T, Iversen I, Kotchoubey B, Kübler A, Perelmouter J, Taub E, Flor H (1999) A spelling device for the paralysed. Nature, 398:97–298
16. Birbaumer N, Veit R, Lotze M, Erb M, Hermann C, Grodd W, Flor H (2005) Deficient fear conditioning in psychopathy: A functional magnetic resonance imaging study. Archives of General Psychiatry, 62:799–805
17. Breitbart W, Rosenfeld B, Penin H (2000) Depression, hopelessness, and desire for hastened death in terminally ill patients with cancer. Journal of American Medical Association, 284:2907–2911
18. Caria A, Veit R, Sitaram R, Lotze M, Weiskopf N, Grodd W, Birbaumer N (2007) Regulation of anterior insular cortex activity using real-time fMRI. NeuroImage, 35:1238–1246
19. Cohen H, Kaplan Z, Kotler M, et al. (2004) Repetitive transcranial magnetic stimulation of the right dorsolateral prefrontal cortex in posttraumatic stress disorder: A double-blind, placebo-controlled study. American Journal of Psychiatry, 161:515–524
20. Cuthbert B, Kristeller J, Simons R, Hodes R, Lang PJ (1981) Strategies of arousal control: Biofeedback, meditation, and motivation. Journal of Experimental Psychology: General, 110(4):518–546
21. DeCharms RC, Maeda F, Glover GH, Ludlow D, Pauly JM, Soneji D, Gabrieli JD, Mackey SC (2005) Control over brain activation and pain learned by using real-time functional MRI. Proceedings of the National Academy of Sciences, USA, 102(51):18626–18631

22. Donchin E (1981) Surprise!...Surprise? Psychophysiology, 18:493–513
23. Donoghue JP (2002) Connecting cortex to machines: Recent advances in brain interfaces. Nature Neuroscience, 5:1085–1088
24. Dworkin BR, Miller NE (1986) Failure to replicate visceral learning in the acute curarized rat preparation. Behavioral Neuroscience, 100:299–314
25. Dworkin B, Miller NE, Dworkin S, Birbaumer N, Brines M, Jonas S, Schwentker E, Graham J (1985) Behavioral method for the treatment of idiopathic scoliosis. Proceedings of the National Academy of Sciences, USA, 82:2493–2497
26. Elbert T, Rockstroh B, Lutzenberger W, Birbaumer N (Eds.) (1984) Self-Regulation of the Brain and Behavior. New York: Springer
27. Engel BT (1981) Clinical biofeedback: A behavioral analysis. Neuroscience and Biobehavioral Reviews, 5(3):397–400
28. Farwell LA, Donchin E (1988) Talking off the top of your head: Toward a mental prosthesis utilizing event-related brain potentials. Electroencephalography and Clinical Neurophysiology, 70:510–523
29. Flor H, Birbaumer N (1993) Comparison of the efficacy of EMG biofeedback, cognitive behavior therapy, and conservative medical interventions in the treatment of chronic musculoskeletal pain. Journal of Consulting & Clinical Psychology, 61(4):653–658
30. Fregni F, Boggio PS, Mansur CG, Wagner T, Ferreira MJ, Lima MC, Rigonatti SP, Marcolin MA, Freedman SD, Nitsche MA, Liebetanz D, Antal A, Lang N, Tergau F, Paulus W (2003) Modulation of cortical excitability by weak direct current stimulation–technical, safety and functional aspects. Supplements of Clinical Neurophysiology, 56:255–276
31. Fregni F, Simon DK, Wu A, Pascual-Leone A (2005) Non-invasive brain stimulation for Parkinson's disease: A systematic review and meta-analysis of the literature. Journal of Neurology, Neurosurgery, and Psychiatry, 76:1614–1623
32. Fuchs T, Birbaumer N, Lutzenberger W, Gruzelier JH, Kaiser J (2003) Neurofeed back training for attention-deficit/hyperactivity disorder in children: A comparison with methylphenidate. Applied Psychophysiology and Biofeedback, 28(1):1–12
33. Gallese V, Keysers C, Rizzolatti G (2004) A unifying view of the basis of social cognition. Trends in Cognitive Sciences, 8(9):396–403
34. Gastaut H (1952) Electrocorticographic study of the reactivity of rolandic rhythm. Review Neurologique (Paris), 87(2):176–182
35. Gastaut H, Terzian H, Gastaut Y (1952) Study of a little electroencephalographic activity: Rolandic arched rhythm. Marseille Medical, 89(6):296–310
36. Giacobbe P, Kennedy SH (2006) Deep brain stimulation for treatment-resistant depression: A psychiatric perspective. Current Psychiatry Reports, 8:437–444
37. Henderson JM, Lad SP (2006) Motor cortex stimulation and neuropathic facial pain. Neurosurgical Focus, 15(21):E6
38. Hinterberger T, Veit R, Strehl U, Trevorrow T, Erb M, Kotchoubey B, Flor H, Birbaumer N (2003) Brain areas activated in fMRI during self regulation of slow cortical potentials (SCPs). Experimental Brain Research, 152:113–122
39. Hinterberger T, Weiskopf N, Veit R, Wilhelm B, Betta E, Birbaumer N (2004) An EEG-driven brain–computer-interface combined with functional magnetic resonance imaging (fMRI). IEEE Transactions on Biomedical Engineering, 51(6):971–974

40. Hinterberger T, Birbaumer N, Flor H (2005a) Assessment of cognitive function and communication ability in a completely locked-in patient. Neurology, 64:1307–1308
41. Hinterberger T, Veit R, Wilhelm B, Weiskopf N, Vatine J-J, Birbaumer N (2005b) Neuronal mechanisms underlying control of a brain–computer-interface. European Journal of Neuroscience, 21:3169–3181
42. Hochberg LR, Serruya MD, Friehs GM, Mukand JA, Saleh M, Caplan AH, Branner A, Chen D, Penn RD, Donoghue JP (2006) Neural ensemble control of prosthetic devices by a human with tetraplegia. Nature, 442:164–171
43. Hoelzl R, Whitehead W (Eds) (1983) Psychophysiology of the Gastrointestinal Tract. New York: Plenum Press
44. Hoffman RE, Gueorguieva R, Hawkins KA, et al. (2005) Temporoparietal transcranial magnetic stimulation for auditory hallucinations: Safety, efficacy and moderators in a fifty patient sample. Biological Psychiatry, 58, 97–104
45. Hummel FC, Cohen LG (2006) Non-invasive brain stimulation: A new strategy to improve neuro-rehabilitation after stroke? Lancet Neurology, 5, 708–712
46. Hummel F et al. (2005) Effects of non-invasive cortical stimulation on skilled motor function in chronic stroke. Brain, 128:490–499
47. Jackson A, Mavoori J, Fetz EE (2006) Long-term motor cortex plasticity induced by an electronic neural implant. Nature, 444:56–60
48. Jorgensen HS, Nakayama H, Raaschou HO, Vive-Larsen, J, Stoier M, Olsen TS (1995a) Outcome and time course of recovery in stroke. Part II: Time course of recovery. The Copenhagen Stroke Study. Archives of Physical Medicine and Rehabilitation, 76:406–412. Doi: 10.1016/S0003-9993(95)80568-0
49. Jorgensen HS, Nakayama H, Raaschou HO, Vive-Larsen J, Stoier M, Olsen TS (1995b) Outcome and time course of recovery in stroke. Part I: Outcome. The Copenhagen Stroke Study. Archives of Physical Medicine and Rehabilitation, 76:399–405
50. Karim AA, Kammer T, Lotze M, Nitsche MA, Godde B, Hinterberger T, Cohen LG, Birbaumer N (2004) Effects of TMS and tDCS on the physiological Regulation of cortical excitability in a Brain-Computer Interface. Biomedical Engineering 49(1):55–57
51. Kennedy PR, Kirby MT, Moore MM, King B, Mallory A (2004) Computer control using human intracortical local field potentials. IEEE Transactions on Neural Systems and Rehabilitation Engineering, 12(3):339–344
52. Khedr EM, Ahmed MA, Fathy N, Rothwell JC (2005) Therapeutic trial of repetitive transcranial magnetic stimulation after acute ischemic stroke. Neurology, 65:466–468
53. Kotchoubey B, Strehl U, Uhlmann C, Holzapfel S, König M, Fröscher W, Blankenhorn V, Birbaumer N (2001) Modification of slow cortical potentials in patients with refractory epilepsy: A controlled outcome study. Epilepsia, 42(3):406–416
54. Kübler A, Kotchoubey B, Kaiser J, Wolpaw J, Birbaumer N (2001a) Brain–computer communication: Unlocking the locked-in. Psychological Bulletin, 127(3):358–375
55. Kübler A, Neumann N, Kaiser J, Kotchoubey B, Hinterberger T, Birbaumer N (2001b) Brain–computer communication: Self-regulation of slow cortical potentials for verbal communication. Archives of Physical Medicine and Rehabilitation, 82:1533–1539

56. Kübler A, Winter S, Ludolph AC, Hautzinger M, Birbaumer N (2005a) Severity of depressive symptoms and quality of life in patients with amyotrophic lateral sclerosis. Neurorehabilitation and Neural Repair, 19(3):182–193
57. Kübler A, Nijboer F, Mellinger J, Vaughan TM, Pawelzik H, Schalk G, McFarland DJ, Birbaumer N, Wolpaw JR (2005b) Patients with ALS can use sensorimotor rhythms to operate a brain–computer interface. Neurology, 64:1775–1777
58. Lang P, Bradley M, Cuthbert B (1999) International Affective Picture System. Gainesville, Fl: The Center for Research in Psychophysiology, University of Florida
59. Lefaucheur JP (2004) Transcranial magnetic stimulation in the management of pain. Clinical Neurophysiology (Supplement), 57:737–748
60. Logothetis N, Pauls J, Augath M, Trinath T, Oeltermann A (2001) Neurophysiological investigation of the basis of the fMRI signal. Nature, 412:150–157
61. Lulé D, Kurt A, Jürgens R, Kassubek J, Diekmann V, Kraft E, Neumann N, Ludolph AC, Birbaumer N, Anders S (2005) Emotional responding in amyotrophic lateral sclerosis. Journal of Neurology, 252:1517–1524
62. Lutzenberger W, Birbaumer N, Elbert T (1980) Self-regulation of slow cortical potentials in patients with frontal lobe lesions. In: Kornhuber H, Deecke L (eds) Motivation, Motor and Sensory Processes of the Brain, Elsevier, Amsterdam
63. Mansur CG, Fregni F, Boggio PS et al. (2005) A sham stimulation-controlled trial of rTMS of the unaffected hemisphere in stroke patients. Neurology, 64:1802–1804
64. McGrady A, Olson P, Kroon J (1995) Biobehavioral treatment of essential hypertension. In M. Schwartz (Ed.) Biofeedback (2nd ed), New York: Guilford
65. Miller N (1969) Learning of visceral and glandular responses. Science, 163:434–445
66. Miniussi C, Bonato C, Bignotti S et al. (2005) Repetitive transcranial magnetic simulation (rTMS) at high and low frequency: An efficacious therapy for major drug-resistant depression? Clinical Neurophysiology, 116:1062–1071
67. Nicolelis MA (2003) Brain–machine interfaces to restore motor function and probe neural circuits. Nature Reviews Neuroscience, 4(5):417–422
68. Pfurtscheller G, Neuper C, Birbaumer N (2005) Human brain–computer interface (BCI). In A. Riehle and E. Vaadia (Eds.), Motor Cortex in Voluntary Movements. A Distributed System for Distributed Functions (pp. 367–401). Boca Raton: CRC Press
69. Plewnia C, Reimold M, Najib A, Reischl G, Plontke SK, Gerloff C (2007) Moderate therapeutic efficacy of positron emission tomography-navigated repetitive transcranial magnetic stimulation for chronic tinnitus: A randomised, controlled pilot study. Journal of Neurology, Neurosurgery, and Psychiatry. 78:152–156
70. Quill TE (2005) ALS, depression, and desire for hastened death: (How) are they related? Neurology, 65:1
71. Rockstroh B, Elbert T, Birbaumer N, Lutzenberger W (1989) Slow Brain Potentials and Behavior (2. Aufl.). Baltimore, MD: Urban & Schwarzenberg
72. Rockstroh B, Elbert T, Birbaumer N, Wolf P, Düchting-Röth A, Reker M et al. (1993) Cortical self-regulation in patients with epilepsies. Epilepsy Research, 14:63–72
73. Scherberger H et al. (2005) Cortical local field potentials encodes movement intentions in the posterior parietal cortex. Neuron, 46:347–354

74. Schneider F, Rockstroh B, Heimann H, Lutzenberger W, Mattes R, Elbert T, Birbaumer N, Bartels M (1992) Self-regulation of slow cortical potentials in psychiatric patients: Schizophrenia. Biofeedback & Self-Regulation, 17(4):277–292
75. Schwartz AB (2007) Useful signals from motor cortex. Journal of Physiology, 579:581–601
76. Sellers EW, Donchin EA (2006) A P300 based brain–computer interface: Initial tests by ALS patients. Clinical Neurophysiology, 117(3):538–548
77. Skinner F (1953) Science and Human Behavior. New York: Macmillan
78. Sterman MB (1977) Sensorimotor EEG operant conditioning: Experimental and clinical effects. The Pavlovian Journal of Biological Science, 12(2):63–92
79. Sterman MB (1981) EEG biofeedback: Physiological behavior modification. Neuroscience and Biobehavioral Reviews, 5:405–412
80. Sterman MB, Clemente CD (1962a) Forebrain inhibitory mechanisms: Cortical synchronization induced by basal forebrain stimulation. Experimental Neurology, 6:91–102
81. Sterman MB, Clemente CD (1962b) Forebrain inhibitory mechanisms: Sleep patterns induced by basal forebrain stimulation in the behaving cat. Experimental Neurology, 6:103–117
82. Sterman MB, and Friar L (1972) Suppression of seizures in an epileptic following sensorimotor EEG feedback training. Electroencephalography and Clinical Neurophysiology, 33(1):89–95
83. Strehl U, Leins U, Goth G, Klinger C, Hinterberger T, Birbaumer N (2006) Self-regulation of slow cortical potentials: A new treatment for children with attention-deficit/hyperactivity disorder. Pediatrics, 118:1530–1540
84. Taylor DM, Tillery SI, Schwartz AB (2002) Direct cortical control of 3D neuroprosthetic devices. Science, 296:1829–1832
85. Theodore WH (2003) Transcranial magnetic stimulation in epilepsy. Epilepsy Currents, 3:191–197
86. Weiskopf N, Veit R, Erb M, Mathiak K, Grodd W, Goebel R, Birbaumer N (2003) Physiological self-regulation of regional brain activity using real-time functional magnetic resonance imaging (fMRI): Methodology and exemplary data. NeuroImage, 19:577–586
87. Weiskopf N, Scharnowsi F, Veit R, Goebel R, Birbaumer N, Mathiak K (2005) Self-regulation of local brain activity using real-time functional magnetic resonance imaging (fMRI). Journal of Physiology, Paris, 98:357–373
88. Weiskopf N, Sitaram R, Josephs O, Veit R, Scharnowski F, Goebel R, Birbaumer N, Deichmann I, Mathiak K (2007) Real-time functional magnetic resonance imaging: Methods and applications. Magnetic Resonance Imaging 25:898–1003
89. Wilhelm B, Jordan M, Birbaumer N (2006) Communication in locked-in syndrome: Effects of imagery on salivary pH. Neurology, 67:534–535
90. Wolpaw JR (2007) Brain–computer interfaces as new brain output pathways. Journal of Physiology, 579:613–619
91. Wolpaw JR, McFarland DJ (2004) Control of a two-dimensional movement signal by a noninvasive brain–computer interface in humans. Proceedings of the National Academy of Science USA, 101(51):17849–17854
92. Yoo SS, Fairneny T, Chen NK, Choo SE, Panych LP, Park H, Lee SY, Jolesz FA (2004) Brain–computer interface using fMRI: spatial navigation by thoughts. Neuroreport, 15:1591–1595

Sleep, Consciousness and the Brain: A Perturbational Approach

Giulio Tononi and Marcello Massimini

Department of Psychiatry, University of Wisconsin, Madison, WI 53719, USA

Sleep reminds us every day that consciousness is something that can come and go, grow and shrink, depending on how our brain is functioning. Everyone is familiar with the impression of nothingness that lies in between our falling into and awakening from dreamless sleep. Of course, blank reports upon awakening from sleep are not the rule and many awakenings, especially from rapid eye movement (REM) sleep, yield dream reports, and dreams can be at times as vivid and intensely conscious as waking experiences. Dream-like consciousness also occurs during various phases of slow wave sleep, especially at sleep onset and during the last part of the night. Nevertheless, there are always a certain proportion of awakenings that do not yield any dream report, suggesting a marked reduction of consciousness. Such "empty" awakenings typically occur during the deepest stages of NREM sleep (stages 3 and 4), especially during the first half of the night. Understanding why consciousness fades during certain phases of sleep is important not just with respect to brain function during sleep, but first and foremost because it can help us in identifying what is really necessary and sufficient for the brain to give rise to conscious experience.

The relationships between sleep and consciousness are indeed interesting and puzzling. It was first thought that the fading of consciousness during sleep was due to the brain shutting down. However, while metabolism is reduced, the thalamocortical system remains active also during stages 3 and 4, with mean firing rates comparable to those of quiet wakefulness [9]. It was also hypothesized that sensory inputs are blocked during sleep and that they are necessary to sustain conscious experience. However, we now know that, even during deep sleep, sensory signals continue to reach the cerebral cortex where they are processed subconsciously [5]. Gamma activity and synchrony have been viewed as possible correlates of consciousness and they were found to be low in NREM sleep [2]. However, they were equally low in REM sleep, when conscious experience is usually vivid, and they can be high in anesthesia [11]. Moreover, intracellular recordings show that gamma activity persists during NREM sleep, and other studies report that gamma coherence is a local phenomenon that does not change between wakefulness and sleep [1]. Large-scale synchrony in the alpha and theta bands may also correlate with conscious perception during wakefulness, but synchrony in these frequency bands actually increases during NREM sleep [3]. Why, then, does consciousness fade?

According to the Integrated Information Theory of Consciousness (IITC, [10]), what is critical for consciousness are not firing rates, sensory input, specific frequency bands, or synchronization per se, but rather a system's capacity for integrated information. If consciousness is integrated information, then the brain substrate of consciousness should be a complex of neural elements, presumably located within the thalamocortical system, that has a large repertoire of available states (*information*), yet cannot be decomposed into a collection of causally independent subsystems (*integration*). An exhaustive measure of complexes and the associated value of integrated information is currently feasible in simple artificial networks [10] but it is a daunting proposition in a complex biological system such as the human brain. Nonetheless, the IITC makes clear-cut predictions that can be addressed experimentally at least at a gross level. According to the IITC, the fading of consciousness during early NREM sleep should be associated with either a reduction of integration within the main thalamocortical complex (for example, it could break down into causally independent modules) or a reduction of information (the repertoire of available states might shrink), or both.

The theory also suggests that, to evaluate integrated information, it is not enough to observe activity levels or patterns of temporal correlations among distant brain regions (*functional connectivity*). Instead, the ability to integrate information among distributed cortical regions must be examined from a *causal* perspective: one must employ a perturbational approach (*effective connectivity*) and examine to what extent cortical regions can interact causally (*integration*) and produce differentiated responses (*information*). One should probe effective connectivity by directly stimulating the cortex, to avoid possible subcortical filtering and gating, and ideally one should do so in humans, as only with humans we know for sure that consciousness fades during early NREM sleep. The problem, of course, is that perturbing directly and non-invasively one particular cortical region while recording the overall response of the rest of the human brain is technically challenging.

In a series of recent experiments [6] we employed a combination of navigated transcranial magnetic stimulation (TMS) and high-density EEG (hd-EEG) to measure non-invasively and with good spatiotemporal resolution the brain response to the direct perturbation of a chosen brain region. Using a 60-channel TMS-compatible EEG amplifier, we recorded TMS-evoked brain responses while subjects, lying with eyes closed on a reclining chair, progressed from wakefulness to deep NREM sleep. Thanks to noise masking and other procedures, subjects were unaware of TMS.

The left panel of Fig. 1 displays the single-trial responses recorded from one electrode located under the stimulator during a transition from wakefulness through stage 1 to NREM (stages 3 and 4) sleep. In the right panel the averages calculated from the single trials collected in these three vigilance

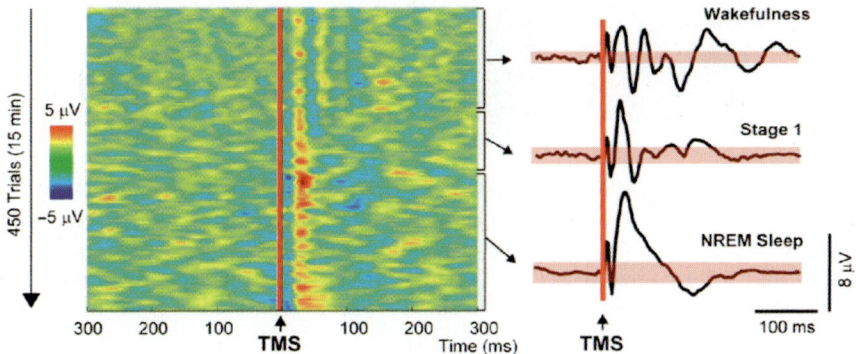

Fig. 1. Changes in the TMS-evoked response during shifts in the state of vigilance. *Left*: Single trials recorded from one channel located under the stimulator while the subject transitioned from wakefulness through stage 1 to NREM sleep. Single-trial EEG data (filtered 4–100 Hz) are color-coded for voltage. *Right*: Averaged TMS-evoked responses (filtered 1–100 Hz) obtained during the three states of vigilance. The *horizontal pink bands* indicate the significance level (three SDs from the mean prestimulus voltage)

states are shown. During wakefulness, TMS induced a sustained response made of recurrent waves of activity. Specifically, a sequence of time-locked, high-frequency (20–35 Hz) oscillations occurred in the first 100 ms and was followed by a few slower (8–12 Hz) components that persisted until 300 ms. As soon as the subjects transitioned into stage 1, the TMS-evoked response grew stronger at early latencies but became shorter in duration due to dampening of later fast waves. With the onset of NREM sleep, the brain response to TMS changed markedly. The initial wave doubled in amplitude and became slower. Following this large wave, no further TMS-locked activity could be detected, except for a negative rebound between 80 and 140 ms. Specifically, fast waves, still visible during stage 1, were completely obliterated, and all TMS-evoked activity had ceased by 150 ms.

To reveal the pattern of effective connectivity associated with each cortical perturbation, we performed source modeling and calculated the spatiotemporal dynamics of the currents induced by TMS. Figure 2 shows the patterns of activation evoked by TMS over premotor cortex during wakefulness and NREM. Black traces represent the global mean field powers and the yellow lines indicate significance levels. For each significant time sample, maximum current sources are plotted and color-coded according to their latency of activation (light blue, 0 ms; red, 300 ms). The yellow cross marks the TMS target on the cortical surface. During wakefulness, the initial response to TMS was followed until about 300 ms by multiple waves associated with spatially and temporally differentiated patterns of activation. During NREM sleep, by contrast, the activity evoked by TMS did not propagate in space and time. Thus, although TMS during sleep elicits an initial response that is even stronger

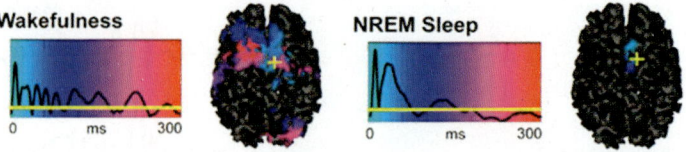

Fig. 2. Spatiotemporal cortical current maps during wakefulness and NREM sleep. *Black traces* represent the global mean field powers and the *yellow lines* indicate significance levels. For each significant time sample, maximum current sources were plotted and color-coded according to their latency of activation (light blue, 0 ms; red, 300 ms). The yellow cross marks the TMS target on the cortical surface

than during wakefulness, this response remains localized, does not propagate to connected brain regions, dissipates rapidly and is stereotypical regardless of stimulation site.

These findings suggest that the fading of consciousness during certain stages of sleep may be related to a breakdown in cortical effective connectivity. What prevents the emergence of a specific long-range pattern of activation during sleep? During NREM sleep, cortical neurons are depolarized and fire tonically just as in quiet wakefulness, but these depolarized up-states are interrupted by short, hyperpolarized down-states when neurons remain silent [7]. This alternation involves large populations of cortical neurons and is reflected in the EEG as high-amplitude slow oscillations. The transition from up- to down-states appears to be due to depolarization-dependent potassium currents that increase with the amount of prior activation [7]. Perhaps, because of this bistability of cortical networks during NREM sleep [4], any local activation, whether occurring spontaneously or induced by TMS, will eventually trigger a local down-state that prevents further propagation of activity.

The role of bistability in altering information processing during sleep has been corroborated in subsequent experiments in which TMS, applied with the appropriate parameters, triggered full-fledged slow oscillations. In Fig. 3 few cycles of the slow oscillation evoked by TMS in humans are compared to the spontaneous slow oscillations recorded in cats (field potential and membrane potential). Each TMS pulse, delivered with high intensity over the sensorimotor cortex, triggered a negative wave associated with global down-state in vast regions of the brain, followed by a positive component crowned by spindles. Spatially, the TMS-evoked slow oscillation was associated with a broad and aspecific response: cortical currents spread, like an oil-spot, from the stimulated site to the rest of the brain.

Altogether, these TMS–EEG measurements suggest that the sleeping brain, despite being active and reactive, loses its ability of entering states that are both integrated and differentiated: it either breaks down in causally independent modules or it burstsinto an explosive and aspecific response. In

Human

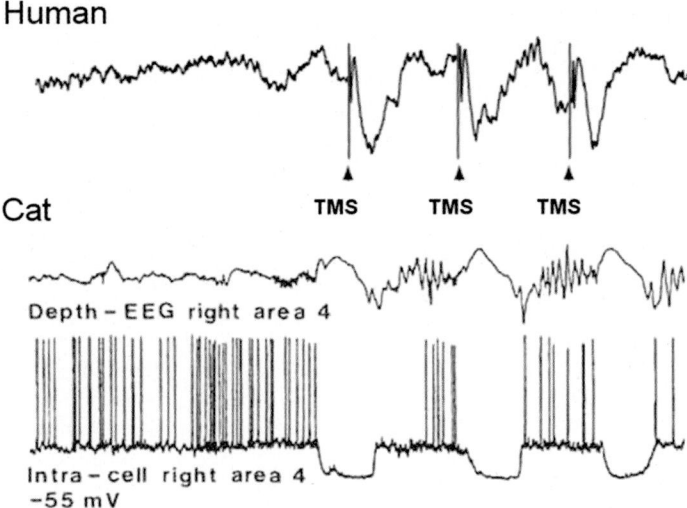

Fig. 3. TMS triggering of full-fledged sleep slow oscillations. The upper trace is from one channel located under the stimulator. Each TMS pulse, delivered at high intensity (70% of maximum output), triggers a typical slow oscillation: a negative wave (more negative than $-80\,$mV) followed by a positive rebound crowned by spindles. As highlighted by the comparison with recordings in cats (modified from [8]), the surface negative wave recorded in human (depth positive in cats) corresponds to the intracellular down-state of the slow oscillation

no case, during NREM sleep, did TMS result in a long-range differentiated pattern of activation. The TMS–EEG perturbational approach also suggests that intrinsic bistability in thalamocortical networks may represent not only the key mechanism responsible for the occurrence of the spontaneous slow oscillations of sleep, but also the reason why information integration is impaired in early NREM sleep. In this perspective, the inescapable occurrence of a down-state after a second or so in an up-state, be it a local or global phenomenon, suggests that due to bistability the availability of a large repertoire of neural activation patterns is much reduced. Moreover, this bistability and consequent restriction of neural repertoire may be present, but covert, even during periods of stable ongoing EEG with no slow oscillations: to uncover the network's bistability, it may be necessary to perturb it directly. And indeed, in our experiments, TMS applied to the cortex during NREM sleep invariably resulted in a stereotypical down-state, even during periods characterized by low-amplitude EEG. Observing a system's ongoing dynamic is usually a good indicator of a system repertoire of potential states. Sometimes, however, observation is not enough, and perturbation is in order.

References

1. Bullock TH, McClune MC, Achimowicz JZ, Iragui-Madoz VJ, Duckrow RB, Spencer SS (1995) Temporal fluctuations in coherence of brain waves. Proc Natl Acad Sci USA 92:11568–11572
2. Cantero JL, Atienza M, Madsen JR, Stickgold R (2005) Precise timing accounts for posttraining sleep-dependent enhancements of the auditory mismatch negativity. Neuroimage 26:628–634
3. Duckrow RB, Zaveri HP (2005) Coherence of the electroencephalogram during the first sleep cycle. Clin Neurophysiol 116:1088–1095
4. Hill S, Tononi G (2005) Modeling sleep and wakefulness in the thalamocortical system. J Neurophysiol 93:1671–1698
5. Kakigi R, Naka D, Okusa T, Wang X, Inui K, Qiu Y, Tran TD, Miki K, Tamura Y, Nguyen TB, Watanabe S, Hoshiyama M (2003) Sensory perception during sleep in humans: A magnetoencephalograhic study. Sleep Med 4:493–507
6. Massimini M, Ferrarelli F, Huber R, Esser SK, Singh H, Tononi G (2005) Breakdown of cortical effective connectivity during sleep. Science 309: 2228–2232
7. Sanchez-Vives MV, McCormick DA (2000) Cellular and network mechanisms of rhythmic recurrent activity in neocortex. Nat Neurosci 3:1027–1034
8. Steriade M (2000) Corticothalamic resonance, states of vigilance and mentation. Neuroscience 101:243–276
9. Steriade M, Timofeev I, Grenier F (2001) Natural waking and sleep states: A view from inside neocortical neurons. J Neurophysiol 85:1969–1985
10. Tononi G (2004) An information integration theory of consciousness. BMC Neurosci 5:42
11. Vanderwolf CH (2000) Are neocortical gamma waves related to consciousness? Brain Res 855:217–224

Part III

Social Dynamics

Skilled Operating

Language Use, Coordination, and the Emergence of Cooperative Action

Carol A. Fowler[1,2] and Michael J. Richardson[3], Kerry L. Marsh[4], and Kevin D. Shockley[5]

[1] Haskins Laboratories, New Haven, CT, USA
[2] University of Connecticut, Storrs, CT 06511, USA
[3] Department of Psychology, Colby College, Waterville, ME 04901-8885, USA
[4] University of Connecticut, West Hartford, CT 06117, USA
[5] University of Cincinnati, Cincinnati, OH 45221-3124, USA

1 Introduction

Over the last three decades, two major theoretical developments within cognitive science have enriched our understanding of perceptual function and coordinated action in real-world environments.

First is the development of an "ecological" approach to perception pioneered by James Gibson ([20]; see also [50]). Hallmarks of the ecological approach include its focus away from mental processes that may support perception and toward perceptual function in the environment. The relevant parts of the environment, the animal's ecological niche, provide, among other things, nourishment, shelter, and surfaces that support activity. The possibilities that niches provide for action are "affordances". Informational support for perception of affordances is provided by invariant structure in reflected light, air, and the surfaces of the body among others. Invariants lawfully entail affordances and are specific to them; their specificity to affordances underlies their service as informational support. The relation of animals to their environment is one of "mutuality"; that is, the possibilities for action for an actor-perceiver are emergent consequences of properties of the environment in relation to those of the actor-perceiver.

Second is the development of an understanding that animals are dynamical systems. That is, they are self-organizing systems in which "contents and representations emerge from the systemic tendency of open, nonequilibrium systems to form patterns" [31]. In the case of actor-perceivers, as Kelso [31] notes, intelligent behaviors emerge in the absence of agents directing the behaviors. The dynamical systems approach to perceptually guided action offers an account of both the stability and the flexibility of coordinated action. Pioneers in this domain include Kelso [31, 32] and Kugler and Turvey [34].

Recently, Warren [71] has proposed an integration of the ecological and dynamical systems approaches in a theory that individuals and their environments are properly considered coupled dynamical systems, with coupling being both informational and mechanical in nature. Interactions between the systems give rise to a behavioral layout having attractors corresponding to stable behaviors that achieve intended goals and bifurcations that lead to transitions between behaviors.

Our research has focused on situations in which coupled dynamical systems that form an overarching dynamical system are minimally three in number, that is, two individuals and the environment. It is, after all, a critical characteristic of the ecological niches of humans and of many other animals that it is populated by conspecifics, who are sources of some of the affordances for the coupled systems (cf. [42]). In addition, people frequently engage in "joint actions" [13] in which, sometimes, affordances are only afforded because people are able to coordinate their behaviors. For example, two people together may wrestle a large, heavy piece of furniture through a narrow doorway, an achievement not possible for either one alone.

We are interested in the formation of dynamical systems that encompass more than one person, and we have been especially interested in the role that language use plays in it. Clark [13] remarks that language can serve as a "coordination device" – a means, that is, by which coordination of two or more individuals can be achieved. So, for example, our furniture movers above might talk back and forth about how to tilt or rotate a large, heavy desk to get it through a doorway so that not only does it get through the doorway but, in addition, the desk and the desk movers emerge through the undamaged doorway themselves unscathed.

Clark [13] provides an example that illustrates the role that language may play in coordinating activities involving two or more individuals. He recorded his speech and that of a cashier as he bought some items in a pharmacy. He brought the items over to the counter on which a cash register sat. The cashier, doing an inventory, caught his eye and said "I'll be right there." In a minute, she came over and, pointing, Clark said "These two things over here." She totaled the amount using the cash register and said "$12.77." Clark repeated the amount and, after a brief interaction with the cashier who made change for a $20.00 bill, left the store. This is an informative example not only because it shows language serving as a coordination device but also because it shows just how spare the language use can be. In ordinary cross-person interactions, utterances may be interjected minimally, serving the role of moving a joint activity forward.

Following a brief review of the literature showing some ways in which language underlies interpersonal coordination or entrainment, we will review our recent research on the formation of cross-person dynamical systems, beginning with a series of experiments in which, although language use occurred, it was not the focus of attention. Then we turn to our experiments on language use.

2 Language Use and Cross-Person Entrainment

As noted, Clark [13] has suggested that a primary role of language is to serve as a medium for the achievement of common goals that he calls "joint actions". The implication is that language use is inherently cooperative in nature. This notion is evidenced by a variety of studies that have demonstrated the spontaneous convergence in speaking patterns in social settings. For example, verbal interactions among individuals demonstrate convergence in dialect [22], speaking rate [65], vocal intensity [46], and pausing frequency [8] (see [23] for a review).

These tendencies to coordinate speech patterns likely have a social, or cooperative, motivation (cf. [7]). However, they appear to rest on a more fundamental disposition to imitate speech whether or not the setting is a social one. For example, Goldinger [24] and Shockley et al. [63] elicited utterances by participants that, in baseline conditions, could not be imitations of a model's speech (words read aloud from a computer screen) and in another repetition condition could be imitations of a model's speech (utterances of a model shadowed by the listener). Both studies found that listeners judged the shadowed repetitions to be better imitations of the model than were the baseline productions. Moreover, Shockley et al. [63] modified a phonetic property of model-produced words – the voice onset times of syllable-initial voiceless stops – and found that, in the absence of instructions to imitate, listeners modified their voice-onset times in the direction of those of the model.

Indeed, imitation is generally pervasive among humans. Imitation of facial gestures (e.g., tongue or lip protrusion) is observed in neonates [44], and by 12 weeks of age, infants imitate vocalic sounds [35]. The tendency to imitate persists into adulthood. For example, adults imitate facial expressions [40]. Interlocutors do not limit their convergences to speech. They are said to move in synchrony with each other's speech rhythms [14] and to match one another's postures [15, 33, 37, 64] in the absence of an intention to entrain. For example, LaFrance [37] demonstrated that listeners to a speaker whom they find engaging tend to mirror the speaker's postures. Listeners are also reported to move in time with the rhythms of a speaker's speech (exhibiting "interactional synchrony" [16, 49]). Shockley et al. [64] evaluated the degree of shared postural sway activity in the context of a cooperative verbal task. They, likewise, found that there was more shared postural activity between two interlocutors when they were speaking with each other than when they were speaking to someone else.

3 Joint Action and the Formation of Dynamical Systems That Encompass More Than One Person

Many everyday tasks require that individuals come together to perform coordinated actions such as when moving large pieces of furniture together, dancing, paddling a canoe, or simply walking and talking with friends. Although language use can facilitate (and be facilitated by) the performance of these activities, joint-action and interpersonal coordination often result from individuals picking up information about another's movements or perceiving information that specifies the constraints of an environment or task in relation to the limitations of their own or co-actors' capabilities [51, 57, 59, 64]. In other words, the stable organization and patterning of joint action emerges from informational couplings that exist between individuals and between individuals and the environment. Moreover, despite the intuition that joint action is qualitatively different from solo action, there is often a similarity in how the coordinated activity of individuals and pairs is constrained and dynamically self-organized [42]. That is, there is an equivalence of the dynamics that underlie joint and solo actions.

Schmidt et al. [56] first demonstrated the dynamic equivalence of joint and solo actions by having pairs of participants visually synchronize the oscillatory motions of their legs. By manipulating the frequency of the coordinated leg oscillations, the authors found that the couplings were constrained (without practice) to inphase and antiphase modes of coordination. Moreover, inphase coordination was more stable than antiphase coordination. That is, as the rate of leg oscillations increased, the individuals spontaneously shifted from an antiphase mode of coordination to an inphase mode. They did not make a shift in the opposite direction when the frequency of inphase cycling was increased. The movement patterns that characterized pairs of participants also characterize rhythmic interlimb coordination within a person (e.g., between a single individual's left and right wrists or arms [26, 30]). Accordingly, Schmidt et al. concluded that the same dynamical processes of entrainment, namely a coupled oscillator dynamic, must also underlie rhythmic between person coordination. Subsequent investigations by Schmidt and colleagues, using the Kugler and Turvey [34] wrist-pendulum paradigm, have further verified that the same coupled oscillator dynamic constrains interlimb coordination within and between people [1, 55, 59].

Additional research by Schmidt and O'Brien [58] and Richardson et al. [52] has demonstrated that a coupled oscillator dynamic also constrains the interpersonal coordination that occurs unintentionally between two interacting individuals. In these studies, individuals swung a wrist-pendulum at their own preferred tempo while visual information about the rhythmic wrist-pendulum movements of a co-actor was available. Despite the fact that the pairs were not instructed to coordinate their movements (and were largely unaware of their movements becoming entrained), the results demonstrated that the movements of the two individuals still became synchronized. Furthermore, like the patterns of intentional interpersonal coordination, the patterns of unintentional coordination were consistent with a coupled oscillator dynamic – the pair's movements were attracted toward an inphase or antiphase mode of coordination.

In addition to demonstrating that interpersonal rhythmic coordination is constrained by the same dynamical entrainment processes as intrapersonal rhythmic coordination, the studies just described indicate that rhythmic interlimb coordination is indifferent to the manner by which component limbs are coupled. That is, the findings demonstrate that the same stable patterns of coordination are observed independent of whether the component limbs are physically coupled by means of the neural muscular tissue of an individual's central and peripheral nervous systems or by means of the non mechanical visual information of the optic array [55]. This suggests that coordination has less to do with the particular anatomical and corporeal substrates of the perceptual-motor system and more to do with the incorporeal lawful relations and interactions that exist between the subcomponents of a system [52, 56]. Such findings also illustrate a fundamental premise of dynamical systems theory, namely that the same lawful properties of the animal–environment system regulate and constrain the behavior of a broad range of systems independent of the components that comprise them [31, 34]. Indeed, it is this premise that leads to an understanding of joint-action systems as nondecomposable units of analysis and, moreover, as self-organized dynamical systems that emerge from the nonlinear interactions and couplings that exist between and among individuals and the environment.

Research by Mottet et al. [45] identified a similitude in the dynamics of one-handed, two-handed, or two-person Fitts-law tasks. This research has also provided evidence for the self-organized and nondecomposable nature of joint-action systems. Here, the authors found that the kinematic and dynamic organization of hand movements remained the same irrespective of whether the participants performed the Fitts-law task with one hand (moving a pointer back and forth between two stationary targets), two hands (one hand moving the pointer and the other hand moving the targets), or with another individual (one individual moving the pointer and the other individual moving the targets). That is, although the kinematic and dynamic organization of hand movements varied with task difficulty, it remained invariant across solo- and joint-action conditions.

More recently, the invariance and similitude of dynamical and informational constraints on joint- and solo-action systems have been shown by Richardson et al. [51] for the task of moving planks of wood. These experiments addressed whether emerging cooperation between individuals could be understood as a shift in action that occurs naturally in response to environmental and individual constraints, paralleling shifts between one- to two-hand lifting. These experiments in cooperative action stand in contrast to approaches in which joint action is understood as mediated by cooperators' cognitive representations of the task [60]. They also contrast with the typical conceptualization of cooperation as a series of strategic decisions, occurring against a backdrop of competition, from a game theory orientation [3, 29]. In such game playing approaches to cooperation, there is a historical dynamic that operates in the sense that the past sequence of decisions has an impact on the likelihood of particular future choices. The primary focus of the investigators, however, is on the cognitive processes brought to bear – the effects of one's prior expectations on how one plays the game, for instance [29] – or on the effects of structural aspects of the game [2]. These games do not require that the competitor be present. In fact, no information about the other person or about the environment is available to the perceiver apart from the payoff matrix for the game and the competitor's choices on each trial. Such game-based situations are substantially different from many of the day-to-day situations in which cooperation lawfully unfolds, online, in dynamic response to information from the other individual and the environmental constraints. Studying these latter situations requires the presence of another person (and the verbal and nonverbal information they emit) and the bodily and environmental dynamics that being copresent in a situation involves. Thus, examining cooperation as an emergent, dynamical process requires embodied methods that focus on how the presence of others can extend the action possibilities (i.e., "affordance" [20]) for an individual.

Accordingly, Richardson et al. [51] investigated the emergence of cooperative action by comparing how participants shifted between using one versus two hands and between solo and joint actions in a task that required that they move lightweight, narrow planks of wood, which could only be touched at the ends (as if they were freshly painted in the middle; see Fig. 1a,b). No other constraints were placed on participants. For instance, in two-person experiments, pairs could talk if they chose; some pairs did and some did not. In experiments involving pairs, individuals might choose to take turns or they might not. In all experiments involving pairs, the typical pattern was turn-taking, and this was something that occurred without discussion. Understood as a dynamical system, cooperation versus autonomous action should be determined by the point at which some critical control parameter value is reached – this control parameter being a relation between an aspect of the actor(s) and an aspect of the environment. That is, the affordances [20, 70] of the situation – planks as movable with one hand, movable with two hands, or movable with two people – are determined by the size of individuals' hands (or arm spans) taken with respect to the length of the plank. In such a paradigm, therefore, the implicit commitment to act as a "plural subject" of action [21], that is, to choose to cooperate, is something that emerges without prior planning or a priori expectations, in response to individual, environmental, and social constraints.

As expected, Richardson et al. [51] found that for individuals in solo experiments, shifts between moving planks with one hand and two hands were determined by the control parameter of a body-scaled ratio [70] of hand span to plank length, replicating other research [9, 47, 48, 69]. Moreover, when planks of wood varied from small enough to move with two hands to those that were impossible to move alone, the point of transition between solo and cooperative action was determined by a comparable value on a body-scaled (arm span to plank length) ratio. When these studies were replicated with

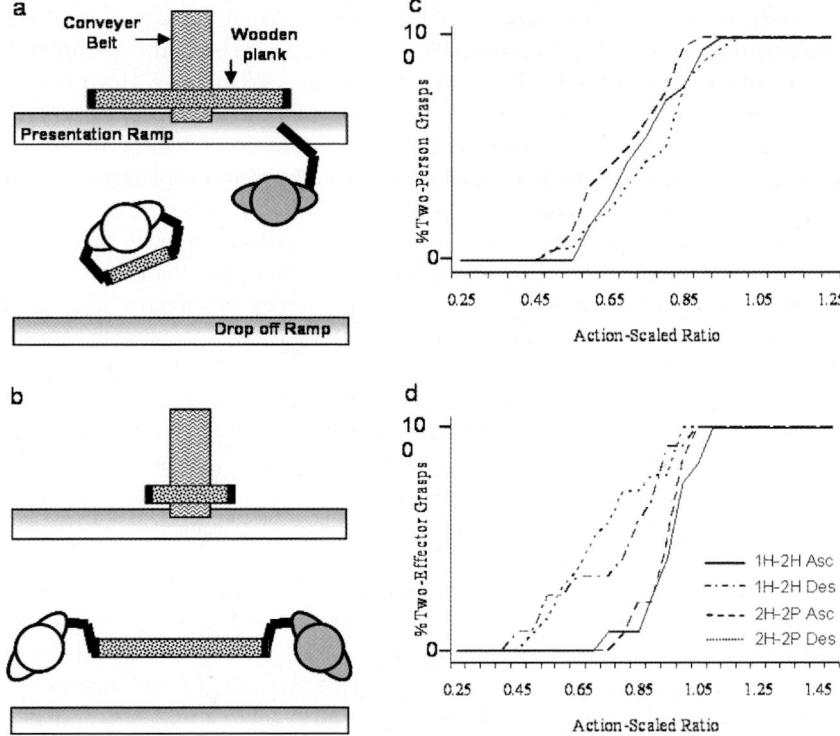

Fig. 1. Experimental setup used by Richardson et al. [51] and Isenhower et al. [27]. (**a**) Example of individual two-hand grasping. (**b**) Example of interpersonal two-person grasping. (**c**) The mean percentage of two-person grasps for participants pairs with the same (small–small, large–large) or different (small–large) size arm spans, as a function of the plank-length/mean arm span. Reproduction of data collected by Isenhower et al. [27]. (**d**) The mean percentage of two-effector (2H or 2P) grasps for the one-hand (1H) to two-hand (2H) transition and the 2H to two-person (2P) transition as a function of action-scaled ratio (plank-length/hand span or plank-length/mean arms span) for the ascending (Asc) or descending (Des) plank-length orders of presentation. Note: From "Judging and actualizing intrapersonal and interpersonal affordances," by M. J. Richardson, K. L. Marsh, and R.M. Baron, Journal of Experimental Psychology: Human Perception and Performance, 33, p. 854 (panels a, b), p. 855 (panel d). Copyright 2007 by the American Psychological Association. Adapted with permission

pairs of participants chosen to have long or short arm spans, or chosen to have arm spans that were mismatched in size, the results were comparable [27]. Participant pairs with longer arm spans picked up metrically larger pieces of wood alone than pairs with shorter arm spans, but there were no differences in these pairs in terms of the body-scaled ratio at which they shifted to joint action (see Fig. 1b). For pairs with mismatched arm spans, the situation was constrained by the inability of the individual with the shorter arm span to pick up larger pieces alone, as well as by social constraints – namely the (nonverbalized) social norms of turn – taking and of sharing the work load. Thus, the body-scaled ratio at which mismatched pairs shifted to cooperative action was, on average, a somewhat smaller proportion of the pairs' average arm span (0.68, compared to 0.73 and 0.75 for the small and large arm-span pairs, respectively), meaning that when the plank reached a size that the shorter participant could not move, participants usually moved the wood together. For a few pairs, the participant with the larger arm span instead moved the wood alone, if that was possible. For coders of the videotape, whether this evoked negative impressions of the individual with the larger arm span (the "what a jerk" effect) was determined by whether it involved a rejection of the implied joint commitment to be a plural subject of action (e.g., ignoring that the waiting partner had first attempted to move it, and was now grasping one end of the plank that rested on the table) or whether it was congruent with this plural subjecthood (e.g., the partner first stepped back from the table where the plank rested, inviting the individual to step forward).

One important feature of a dynamical system that was studied in the embodied cooperation experiments is the feature of multistability. That is, for some values of the control parameter, multiple attractors of meaning should be present in the system – attractors should exist simultaneously for both autonomous and cooperative action. That is, there should be regions where both solo and joint actions are comfortable modes; there should be planks that can readily be moved by either means. In such regions, the past trajectory of action through which partners reached the point at which multiple attractors became present will determine whether cooperation or solo action occurs. In particular, the most common pattern of multistability in action would be hysteresis such that the past history of action propels partners into that region where both actions are comfortable. In such a case, partners will continue for a time in the current mode before shifting action modes.

To test for multistability, participants moved planks of wood arrayed in ascending lengths, descending lengths, or, for comparison, a random ordering of size. As expected, in all experiments, participants' actions displayed hysteresis – participants in the ascending condition persisted in solo action to plank lengths somewhat longer than they did in the random condition ([27, 51], see Fig. 1d). In the descending conditions, pairs persisted in cooperative action down to significantly smaller pieces of wood than in the ascending conditions. An identical pattern of hysteresis occurred in action experiments that involved solo actors moving smaller pieces of wood with one or two hands. Thus, as in the research on rhythmic interpersonal coordination described earlier, the present results reveal an equivalence of the dynamics that underlie joint and solo actions.

4 Language as a Coordination Device

In the research of Richardson et al. [51] just described, participants were free to talk back and forth or not as they chose. The task did not require discussion, and some participant pairs performed the task without using language as a coordination device. However, there are kinds of coordinative activities in which language either use fosters coordination (as in the example of Clark and the cashier described above) or is essential to it. Some of our recent research has examined joint actions of the latter sort.

Shockley et al. [64] had pairs of participants, or individual participants paired with a confederate play a game that required that they talk back and forth. Each participant or confederate had a picture that they, but not their partner, could see. Their partner had a similar picture. The two pictures differed in a number of fairly subtle ways (e.g., a person might be holding four balloons in one picture, but just three in the other). Individuals were instructed to talk back and forth to uncover the differences between the pictures. The experiment had a 2×2 design. One independent variable was whether the two participants were performing the talking task with each other or whether each was performing it with a confederate. The second was whether participants were facing each other or were back to back (see Fig. 2a). The dependent measure was postural sway measured at the waist in the anterior–posterior direction.

Our question was whether between-participant talk and/or visual contact fostered interpersonal coordination. In particular, was between-participant coordination more likely to be present when participants were talking back and forth to each other than when each participant was performing the task with a confederate? Likewise, was between-person coordination more present when participants could see one another than when they could not?

Our measures of coordination were provided by "cross-recurrence quantification analysis" (CRQ) [62, 74] performed on the time series of samples of anterior–posterior positions of the waist of each participant. In this analysis, the multidimensional space in which the sway activity is taking place is reconstructed, and for each pairing of sampled waist positions of the two participants, it is determined whether the positions are close enough in the multidimensional space to count as coincident or recurrent. We take the recurrent points as indications of entrainment or coordination. We focused on two measures of recurrence, percent recurrence (that is, the percentage of all possible recurrent points that are recurrent), and "maxline" (see Fig. 2b). When $x(i)$ (participant x at time i), $y(j)$ (participant y at time j) points are found to be recurrent in reconstructed space, and then $x(i+1)$ and $y(j+1)$ are also recurrent, a pair of parallel lines is formed in reconstructed space (meaning over two consecutive temporal samples, recurring locations are found). If $x(i + 2)$ and $y(j + 2)$ are also recurrent, the duration of this parallelism is longer. When parallel lines form, it means that participants' postural sway patterns are coevolving in a similar way and that these sway patterns continue

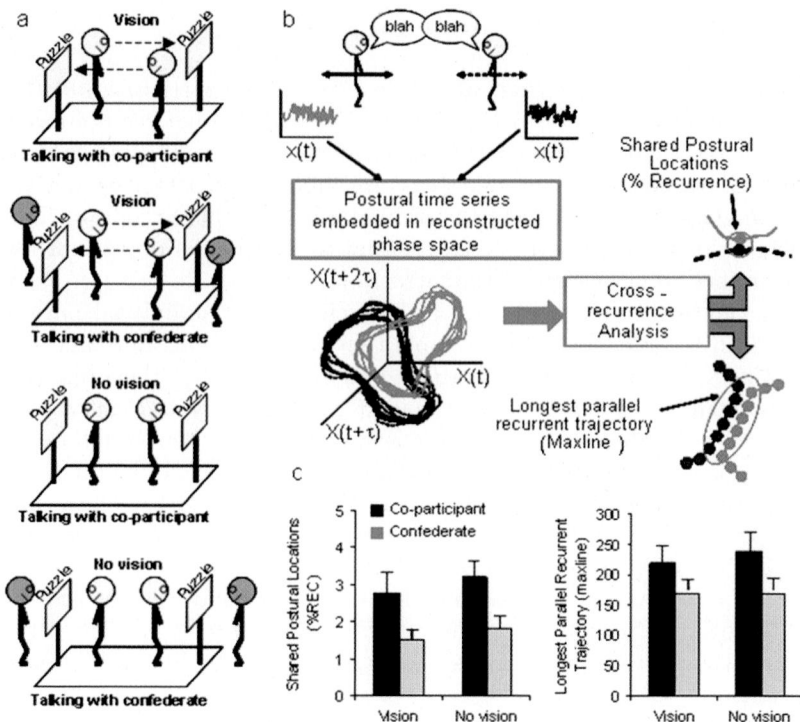

Fig. 2. (a) Experimental setup used by Shockley et al. [64]. (b) Illustration of cross-recurrence analysis. (c) Mean percent recurrence (%REC) and Maxline for the different experimental conditions. Note: From "Mutual interpersonal postural constraints are involved in cooperative conversation," by K. Shockley, M.-V. Santana and C.A. Fowler, 2003 Journal of Experimental Psychology: Human Perception and Performance, 29, p. 329 (panel a), 330 (panel c). Copyright 2003 by the American Psychological Association. Adapted with permission

to coevolve similarly over some time period reflected by how long the lines (x and y) remain parallel. Maxline is the length of the longest of these parallel lines in the reconstructed space.

Shockley et al. found that both percent recurrence and maxline were significantly greater when participants were performing the talking task with each other than when they were performing the task with a confederate (see Fig. 2c). Remarkably, it did not matter whether they could see each other or not.

This finding can be interpreted as showing that language use, indeed, serves as a coordination device. However, it raises the question of how language use can foster coordination. That is, how can talking back and forth about a pair of pictures, with or without being able to see one's partner, lead to entrainment of postural sway?

One possibility has to do with a special property of the task. The pictures that participants were talking about were very similar. It is likely, therefore, that the participants were producing utterances that shared many content words. The vocal tract movements that occur when speakers talk have an impact on postural movements [17, 18, 28, 53, 72]. Did the shared phonetic content of the participants' utterances underlie entrainment of their postural sway? Among phonetic properties that might affect postural sway, a likely candidate is stress pattern. Multisyllabic words have stronger and weaker syllables, and stronger syllables are produced with greater vocal effort than weaker syllables. Production of stronger syllables, then, might have more of an impact than weaker syllables. If speakers share stress patterns, and so the stress rhythms of speech, because they are producing the same words, possibly they will share the impact of those words on their postural sway.

To test that idea, Shockley et al. [61] had pairs of standing speakers (see Fig. 3a) utter words that shared a strong–weak (e.g., donut, attic) or weak–strong (admit, descend) stress pattern. Alternatively, they produced sequences that differed from each other in stress pattern, with one participant producing strong–weak words and the other producing weak–strong words. Participants exhibited more shared postural activity when the words they spoke shared their stress pattern than when they uttered words that differed in stress pattern (Fig. 3b). Importantly, however, the coordinated postural activity could not be attributed to speech patterns alone. We compared the shared postural activity of two kinds of participant pairs, copresent pairs and virtual pairs. The copresent pairs produced words that shared or did not share a stress pattern in each other's presence. The virtual pairs produced the same word sequences as the copresent pairs; however, they were recorded separately, not

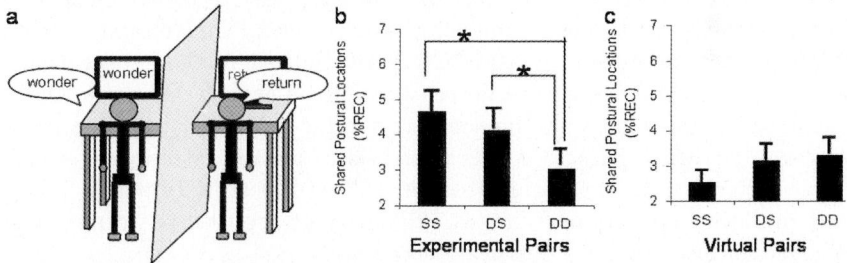

Fig. 3. (a) Experimental setup used by Shockley et al. [61]. (b) Mean percent recurrence (%REC) for experimental pairs who uttered the same words (SS), different words with the same stress patterning (SD) or different words with a different stress patterning (DD). (c) Mean %REC for "virtual pairs", who did not perform the task together. Note: From "Articulatory constraints on interpersonal postural coordination," by K. Shockley, A. Baker, M. J. Richardson, & C.A. Fowler, Journal of Experimental Psychology: Human Perception and Performance, 33, pp. 201–208. Copyright 2007 by the American Psychological Association. Adapted with permission

in one another's presence. (They were participants who had been copresent with someone else.) Virtual pairs exhibited less shared postural activity than the copresent pairs (Fig. 3c). By implication, the presence of another individual fosters interpersonal coordination beyond the requirements of a given speaking task (cf. [39, 73]).

The findings of Shockley et al. [61] indicate that interpersonal postural coordination is mediated, in part, by the effects of articulatory processes on an individual's postural activity. It is possible, however, that other linguistic processes might play a role in coupling the postural activity of conversing individuals. For one, interpersonal entrainment could result from an individual's postural movements entraining to the vocal patterns or rhythms latent within someone else's speech signal. Indeed, a number of previous experiments have demonstrated that acoustic patterns can influence postural activity. Santana et al. [54] asked participants to listen to audio recordings of words or to read visually presented words while standing. For the listening condition, they found a greater correlation between head and waist movements in the anterior–posterior direction within an individual when the participant was listening to rhyming words as compared to nonrhyming words. They found no such difference, however, when participants were reading the same words.

Baker et al. [4] explicitly tested whether interpersonal postural entrainment is (at least partially) a result of an individual's postural movements entraining to the vocal patterns or rhythms latent within a speech signal. Pairs of participants listened to audio recordings of two-syllable words presented through headphones. The words were either (a) the same as the words presented to their partner, (b) different but had the same stress patterning as the words presented to their partner, or (c) different and had a different stress patterning to the words presented to their partner. As in the studies of Shockley et al. [61, 64], postural activity was recorded using motion trackers attached at the waist of each participant and the amount of postural entrainment was determined using CRQ. In contrast to the expectation that the amount of shared postural activity would be greater for pairs who were listening to words that had the same or similar stress patterns compared to pairs who were listening to words that had a different stress pattern, there was no difference in the amount of shared postural activity across the three word similarity conditions. That is, similarity of acoustic patterning had no influence on interpersonal postural entrainment.

This lack of acoustic influence may, however, reflect a lack of engagement in the task rather than a lack of sensitivity of postural activity to acoustic patterns. Specifically, there was no imperative for the participants to listen to the words presented. In a normal conversation, listeners must understand the words spoken by others in order to successfully achieve the goal(s) of the interaction. Thus, if participants in the Baker et al. study were required to pay more careful attention to the words they heard (e.g., report how many words were presented, classify the words in some way, or be told that their memory will be tested at the end), the acoustic influence on the evolution of an individual's postural trajectory may be more apparent. In addition, natural conversation entails more than a separate presentation of words. Rather, natural conversation involves a continuous flow of acoustic sounds, in which the rhythms latent in a speech signal emerge from the structure of spoken sentences and phrases, as well as from a speaker's dialect, pausing frequency, rate of speech, and respiratory activity. Hence, the acoustic rhythms of language might influence postural sway, but only in the context of a more natural, ongoing monologue.

To test these possibilities, we recently conducted a study in which the amount of shared postural activity was examined between two participants, each of whom listened to a 4-min monologue. The monologues were spoken either by the same speaker or by different speakers. Two different monologues were used. The monologues were produced by the same author (Garrison Keillor) and had a similar content – both monologues were comic tales about a day in the life of people who lived in a small American town. Audio recordings of these monologues were made by three different speakers (each speaker was recorded reading each monologue). The monologues were read by the speakers while they viewed a teleprompter so that the length in seconds of each recording was similar across monologue and speaker. Thirty-six participants were instructed to listen to one of the two monologues and then complete a short comprehension test. By recording the postural activity of participants while they listened to the monologues, the amount of shared postural activity was determined between participants who listened to (a) the same monologue spoken by the same speaker, (b) the same monologue spoken by a different speaker, (c) different monologues spoken by the same speaker, and (d) different monologues spoken by different speakers. Despite the expectation that pairs would exhibit more shared postural activity when listening to the same monologue and/or the same speaker, pairs exhibited the same amount of shared postural activity irrespective of the monologue or speaker heard. That is, pairs who listened to the same speaker and/or monologue exhibited the same amount of shared postural activity (or lack thereof) as pairs who listened to different monologues and/or speakers. There was also no relationship between performance on the comprehension test and the amount of shared postural activity – pairs who performed well on the comprehension test exhibited the same amount of shared postural activity as pairs who performed poorly.

The results of this latter experiment and those of Baker et al. [4] suggest that the vocal patterns or rhythms latent within a speech signal have little effect on the strength of interpersonal postural entrainment and that such coordination, as demonstrated by Shockley et al. [61], is likely to be the result of the effects of articulatory processes on postural sway. Another study conducted in our lab, in which pairs of participants stood and completed an interpersonal puzzle task cooperatively or competitively, provides further evidence that interpersonal postural entrainment is a result of the effects of articulatory processes on postural activity. In this study, pairs who completed the puzzle task cooperatively were found to exhibit the same amount of shared postural activity as pairs who completed the puzzle task competitively. Given that an analysis of the conversational transcripts revealed a similarity in the number of utterances and words produced by participants in the cooperative and competitive pairs, one can conclude that interpersonal postural entrainment will occur between individuals as long as they are conversing.

It is important to appreciate, however, that this latter conclusion does not suggest that there is no relationship between interpersonal postural entrainment and cooperative intent. Other research has demonstrated a clear relationship between similarity of movement (e.g., mimicry) and cooperative intent ([5], see [11, 19] for reviews), as well as feelings of rapport et al. [6, 10, 25, 36]. Still others have demonstrated that mimicking another's movement can reflect a relational or other-directed focus [12, 38, 68]. That is, movement similarity not only is a concomitant of rapport but also causes individuals to be more helpful and generous to others [66, 67]. Therefore, although interpersonal postural entrainment is mediated by the articulatory processes inherent in conversational language, such entrainment should still facilitate social interaction by placing individuals within a common frame of reference. Being in a common frame of reference with another may be the basis for individuals forming a comfortable, smoothly operating social unit, something that should be indicated by feelings of "harmony", or greater perceptual and cognitive fluency. Recent findings support this. In the plank-moving studies, pairs for whom there was a stronger cooperation attractor reported greater feelings of comfort with the other person [41]. Similarly, creating environmental circumstances that made movement entrainment more difficult in the wrist-pendulum paradigm led pairs to report lower feelings of interpersonal harmony and environmental smoothness than when such entrainment was easy to achieve [43].

Thus, the extent to which articulatory processes are able to engage entrainment should be associated with feelings of social connectedness and environmental harmony. But the research on articulatory processes thus far discussed illustrates how language operates as a coordination device in interpersonal postural sway in particular. What about other forms of interpersonal coordination? For instance, can language use couple the rhythmic limb movements of interacting individuals? Richardson et al. [52] examined this question by testing whether a conversational exchange between two participants swinging handheld pendulums resulted in (a) the movements of the participants becoming unintentionally coordinated without vision and/or (b) increased the amount of unintentional coordination observed between interacting participants who could see one other. In contrast to the expectation that verbal conversation would, at a minimum, operate to strengthen the unintentional coordination that occurs between visually coupled individuals, the results demonstrated that a conversational exchange actually decreased the amount of unintentional coordination that occurs between visually coupled participants. That is, there was less unintentional coordination observed when the participants were looking at each and conversing than when they were simply looking at each other (not conversing).

Together, the above findings suggest that a linguistic, but not a visual, coupling provides a basis for interpersonal postural entrainment [61, 64] and that a visual, but not a linguistic, coupling provides a basis for unintentional interpersonal rhythmic limb coordination [52, 58]. This appears to relate to the fact that the verbal entrainment of posture does not occur on the basis of perceived speech signals (as indicated above) but rather is dependent upon the interpersonal coordination of speech productions. In other words, when people are copresent and producing utterances simultaneously, the speech rhythms and articulatory processes of two people become coordinated in a natural conversation (i.e., individuals speak the same words and at a similar rate) and these rhythms and processes affect postural activity, postural sway becomes coordinated [61]. In this respect, interpersonal postural coordination can be said to be mediated by an intrapersonal coupling, that between the vocal and postural machinery of an individual. This latter conclusion leads to a better understanding of why language does not couple the rhythmic limb movements of interacting individuals, as the influence of articulatory processes on locally controlled limb movements (e.g., of the arms and legs) would be very small. In fact, it is more likely that vocal gestures and articulatory processes influence rhythmic limb movements, but as small stochastic perturbations. This may explain why a conversational exchange decreases the amount of visually mediated unintentional interlimb coordination between people [52].

5 Conclusions

The limbs of an individual couple in ordinary actions to serve jointly as a dynamical system, the attractor states of which serve the action goals of an individual. By the same token, pairs of individuals, and doubtless larger groups, couple with relevant components of their environment to achieve "joint actions." In some settings, language use serves directly to foster the coordinations that serve these actions as when, for example, individuals talk back and forth as they jockey a large piece of furniture through a narrow doorway. Our research suggests that language use can serve as a coordination device less directly as when copresent individuals produce speech that serves to entrain postural sway. Speech exchanges between copresent individuals do not always lead to entrainment however. Our findings suggest that entrainment occurs, in part, when similar actions (e.g., producing speech with similar stress patterns) have similar impacts on postural sway. This may help to explain why entrainment of postural sway is observed, but not when entrainment of oscillatory activity (pendulum swinging) is observed. However, that is not a complete explanation of our findings because our comparison of copresent and virtual pairs in our postural sway experiments shows that entrainment requires not just that similar speech be produced but that individuals be copresent. We plan to pursue these intriguing sets of findings to uncover the role that language use plays in the emergence of cross-person dynamical systems in the service of joint action.

Acknowledgments

Preparation of the manuscript was supported by NIDCD grant DC-03782 to Haskins Laboratories and grants from National Science Foundation to the University of Connecticut (BSC-0240277 and BSC-0342802), to the College of the Holy Cross (BCS-0240266), and to University of Cincinnati (CMS-0432992). The authors thank Annie Olmstead and Miguel Moreno for help in collecting data for the experiment examining the shared postural activity that occurred between participants who listened to the same or different monologues, spoken by the same speaker or by a different speaker.

References

1. Amazeen PG, Schmidt RC, Turvey MT (1995) Frequency detuning of the phase entrainment dynamics of visually coupled rhythmic movements. Biol Cybern 72:511–518
2. Axelrod R, Dion D (1988) The further evolution of cooperation. Science 242:1385–1390
3. Axelrod R, Riolo RL, Cohen MD (2002) Beyond geography: Cooperation with persistent links in the absence of clustered neighborhoods. Pers Soc Psychol Rev 6:341–346
4. Baker AA, Shockley KD, Richardson MJ, Fowler CA (2005) Verbal constraints on interpersonal postural coordination. In Heft H, Marsh KL (eds) Studies in perception and action VIII, pp. 41–44. Lawrence Erlbaum Associates, Mahwah, NJ
5. Bargh JA, Chartrand TL (1999) The unbearable automaticity of being. Am Psychol 54:462–479
6. Bernieri FJ, Reznick JS, Rosenthal R (1988) Synchrony, pseudosynchrony, and dissynchrony: Measuring the entrainment process in mother-infant interactions. J Pers Soc Psychol 54:243–253
7. Bourhis RY, Giles H (1977) The language of intergroup distinctiveness. In Giles H (ed) Language, ethnicity and intergroup relations, pp. 119–135. Academic Press, London
8. Capella J, Planalp S (1981) Talk and silence sequences in informal conversations, III: Interspeaker influence. Hum Commun Res 7:117–132
9. Cesari P, Newell KM (1999) The scaling of human grip configurations. J Exp Psychol Hum Percept Perform 25:927–935
10. Charney EJ (1966). Psychosomatic manifestation of rapport in psychotherapy. Psychosom Med 28:305–315
11. Chartrand TL, Jefferis VE (2003) Consequences of automatic goal pursuit and the case of nonconscious mimicry. In: Forgas JP, Williams, KD, von-Hippel W (eds) Social judgments: Implicit and explicit processes, pp. 290–305. Cambridge University Press, New York

12. Cheng CM, Chartrand, TL (2003) Self-monitoring without awareness: Using mimicry as a nonconscious affiliation strategy. J Pers Soc Psychol 85:1170–1179
13. Clark H (1996) Using language. Cambridge University Press, Cambridge, UK
14. Condon W (1976) An analysis of behavioral organization. Sign Lang Stud 13:285–318
15. Condon W, Ogston W (1966). Sound film analysis of normal and pathological behavior patterns. J Nerv Ment Dis 143:338–347
16. Condon W, Ogston W (1971) Speech and body motion synchrony of the speaker–hearer. In: Horton D, Jenkins J (eds) The perception of language, pp. 150–184. Charles E. Merrill, Columbus, OH
17. Conrad B, Schonle P (1979) Speech and respiration. Arch Psychiat Nerven 226:251–268
18. Dault MC, Yardely L, Frant JS (2003) Does articulation contribute to modifications of postural control during dual-task performance? Cognitive Brain Res 16:434–440
19. Dijksterhuis A, Bargh JA (2001) The perception-behavior expressway: Automatic effects of social perception on social behavior. In: Zanna MP (ed) Advances in experimental social psychology, Vol. 33, pp. 1–40. Academic Press Inc., San Diego, CA
20. Gibson JJ (1979) The ecological approach to visual perception. Houghton Mifflin, Boston, MA
21. Gilbert M (1996) Living together: Rationality, sociality, and obligation. Rowman & Littlefield, Lanham, MD
22. Giles H (1973) Accent mobility: A model and some data. Anthropol Linguist 15:87–105
23. Giles H, Coupland N, Coupland J (1991) Accommodation theory: Communication, context, and consequence. In: Giles H, Coupland J, Coupland N (eds) Contexts of accommodation: Developments in applied sociolinguistics, pp. 1–68. Cambridge University Press, Cambridge, UK
24. Goldinger S (1998) Echoes of echoes: An episodic theory of lexical access. Psychol Rev 105:251–279
25. Grahe JE, Bernieri FJ (1999) The importance of nonverbal cues in judging rapport. J Nonverbal Behav 23:253–269
26. Haken H, Kelso JAS, Bunz H (1985) A theoretical model of phase transitions in human hand movements. Biol Cybern 51:347–356
27. Isenhower RW, Richardson MJ, Carello C, Baron RM, Marsh KL (in preparation). Individual armspan variation and invariants in embodied cooperative action
28. Jeong BY (1991) Respiration effect on standing balance. Arch Phys Med Rehab 72:642–645
29. Kelley HH, Stahelski AJ (1970) The social interaction basis of cooperators' and competitors' beliefs about others. J Pers Soc Psychol 16:66–91
30. Kelso JAS (1984) Phase transitions and critical behavior in human bimanual coordination. Am J Physiol 246:R1000–R1004
31. Kelso JAS (1995) Dynamic patterns: The self-organization of brain and behavior. MIT Press, Cambridge, MA
32. Kelso JAS, Holt K, Rubin P (1981) Patterns of interlimb coordination emerge from the properties of nonlinear limit cycle oscillatory processes: Theory and data. J Motor Behav 13:226–261

33. Kendon A (1970). Movement coordination in social interaction: Some examples. Acta Psychol 32:1–25

34. Kugler P, Turvey MT (1987) Information, natural law, and the self-assembly of rhythmic movement. Lawrence Erlbaum Associates, Hillsdale, NJ

35. Kuhl P, Meltzoff A (1996) Infant vocalizations in response to speech: Vocal imitation and developmental change. J Acoust Soc Am 100:2425–2438

36. LaFrance M (1979) Nonverbal synchrony and rapport: Analysis by the cross-lag panel technique. Soc Psychol Quart 42:66–70.

37. LaFrance M (1982) Posture mirroring and rapport. In: Davis M (ed) Interaction rhythms: Periodicity in communicative behavior, pp. 279–298. Human Sciences Press, New York

38. Lakin JL, Chartrand TL (2003) Using nonconscious behavioral mimicry to create affiliation and rapport. Psychol Sci 14:334–339

39. Latane B (1981) The psychology of social impact. Am Psychol 36:343–56

40. McHugo G, Lanzetta J, Sullivan D, Masters R, Englis B (1985) Emotional reactions to a political leader's expressive displays. J Pers Soc Psychol 49:1513–1529

41. Marsh KL, Richardson MJ, Baron RM (in preparation) Interpersonal consequences of embodied cooperation

42. Marsh KL, Richardson MJ, Baron RM, Schmidt RC (2006) Contrasting approaches to perceiving and acting with others. Ecol Psychol 18:1–37

43. Marsh KL, Richardson MJ, Schmidt RC (in preparation) Interpersonal synchrony and perceptions of harmony

44. Meltzoff A, Moore M (1983) Newborn infants imitate adult facial gestures. Child Dev 54:702–709

45. Mottet D, Guiard Y, Ferrand T, Bootsma RJ (2001) Two-handed performance of a rhythmical Fitts task by individuals and dyads. J Exp Psychol Hum Percept Perform 27:1275–1286

46. Natale M (1975) Convergence of mean vocal intensity in dyadic communications as a function of social desirability. J Pers Soc Psychol 32:790–804

47. Newell KM, McDonald PV, Baillargeon R (1993) Body scale and infant grip configurations. Dev Psychol 26:195–205

48. Newell KM, Scully DM, Tenenbaum F, Hardiman S (1989) Body scale and the development of prehension. Dev Psychobiol 22:1–13

49. Newtson D (1994) The perception and coupling of behavior waves. In: Vallacher R, Nowak A (eds) Dynamical systems in social psychology, pp. 139–167. Academic Press, San Diego, CA

50. Reed E (1995) Encountering the world: Toward an ecological psychology. Oxford University Press, New York

51. Richardson MJ, Marsh KL, Baron RM (in press). Judging and actualizing intrapersonal and interpersonal affordances. J Exp Psychol Hum Percept Perform 33:201–208

52. Richardson MJ, Marsh KL, Schmidt RC (2005) Effects of visual and verbal information on unintentional interpersonal coordination. J Exp Psychol Hum Percept Perform 31:62–79

53. Rimmer KP, Ford GT, Whitelaw WA (1995) Interaction between postural and respiratory control of human intercostals muscles. J Appl Physiol 79:1556–1561

54. Santana MV, Shockley K, Fowler CA (1999) Physiological reflections of speech and communication. In: Santana MV, Fowler CA, Schmidt RC (chairs) Multiple

levels of constraits on interpersonal coordination. Symposium conducted at the 10th international conference on perception and action, Edimburgh, Scotland

55. Schmidt RC, Bienvenu M, Fitzpatrick PA, Amazeen PG (1998) A comparison of intra- and interpersonal interlimb coordination: Coordination breakdowns and coupling strength. J Exp Psychol Hum Percept Perform 24:884–900
56. Schmidt RC, Carello C, Turvey MT (1990) Phase transitions and critical fluctuations in the visual coordination of rhythmic movements between people. J Exp Psychol Hum Percept Perform 16:227–247
57. Schmidt RC, Christianson N, Carello C, Baron RM (1994) Effects of social and physical variables on between-person visual coordination. Ecol Psychol 6:159–183
58. Schmidt RC, O'Brien B (1997) Evaluating the dynamics of unintended interpersonal coordination. Ecol Psychol 9:189–206
59. Schmidt RC, Turvey MT (1994) Phase-entrainment dynamics of visually coupled rhythmic movements. Biol Cybern 70:369–376
60. Sebanz N, Bekkering H, Knoblich G (2006) Joint action: Bodies and minds moving together. Trends Cogn Sci 10:71–76
61. Shockley KD, Baker AA, Richardson MJ, Fowler CA (in press) Articulatory constraints on interpersonal postural coordination. J Exp Psychol Hum Percept Perform
62. Shockley K, Butwill M, Zbilut J, Webber C (2002) Cross recurrence quantification of coupled oscillators. Phys Lett A 305:59–69
63. Shockley K, Sabadini L, Fowler CA (2004) Imitation in shadowing words. Percept Psychophys 66:422–429
64. Shockley K, Santana MV, Fowler CA (2003) Mutual interpersonal postural constraints are involved in cooperative conversation. J Exp Psychol Hum Percept Perform 29:326–332
65. Street RL (1984) Speech convergence and speech evaluation in fact-finding interviews. Hum Commun Res 11:149–169
66. van Baaren RB, Holland RW, Kawakami K, van Knippenberg A (2003) Mimicry and prosocial behavior. Psychol Sci 15:71–74
67. van Baaren RB, Holland RW, Steenaert B, van Knippenberg A (2003) Mimicry for money: Behavioral consequences of imitation. J Exp Soc Psychol 39:393–398
68. van Baaren RB, Maddux WW, Chartrand TL, de Bouter C, van Knippenberg A (2003) It takes two to mimic: Behavioral consequences of self-construals. J Pers Soc Psychol 84:1093–1102
69. van der Kamp J, Savelsbergh GJP, Davis WE (1998) Body-scaled ratio as a control parameter for prehension in 5- to 9-year-old children. Dev Psychobiol 33:351–361
70. Warren WH (1984) Perceiving affordances: Visual guidance of stair climbing. J Exp Psychol Hum Percept Perform 10:683–703
71. Warren W (2006) The dynamics of perception and action. Psychol Rev 113:358–389
72. Yardely L, Gardner M, Leadbetter A, Lavie N (1999). Effect of articulation and mental tasks on postural control. Neuroreport 10:215–219
73. Zajonc RB (1965) Social facilitation: A solution is suggested for an old social psychological problem. Science 149:269–274
74. Zbilut JP, Giuliani A, Webber CL Jr. (1998) Detecting deterministic signals in exceptionally noisy environments using cross-recurrence quantification. Phys Lett A 246(1–2):122–128

Dynamics of Interpersonal Coordination

Richard C. Schmidt[1] and Michael J. Richardson[2]

[1] Department of Psychology, College of the Holy Cross and University
   of Connecticut, Worcester, MA 01610, USA
[2] Department of Psychology, Colby College and University of Connecticut,
   Waterville, ME 04901-8885, USA

Abstract. Everyday human actions often occur in a social context. Past psychological research has found that the motor behavior of socially situated individuals tends to be coordinated. Our research performed over the last 20 years has sought to understand how the mutuality, accommodation, and synchrony found in everyday interactional coordination can be understood using a dynamical theory of behavioral order, namely coordination dynamics. Using laboratory interpersonal tasks, we have demonstrated that when two people are asked to rhythmically coordinate their limbs they show behavioral phenomena identical to those found in bimanual interlimb coordination, which has been mathematically modeled as a dynamical process. Research has demonstrated that these same dynamical organizing principles can coordinate the rhythmic movements of two people unintentionally and that the weaker, intermittent coordination that ensues is affected by both perceptual (e.g., attentional focus and information pickup activity of the visual system) and dynamical constraints (e.g., intrapersonal rhythmic synergies and period basin of entrainment). Other research has investigated how traditional social and personality properties of a dyad (rapport, social competence) relate to dynamical properties of a dyad's coordinated movements and how the stability of coordinated movements mirrors the stability of mental connectedness experienced in social interactions.

1 Dynamics of Interpersonal Coordination

Psychological theory over the past few decades has had a growing appreciation that mind and knowing are embodied and that action not only reflects the mind (the cornerstone of all cognitive experiments) but is itself imbued with significance, that is, it is the vehicle by which we realize our mind, our thoughts, our values [12, 94]. There has also been for some time an appreciation that many natural actions are performed in a social context [56] and that social and physical environments are nested with their meanings existing side by side in natural perception and action [69]. Simple actions such as reaching, locomotion, or even standing still are often performed to achieve a social goal and can then be socially defined as, for example, a caress, an escape or opportunity for a conversation. Generic motor movements are nested within a higher-order definition of the actions that are defined intersubjectively in terms of another person, in terms of an interpersonal situation that generally occurs over a longer time scale than the motor gesture itself and is used to establish and maintain social relationships as well as achieve mutual action goals. But how does one begin to understand this social level of perception and action [58]? What are our expectations for how to understand the structure of such social behaviors? What taxonomies should we use to define the various kinds of embodied social interactions? One way is to define social actions in terms of cultural rituals as a cultural psychologist or social anthropologists would [3]. Another way is define them in terms of different kinds of social relationships (dyads, strangers) or in terms of different kinds of social properties (gender, rapport) or personality characteristics (expressivity, dominance), from the perspective of a social psychologist [4, 59, 88]. But both of these domains, as important as they are, deal mainly with the 'what' rather than the 'how' of social actions. They focus on the 'semantics' rather than the 'syntax' of social interactions, the 'content' rather than the 'process'.

Focusing on the 'process' of social interactions, we notice two general kinds of social interactions. We all have had the pleasure of interpersonal interactions that have just 'flowed' [21]: that effortless soccer goal or tennis match where you could just read others' minds or that first date that you did not want to end. We also have been victims of untoward interpersonal events, from the mundane, like the awkward 'dance' two people do when they are trying to walk in the same space, to the more intimate, like 'dates from hell' and other failed personal interactions. We can characterize these positive and negative social interactions in terms of the efficiency and stability with which the component actions achieve an intended social goal. In order to understand this continuum of stabilities, the "how well it is going-ness", that characterize these processes of social coordination, we need to turn to the universal logic of stability of natural systems: dynamical processes of self-organization. To understand the process of interpersonal interactions, we need to study the dynamics of interpersonal coordination. The research that we have performed over the past two decades has been dedicated to understanding the extent to which principles of coordination dynamics that govern intrapersonal movements can be used to understand the array of stabilities that arise in interpersonal coordination. In what follows, we will review this work with an eye toward how it is laying a foundation for a more general theory of social perception and action [58].

2 Universal Principles: Dances We Can and Cannot Do

Haskins Labs in the 1980s was an exciting place to study the coordination and control of human movement. The theory of behavioral dynamics was being born and taking its first steps. Seminal theoretical papers written in the early years of the decade by Kelso, Kugler, and Turvey [47, 51, 52] identified ideas empirically investigated in the later years of the decade. The ideas are as big now as they were 25 years ago. In order to answer how the many degrees of freedom of the perceptual-motor system are regulated, the traditional machine (i.e., neural or symbolic computational) theory answer was being replaced with a dynamical conception of order based on theories of physical biology. Neural wiring conceptions, such as the afference and efference distinction, were replaced by synergistic linkages between muscles that were dynamically constrained. Cognitive programs were replaced by equations of constraint that channel and guide a dynamic unfolding of behavior in a non-prescriptive manner.

Key to empirically investigating how dynamics – namely "the free interplay of forces and mutual influences among components tending toward equilibrium or steady states [51] (pg. 6)" – organizes behavior was being able to conceive, model, and measure behavioral equilibrium or attractor states. While some researchers investigated the lawful nature of minimal energy states (comfort or optimal modes [53, 92]), Kelso and his students began to explore how the behavioral patterns associated with coordinated rhythmic limb movements can be modeled as representing the dynamics of periodic attractors or limit cycles [48]. They further explored how the behavioral transitions between oscillatory phase modes (like in quadruped gait transitions) can be investigated in bimanual tasks [42, 43] and understood as phase transitions or bifurcations – a general nonlinear way that a dynamical system reorganizes itself after destabilization. In particular, the bimanual phenomenon was that the antiphase coordination of wrist or index fingers becomes increasingly less stable as the frequency of oscillation is increased, eventually breaking down and leading to a transition to inphase coordination. The dynamics of this behavioral switching was captured by a mathematical formalism that modeled both the steady state and phase transition behavior of coordinated rhythmic limb movements by capturing the dynamics of the relative phase angle (ϕ) between limbs [35]:

$$\dot{\phi} = a \sin \phi - 2b \sin 2\phi + \sqrt{Q}\, \zeta. \tag{1}$$

In this equation (the HKB model), $\dot{\phi}$ is the rate of change of the relative phase angle formed between the two oscillators, a and b are coefficients whose magnitudes govern the strength of the between-oscillator coupling, and ζ is a Gaussian white noise process dictating a stochastic force of strength Q [81]. Using the terminology of *synergetics*, ϕ is an *order parameter* (i.e., it summarizes the spatial temporal order of the rhythmic units) and the variables (e.g., frequency of oscillation) that influence its stability are *control parameters* [33, 34]. The bimanual phase mode switching phenomenon and model has had an inestimable impact on the dynamical theory of human movement because they provided a specific explication of behavioral attractors as well as how transitions occur between them.

This model of behavioral dynamics has generated an astounding amount of subsequent coordination research including our own on interpersonal coordination. The dynamics modeled in the phase transition phenomenon created a reorganization of behavioral order across the limbs of a single person. However, the question of the universality of these behavioral dynamics could be raised. If indeed, inphase and antiphase are canonical steady states that arise as a consequence of the dynamics of oscillators and their interactions, should not they also be differentially stable steady states in rhythmic movements coordinated across two people, across two neurally based oscillators linked by perceptual information [68]? Is it possible to have functional synergies or coordinative structures written across two individuals using the language of dynamics?

In Schmidt, Carello, and Turvey [71], two seated participants were asked to visually coordinate their lower legs in either inphase or antiphase while oscillating them at a tempo of an auditory metronome pulse that increased in frequency from 0.6 to 2.0 Hz by 0.2 Hz increments every 5 s. In the first experiment, six pairs of undergraduate participants were instructed to keep and return to their original coordination phase mode if they fell out of it. The trials were videotaped and the relative phase angle between the legs of the participants was evaluated frame by frame. Of interest was the relative stability of the two modes of phasing and whether the likelihood of a breakdown in phase locking would grow as the frequency of oscillation was increased. Participants had a hard time maintaining the antiphase coordination of their lower legs at the higher frequencies. As seen in Fig. 1 (left), the variability at the higher frequencies for the antiphase mode is indicative of a breakdown

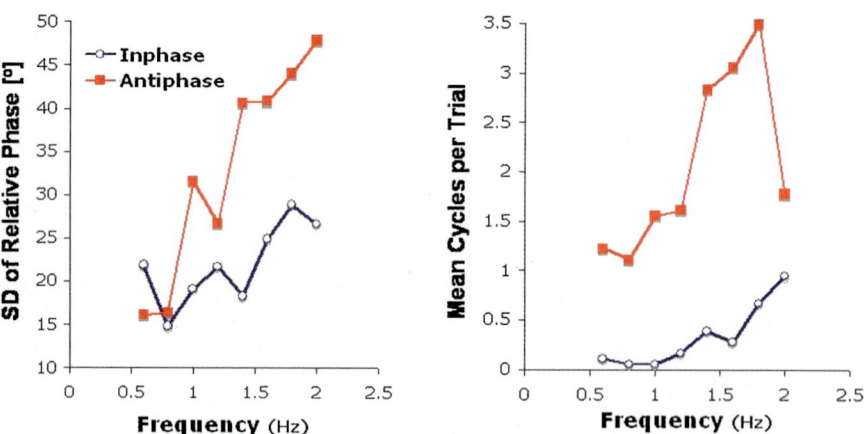

Fig. 1. Data exhibiting the destabilization of relative phase with frequency increase in the interpersonal coordination of lower legs. Both relative phase variability and occurrences of the other phase modes are much greater for antiphase at higher frequencies. Adapted from [71]

in coordination, whereas this variability for inphase is large but still indicative of a stable state. Additionally, Fig. 1 (right) reveals the mean number of cycles that were in the 'other' phase mode was 1.5 times greater for antiphase than inphase for lower frequencies but nearly three times as great for higher frequencies of oscillation. A second experiment in this study analyzed the characteristics of the antiphase breakdown and transition to inphase using "do not intervene" participant instructions to allow the transition to naturally emerge. Results revealed that the transitions had the formal properties of a dynamical reorganization, namely divergence, critical fluctuations, and hysteresis [29]. These visually coordinated interpersonal movements expressed the same differential dynamic stability of the intrapersonal phase modes modeled by Haken et al. [35], namely the inphase mode of interpersonal leg swinging is the globally stable behavioral attractor while the antiphase phase mode is a dynamically unstable local attractor. Scaling the oscillation frequency results in a gradual weakening of the local attractor, followed by a state of criticality (exhibited by increase in relaxation time and amplification of fluctuations), and then leads to a sudden annihilation of the local attractor and a switch to the globally attractive inphase mode (divergent response). A third experiment in this study verified that the critical nature of this phase transition is conditioned by the rate that the control parameter is increased [49]. Increasing, the frequency plateau time (i.e., the length of time the relative phase was observed at each frequency) from 5 to 10 s resulted in an elimination of the state of criticality immediately before the transition because the observation time scale (plateau time = 10 s) was now equivalent to the random walk or first passage time required for such a transition to occur probabilistically due to chance [28, 29].

In summary, the three experiments of Schmidt et al. [71] demonstrated that the self-organizing coordination dynamics that Kelso observed in bimanual coordination of a single individual also occur in interpersonal coordination of movements. The between-person nature of these behavioral attractors demonstrated that the dynamical organizing principles of the HKB equation can operate in neurally based behavioral oscillatory systems that are coupled by perceptual information and, consequently, that these principles represent a universal self-organizing strategy that occurs at multiple scales of nature.

3 Frequency Detuning: The Firefly Dance

An example of inter-organism dynamical synchronization that has been studied and modeled for some time is the mass coordination of firefly flashing in some parts of South East Asia [9, 87]. At dusk, male fireflies of certain species in New Guinea and Malaysia congregate in groups of hundreds or thousands in trees. As the night progresses, a synchronization of their flashing emerges resulting in a large, light-pulsing mass of fireflies. Synchronization has also been observed in certain North American fireflies; however, these species do not congregate and their coupling is apparently weaker, yielding synchronization that is intermittent [20]. Hanson [36] experimentally investigated and dynamically modeled the firefly synchronization process using an experimental method in which he manipulated the rate at which he periodically flashed a light at a firefly. The essential dynamical process in Hanson's model of firefly entrainment is captured by an extension of the HKB equation [45] that contains a $\Delta\omega$ term which specifies the eigenfrequency difference or frequency detuning between the two rhythmic units:

$$\dot{\phi} = \Delta\omega - a\sin\phi - 2b\sin 2\phi + \sqrt{Q}\,\zeta. \qquad (2)$$

Entrainment of the fireflies in this model will occur ($\dot\phi = 0$) when the eigenfrequency difference between the firefly flashing and the experimental flashing of the light ($\Delta\omega$) is small enough to be 'balanced' by the coordination 'forces' of the coupling function ($-a\sin\phi - 2b\sin 2\phi$). If the eigenfrequency difference is too large, $\dot\phi$ will be non-zero and absolute coordination [91] or phase locking will not occur and the firefly will not fully entrain to the light. The dynamical model also predicts that when the balancing of $\Delta\omega$ and the coupling forces does occur and phase locking ensues, a phase lag between the two light flashes will emerge that is proportional in size to the $\Delta\omega$. Furthermore, the model predicts that when the flashing light is inherently faster, it should lead in the cycle, and when it is inherently slower, it should lag in the cycle. Because the phase lag demonstrates that the oscillating rhythmic unit (in this case, the firefly) is retaining some vestige of its original dynamic when in the coordinative state, von Holst referred to it as evidence of a 'maintenance tendency' of the oscillator [91]. As seen in Fig. 2 (left), this increase in phase lag with an increase in frequency detuning is just what Hanson [36] found in his experiment.

Importantly, these phase lag patternings of oscillators with different eigenfrequencies have been observed in a number of other examples of neurally based biological coordinations such as cockroach locomotion [23], the coordination of breathing and sucking in infants [30], and coordination of central pattern generators at the neural scale [84]. Additionally, Turvey, Schmidt, and others [77, 85, 89] using an experimental paradigm developed by Kugler [53] at Haskins Labs have demonstrated that similar detuning patterns are seen in human bimanual coordination. This paradigm allows the manipulation of the detuning by having individuals bimanually swing (using ulnar-radial adduction) handheld pendulums whose lengths (and hence, whose frequencies) can be manipulated: Two pendulums of identical lengths will have a $\Delta\omega$ of 0, whereas two pendulums of different lengths will have a non-zero $\Delta\omega$ whose magnitude depends on the length difference. As reviewed in [80], the intrapersonal, bimanual relative phase patternings found using this wrist-pendulum paradigm beautifully support the predictions of the extended HKB model.

But would the dynamical patterns of frequency detuning predicted by the HKB model and seen in fireflies, cockroaches, and undergraduate UConn

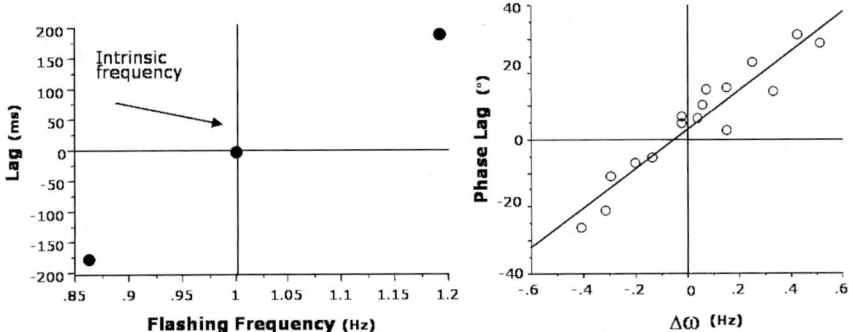

Fig. 2. The effect of frequency detuning ($\Delta\omega$) on the relative phase lag in fireflies (*left*) and interpersonal coordination of wrist pendulums (*right*). Adapted from [36] and [79], respectively. Phase lag increases as the detuning becomes greater (i.e., as inherent frequencies become more different) as predicted by the extended HKB model

students bimanually swinging wrist-pendulums also appear in interpersonal coordination of limbs? An initial study [68, 79] had two participants sitting side by side visually coordinate wrist-pendulum pairs that differed in their $\Delta\omega$'s. Participants could easily coordinate the pendulums in synchrony. But as Fig. 2 (right) depicts, as the length difference between the wrist-pendulums became greater, the phase lag between the pendulums increased such that the person with the shorter pendulum led in the cycle – just what is predicted by the dynamical model. Subsequent interpersonal wrist-pendulum studies investigated effects of phase mode and frequency of oscillation predicted by (2). Visual interpersonal wrist-pendulum coordination was found to be weaker as revealed by exaggerated relative phase lags and fluctuations (a) for antiphase compared to inphase and (b) for higher frequencies of oscillation compared to lower frequencies [2, 70].

These frequency detuning results verified that humans do indeed 'dance' like fireflies – human interpersonal rhythmic coordination is subject to the same dynamical laws as seen elsewhere in nature. They also demonstrate the explanatory power of the explicit modeling of the dynamical self-organizing processes captured by the HKB equation.

4 Coordination Dynamics in the Wild I: Unintentional Social Entrainment

This early research on interpersonal coordination demonstrated that dynamics could constrain social actions in the laboratory during artificial tasks when people *intend* to coordinate. But do such processes of self-organization operate in circumstances where interpersonal coordination arises naturally in the functional context of a social interaction? Social psychologists have been investigating naturalistic behavioral 'entrainment' processes in social interactions since the 1960's (see [7] for a good review). For example, Condon and Ogston [18], who used film analysis ('kinesics') to investigate movement coordination interactions, state

> ...analysis revealed harmonious or synchronous organizations of change between body motion and speech in both intra-individual and inter-actional behavior. Thus the body of the speaker dances in time with his speech. Further, the body of the listener dances in rhythm with that of the speaker (p. 338).

Much social psychological research since that time has been directed at understanding what role this coordination of social behavior plays in human communication processes [7, 16, 50, 88].

Is it possible that the dynamical processes of self-organization captured by the HKB equation are the basis for the basic behavioral phenomenon of interactional synchrony? First pass evidence in support of this claim would be to demonstrate that these dynamical constraints operate even when people do not intend to coordinate. A series of studies we have performed demonstrated that they do. In the initial study [73], the visual wrist-pendulum task was modified to see whether such unintentional entrainment would occur. Participants were told that their task was to swing a pendulum at its comfort mode tempo, one that they could "swing all day". During the first half of a trial, participant pairs were instructed to look straight ahead so that they could not see each other. During the second half of a trial, participants were told to look at the other participant's moving pendulum but maintain their preferred tempo from the first half of the trial. Of interest was (a) whether in the second half of the trials when visual information was available the participants would tend to entrain their movements unintentionally and (b) whether the patterning of relative phase would be indicative of the HKB dynamic. In the first half of the trials, no phase entrainment was observed. The relative phase angle was not constant and all phase angles were equally observed in the coordination of the participants' pendulums. This is not surprising because there was no perceptual coupling and 'phase lapping' occurred between the oscillations. In the second half of the trials when vision was available, the relative phase was also not constant and all phase angles were again observed but this time not equally: Relative phase angles near the attractor regions predicted by the HKB model (near 0 and 180°) tended to dominate. The participants' movements were weakly phase entrained. The coordination was not the absolute coordination (phase locking) exhibited in intentional visual coordination but rather relative coordination – a non-steady state coordination behavior produced by dynamical systems with weak attractor basins and intrinsic noise [46, 91, 96]. Such dynamical systems demonstrate the property of *intermittency*: A constant change in state with an attraction to certain regions of their underlying phase space. We saw further evidence for this by measuring the rate of change of the relative phase angle (ϕ). In the first half of the trials, the average $\Delta\phi$ calculated for nine 20° relative phase regions between 0 and 180° is large and shows no pattern. But in the second half of the trials when vision was available, the average $\Delta\phi$ across the relative phase regions has an inverted u-shape (Fig. 3). The minimum rates of change near 0 and 180° suggests that the system was attracted to these regions. These results demonstrate that the dynamical principles of self-organization in the HKB model can constrain interpersonal coordination unintentionally, outside of participants' awareness in a rather subtle fashion. Participants' behavior was influenced by weak behavioral attractors, mere 'ghosts' of the attractors that we see constrain intentional coordination of rhythmic limb movements.

Subsequent studies have verified this basic conclusion using progressively more naturalistic interaction circumstances. Note that neither components of the Schmidt and O'Brien task [73], neither just looking at another's movements, nor swinging a wrist pendulum is a typical everyday task. To remedy the first concern, Richardson et al. [62] used a dyadic problem-solving task as a foil for investigating whether unintentional entrainment of pendulums would naturally occur. Cartoon faces were attached to the participants' swinging pendulums (in the just looking and looking while talking conditions) or on stands on the side opposite of the other participant (in the no looking-no

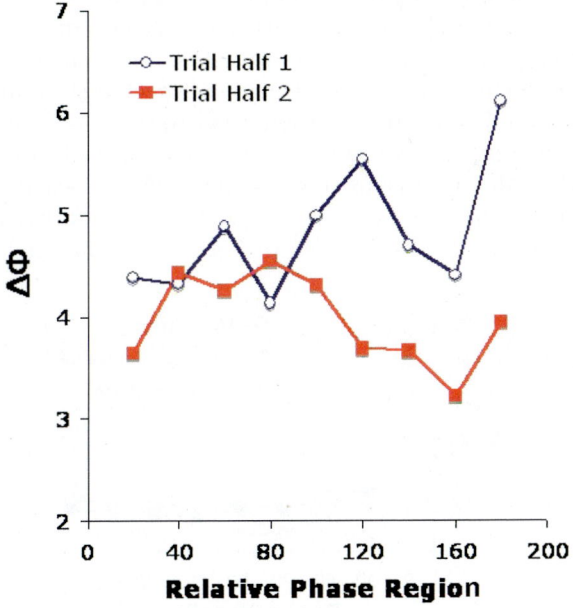

Fig. 3. The rate of change of relative phase ($\Delta\phi$) at different relative phase regions without visual information (trial half 1) and with visual information (trial half 2). With vision available, $\Delta\phi$ tends to decrease near 0 and 180° suggesting unintentional intermittent entrainment. Adapted from [74]

talking control and not looking but talking conditions). Participants were told that the faces were placed on the pendulums in some conditions to make the dyadic problem-solving task more difficult to perform. In truth, they were put there so that participants would necessarily have visual information about each other's movements during the problem-solving task. In each condition, a participant could only see one of the cartoon faces, and the dyad's task was to determine together the differences (i.e., eye color, mouth shape) between the two faces. Debriefing verified that the participants indeed were unaware that the coordination of their pendulum swinging was the true purpose of the experiment. The results for the conditions in which visual information about the other person's pendulum trajectory was available provided evidence similar to Schmidt and O'Brien [73] for unintentional entrainment as predicted by the HKB equation, that is, relative coordination, the intermittent attraction to inphase and antiphase relative phase angles, was observed.

Another recent study [61] eliminated the artificiality of wrist-pendulum swinging by investigating the unintentional coordination of movements between two people rocking in rocking chairs. Participants were told that the experiment was investigating the ergonomics of rocking chair movements and that they were completing the task together because 'testing two people at the same time was a more efficient way of collecting data'. In addition to investigating whether participants would unintentionally entrain their rocking movements, the experiments also manipulated the amount of visual information each participant had of each other's rocking movements. To manipulate the information available, participants were also told that the study was investigating how different postural configurations affected the stability of rocking and that they would have to turn their head to focus on the red target that was located either on the arm rest of their partner's chair, directly in front of them, or on the side away from their partner. These locations corresponded to the participants having focal, peripheral, or no information about their partner's movements. As can be seen in Fig. 4, unintentional entrainment to

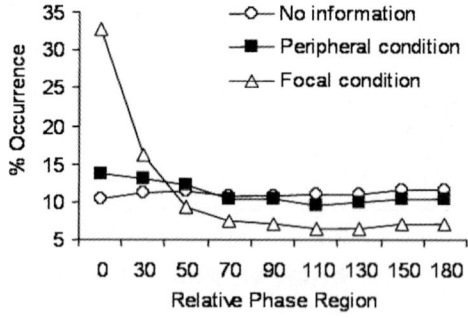

Fig. 4. The distribution of relative phase angles between the rocking chair movements of co-participants as a function of the availability of visual information. More unintentional entrainment near inphase was found for the focal as compared to the peripheral information condition. Adapted from [60]

the inphase mode occurred and the degree of attraction to this mode was influence by the degree of information available – significantly less coordination was observed for the peripheral as compared to the focal information condition. The results of this and the preceding studies thus provide clear evidence that the dynamical processes of self-organization modeled by the HKB equation can constrain the unintentional synchrony in these laboratory tasks and are possibly responsible for the synchrony observed in ordinary, everyday social interactions.

5 The Expanded Interpersonal Synergy: Mediating Intrapersonal Rhythms

In contrast to the coordination paradigms we have used which investigate the interpersonal synchronization of isolated limb movements, natural interpersonal interactions are whole body interactions with nested intrapersonal subtasks. Consider an ordinary interpersonal interaction: Two people have a conversation and walk together while carrying lunch trays in a cafeteria. The interpersonal coordination in their conversation and the interpersonal rhythmic synchrony of their leg movements occur along with other intrapersonally coordinated subtasks, in this case, carrying a tray and navigating through a cluttered environment. The orchestration of these subtasks needs to be coordinated within each person. Consequently, interpersonal interaction occurs in a context of ongoing intrapersonal coordination. Condon and Ogston [17] describe this intrapersonal coordination as being characterized by self-synchrony:

> ...Behavior appears to be composed of flowing "configurations of change", where body elements sustain and change movement together in an ordered fashion. It is this variable and serially-emerging pattern of sustaining and changing together which seems to constitute the "units" of behavior (p. 341).

The idea is that these "configurations of change" define synchronization points for the various movements that we perform across the whole body. As we initiate arm movements to put down our lunch tray, we open our mouth to speak, lean our trunk forward, turn our head to turn right, and use our eyes to guide the tray to the table. In the next moment, we begin to lower our trunk to sit, turn our head toward our interactor to listen, and direct our gaze toward the salt shaker. Many parts of our body change and sustain direction at the same point in time and define the configuration of change for that moment. Condon and Ogsten [17] discussed as well the relation of this self-synchrony to interactional synchrony:

> Subject A, as he listens, displays a similar pattern of "configurations of change" which follows the same principle of change as those of Subject

B. Further, the "configurations of change" of Subject A are in precise
synchrony with those of Subject B (p. 342).

In more dynamical motor control terminology, we see that the interpersonal coordinative structure or synergy that self-organizes across two people in a naturalistic interaction has nested within it speech, bimanual, bipedal, and postural intrapersonal synergies that orchestrate the whole body movements of single individuals. Condon and Ogston's point is that these intrapersonal synergies are coordinated – synchronized – in natural everyday actions and that their total functioning becomes synchronized across two people in interactions. Given this expanded, whole body view of coordination, the question that can be raised is how do the various intrapersonal synergies interact with or mediate the interpersonal entrainment that emerges in an interaction. We have investigated this question by looking at the role of eye movements and speech rhythms in interpersonal interactions.

Our environmental interactions involve the active pickup of visual information that requires eye movements. This is especially true when perceiving animate objects whose movements need to be tracked in order to coordinate with them (e.g., getting on an escalator or entering an automatic revolving door). Much research has been performed investigating the constraints that movements of one effector have on the movements of another [24]. However, because perceptual systems are typically conceived as static (in spite of the work of Gibson [26, 27] and Yarbus [95]), research has tended to ignore the constraints that movements of a visual system have on the coordination of effector movements that are occurring simultaneously. If one is tracking a moving object with one's eyes, the question is whether these eye movements constrain the actions you make. Past research has noted a high degree of intrapersonal coordination between limb and saccade eye movements [37, 38, 90]. Given this intrapersonal synergy, we can ask whether a listener's visual tracking of a speaker's movements (e.g., gesticulation) mediates the interpersonal entrainment of the speaker and listener's arm movements because the listener's arm movements are intrapersonally coordinated with his/her tracking eye movements.

Schmidt et al. [75] used an environmental entrainment paradigm to investigate the role of eye movements on unintentional entrainment. In this paradigm (Fig. 5, left), we were interested in whether participants would unintentionally entrain to a rhythmically moving stimulus. Participants were instructed to read aloud letters that randomly appeared on a projection screen while simultaneously swinging a wrist-pendulum. In addition to the letter stimuli, a sinusoidally oscillating stimulus moved horizontally across the screen. Participants were told that the purpose of the task was to measure the speed and accuracy of their reading and that their wrist movements and the oscillating stimulus were just motor and perceptual distracters. The real purpose of the experiment was to investigate whether participants would entrain their wrist movements to the oscillating stimulus and whether visual tracking the stimulus would facilitate this unintentional entrainment. We manipulated visual

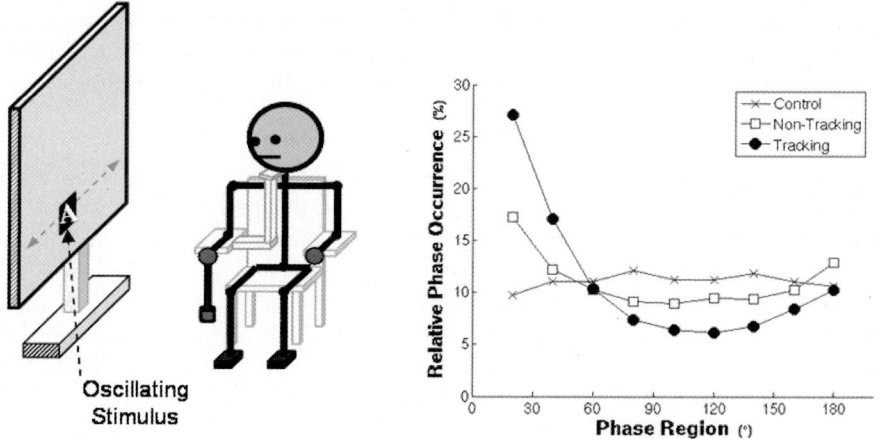

Fig. 5. Environmental entrainment experimental setup used to study effects of visual tracking (*left*) and the distribution of relative phase angles between the visual stimulus and the participant's wrist movements as a function of the visual tracking conditions (*right*). More unintentional entrainment near inphase was found for the tracking compared to the non-tracking condition. Adapted from [75]

tracking by controlling where the letter to be read appeared. When the letters appeared in the center of the screen (above the oscillating stimulus), the participants were required to fix their gaze directly at the center of the screen (non-tracking condition). When the letters appeared on the visual stimulus, the participants needed to track the stimulus with their eyes as it oscillated from side to side in order to read the letters (tracking condition). To measure chance level coordination, trials were performed in which the letters appeared in the center of the screen along with an invisible oscillating stimulus (control condition). Results revealed that unintentional environmental entrainment between the wrist movements and the oscillating visual stimulus occurred. As seen in Fig. 5 (right), the results demonstrated that the visual tracking of the stimulus produced greater unintentional entrainment than the non-tracking condition. However, the non-tracking condition still exhibited some unintentional phase entrainment near 0 and 180°.

But does eye tracking facilitate the unintentional entrainment to the movements of another person as well as an oscillating dot? A follow-up study [60] investigated whether this result generalized to *interpersonal* rhythmic coordination by using the dyadic problem-solving paradigm of Richardson et al. [62] with tracking and non-tracking conditions. In the visual tracking condition, the cartoon pictures were attached to the end of the pendulums such that each participant had to visually track the motion of their co-actor's movements in order to complete the dyadic task. In the non-tracking condition, the pictures were displayed on a floor stand positioned directly behind the motion of their co-actor's pendulum; hence, no tracking movement of the eyes was required to perform the problem-solving task. Results indicated that the wrist movements of the participants became more strongly entrained when visual tracking of their partner's movements was required to complete the dyadic puzzle task.

The results of both the environmental and interpersonal entrainment studies suggest that the dynamical, interpersonal 'coordinative structure' needs to be conceptualized as including the movements of two people coupled via an active, visual information pickup dynamic which is intrapersonally coupled to the individuals' limbs. What we see is that the interaction of separate intrapersonal synergies (rhythmic wrist and eye tracking) constrains the creation of an interpersonal synergy. Future research needs to pay heed to the composite nature of the whole body system in studying environmental and interpersonal interactions and that a full understanding of the rhythmic synergy underlying interpersonal coordination involves understanding the intrapersonal coordination of motor and perceptual rhythms.

Another behavior found in natural interpersonal interactions that needs to be intrapersonally coordinated with other body movements is speech. Speech often serves an explicit coordination function in social interactions [13]. For example, the content of our speech dictates who does what in an interaction with a mutual social goal. (Perhaps most apparent when you are in a foreign country where you do not speak the language.) A more subtle question is whether the verbal information flow in a conversation is a sufficient informational medium to sustain interpersonal synchronization of bodily movements. If just looking at someone's movements causes you to become entrained, will just talking to them also create interpersonal motor entrainment? Shockley et al. [83] investigated whether conversation is sufficient to entrain the *postural* sway of two interacting participants. Using a dyadic puzzle task not unlike that described above for Richardson et al. [62], participants interacted visually and verbally, interacted just verbally (faced opposite directions), interacted just visually (faced each other but conversed with an experimenter), or did not interact (control condition; faced away and conversed with an experimenter). The results demonstrated that postural entrainment occurred when participants were interacting verbally and that having visual information about the other person available by itself did not result in postural entrainment. Richardson et al. [62] using the interpersonal wrist-pendulum task ran similar verbal conditions (not looking but talking and looking while talking conditions; see above) to investigate whether such verbal information would affect interpersonal entrainment of rhythmic *limb* movements. They found that entrainment occurred for conditions where the participants could see each other but just chance level entrainment for the verbal information condition. An explanation for why interpersonal postural entrainment is affected by speech but limb entrainment is not has been suggested by a follow-up study. Using a similar method, Shockley et al. [82] found that verbal entrainment of posture does not occur on the basis of perceived speech signals but rather is dependent upon the rhythmic nature of the speech productions. That is, because speech rhythms of two people in conversation are coordinated and these speech rhythms produce postural changes, the postural sway between the two interactors becomes coordinated. As with the eye-tracking studies above, it appears that the interpersonal synchronization of movements (limbs or posture) seems to be influenced by an interaction of intrapersonal rhythms (i.e., eye-wrist or speech posture).

Why it is that the interpersonal entrainment of limb movements was not influenced by speech rhythms in Richardson et al. [62] needs to be understood as well. Studies of gesticulation have found evidence for intrapersonal synchronization of speech rhythms and hand movements [11], and other research has found that interpersonal gestural synchronization occurs in natural conversation although it depends heavily on the communicative and functional context of a gesture [25]. It may be that the rhythms latent in the between-person conversation may have been too subtle to have any observable influence on the participants' rhythmic limb movements in the Richardson et al. [62] study. There seems to have been a lack of intrapersonal 'intimacy' between the rhythms of a natural conversation and the sinusoidal rhythms of wrist-pendulum swinging. Accordingly, one may expect overtly rhythmic speech to provide a better functional context for finding a coupling between rhythmic limb movements and speech as well as speech-coordinated rhythmic limb movements between two people.

We performed a recent study to address the first of these questions [76]. Participants were required to read letters that appeared rhythmically in the middle of a computer screen while swinging a wrist-pendulum at a comfort mode tempo. Although the participants were told we were interested in their reading speed and accuracy, we were really measuring the degree of unintentional entrainment between the wrist-pendulum swinging and the rhythm of the appearing letters. This entrainment was evaluated under four speech conditions: out loud reading, silent reading (to test whether speech motor movements were necessary), no reading with a rhythmically appearing visual stimulus (to test whether purely visual rather than speech-mediated entrainment occurred), and no reading with no visual stimulus (control condition to test chance entrainment). The tempo of the appearing letters was also manipulated and presented at a period equal to the participant's self-selected comfort mode tempo (determined in pretrials) or at a period that was slightly faster or slower than that tempo. Results provided evidence for both frequency and phase entrainment in the two reading conditions. The difference between the period of the wrist movements and the rhythmic stimulus (Fig. 6, left) decreased when either out loud or silent speech was performed, suggesting a 'magnet effect' [91]. Additionally the relative phase distributions (Fig. 6, right) indicate a tendency to be at the front or back of the cycle ($0°$ arbitrarily defined as the pendulum being away from the screen and $180°$ was defined as the pendulum being toward the screen) when the letter appeared in either the out loud or the silent speech conditions.

These results provide evidence that speech rhythms can entrain limb rhythms. They also provide evidence that if the speech rhythm is synchronized

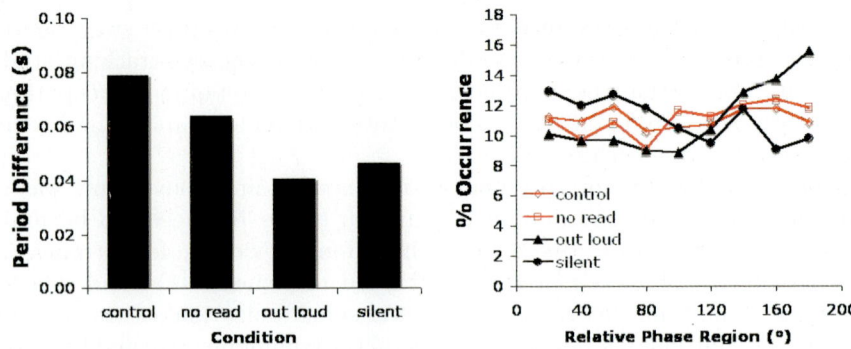

Fig. 6. Period difference (*left*) and distributions of relative phase angles between the letter-to-be-read stimulus and the participant's wrist movements (*right*) for the different rhythmic speech conditions. The decrease in period difference for the reading conditions represents frequency entrainment (magnet effect), whereas the increase in relative phase occurrence near 0 and 180° represents phase entrainment between the wrist movements and the speech rhythm

with an environmental rhythm, speech can mediate the entrainment of a limb rhythm and an environmental rhythm. Moreover, the silent reading results demonstrate that overt motor speech movements are not necessary for speech to play such a mediational role. More important seems to be the attentional rhythm [54] rather than the motor movements themselves. Such an explanation may have implications for why eye-tracking movements facilitate unintentional entrainment with an environmental rhythm. Future research needs to investigate whether the overt eye movements themselves are less important than the attentional attunement that the movements create. Regardless of this particular issue, the implication of this speech study is similar to that of the eye-tracking studies reviewed above: Intrapersonal rhythms and the synergies that create them are themselves coordinated and this self-synchrony can mediate the establishment of synchrony between people.

6 Informational and Dynamical Constraints on Unintentional Interpersonal Entrainment

The research discussed thus far demonstrates that interpersonal entrainment emerges from dynamical laws operating via informational interactions. As much as unintentional interpersonal entrainment is a powerful effect, individual differences observed in our experiments suggest that it does not always occur. Such entrainment occurs only if certain conditions are fulfilled. Additional research we have performed has investigated more specifically the dynamical and informational constraints on unintentional interpersonal entrainment.

Equation (2) suggests that the weak phase entrainment seen in the unintentional interpersonal coordination will occur as long as the difference in inherent tempos of the oscillators (i.e., $\Delta\omega$) is not much larger than the coupling strength. This suggests that if two participants establish comfort mode tempos that are very different (i.e., create a large $|\Delta\omega|$ by virtue of 'personal' tempos), the coupling dynamic of (2) may not be strong enough to parry their respective 'maintenance' tendencies. Empirical results have supported this prediction. In the unintentional coordination experiment of Schmidt and O'Brien [73], $\Delta\omega$ was manipulated by having participants swing pendulums of same or different lengths. A reduced concentration of relative phase angles around the attractors of 0 and 180° was found for conditions in which the participants swung pendulums of different lengths (i.e., different inherent frequencies) compared to conditions in which they swung pendulums of similar length. Furthermore, a reanalysis of the data of Richardson et al. [62] displayed in Fig. 7 suggests that even when participants have the same length pendulums, inherent differences in their selection of a comfort mode tempo will affect the likelihood of entrainment. This result raises not only the issue of what variables effect individual differences in comfort mode tempo (see [78]) but also the question of over what range of $\Delta\omega$'s unintentional visual entrainment will occur. This reanalysis suggests that there is a range of period differences over which interpersonal coordination will occur and that beyond this range the occurrence of unintentional coordination is highly unlikely. In other words, there appears to be a *period basin of entrainment* [86] for unintentional interpersonal coordination. In a recent study [57], we used the environmental coordination paradigm to measure this period basin for unintentional coordination to an environmental rhythm. Participants were

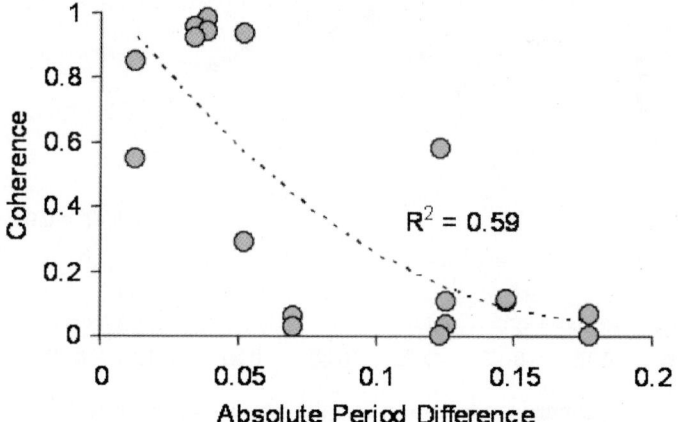

Fig. 7. Cross-spectral coherence as a function of the difference in the natural (self-selected) period of participants' wrist-pendulum movements. The natural period for each participant in a pair was calculated during no visual information control trials. Note that the coherence decreases as the period difference (naturally occurring $\Delta\omega$) increases, suggesting the existence of a period basin of entrainment

required to read letters presented on a sinusoidally oscillating visual stimulus while swinging a pendulum at a comfort mode tempo. This task was the same as the visual tracking condition of Schmidt et al. [75] discussed above in which we found unintentional entrainment in accord with the HKB dynamic. By initially determining the participant's comfort mode tempo and then manipulating the period of the oscillating stimulus in approximately 25 ms increments over a range 200 ms above and below the comfort mode tempo, we demonstrated that rhythmic limb movements become unintentionally entrained to the environmental rhythm when the period of the environmental rhythm was within ±150 ms of the individual's natural period. This range, estimated using entrainment to a non-person rhythm, is similar to that suggested by the interpersonal data in Fig. 7.

A variable that may be related to the size of this basin is the amount of period variability exhibited by an individual. Because, as seen in Fig. 8, the magnitude of coherence between participants' movements and the an oscillating visual stimulus mirrors the mean distribution of participants' cycle-to-cycle periods, we are currently investigating whether participant pairs with more overlap of their distributions of period (due to a sufficiently sized SD of period) will tend to have greater basins of entrainment and thus are more likely to become unintentionally entrained. An additional question concerns

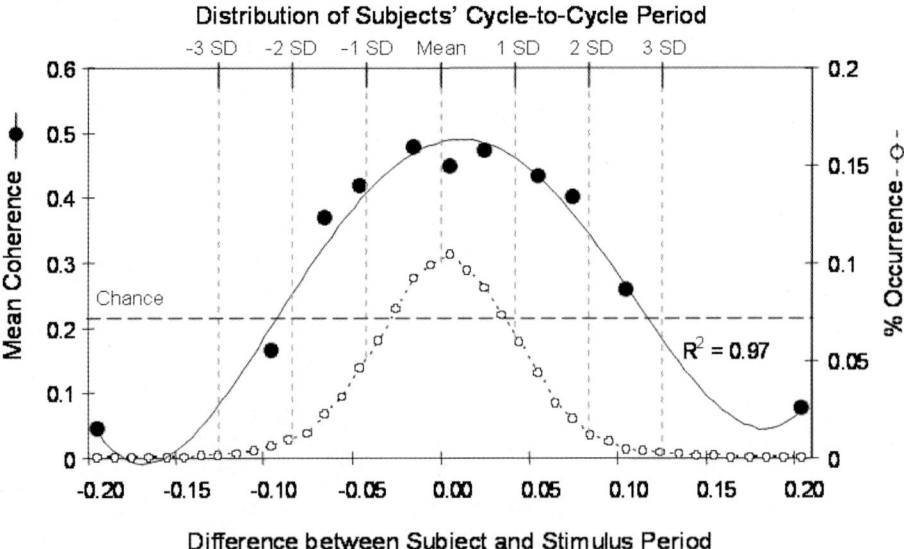

Fig. 8. The cross-spectral coherence between a participant's wrist movements and an oscillating visual stimulus as a function of the difference between the participant's natural period of movements and the stimulus period (*solid line, black dots, bottom x-axis, left y-axis*). Average period distribution of participants swinging a wrist pendulum for a 30-s trial (*dashed line, white dots, top x-axis, right y-axis*)

the constructive effect of movement variability or noise [14, 15, 65] on unintentional interpersonal coordination. Such noise-based enhancement reflects a process known as stochastic resonance, which occurs when the flow of information through a nonlinear system is maximized by the presence of sub-threshold noise [14]. Thus, not only might an individual's movement variability be a parametric constraint on visual entrainment, but it is also possible that such noise may operate to increase the stability of entrainment. We are currently investigating this possibility by using the environmental entrainment paradigm and adding different magnitudes of sub-threshold period variability (e.g., 0, 2, and 5% of the stimulus' target period; see [63] for more details on this method) to the trajectory of the oscillating stimulus. Although participants are expected to become entrained to the visual stimulus when the stimulus is equal to the participant's comfort tempo (no period difference) and contains no variability, more unintentional coordination is expected to be observed for the noisy stimulus when the stimulus' period is less than or greater than the participant's comfort tempo. That is, the basin of entrainment for unintentional coordination is expected to be extended for a 'noisy' stimulus compared to a no-noise stimulus.

In addition to the dynamical constraints of naturally occurring differences in inherent tempos and noise, the stability of visual interpersonal coordination also depends on the availability and pickup of movement information. But what movement information needs to be detected for visual coordination to occur? Will any visual movement information do, or is some information more privileged or relevant for visual entrainment?

Previous research on locomotion [55, 93] and rhythmic catching [1, 67] has demonstrated that the pickup of information during rhythmic tasks is not uniform and that the information available at certain discrete points may be more important for stable coordination. Consistent with these findings, Roerdink et al. [66] have demonstrated how individuals more often fix their gaze on the endpoints of an oscillating visual stimulus during a manually tracking task. Similarly, Byblow et al. [10] have found evidence to suggest that the endpoints of rhythmically moving limbs – namely the peak extension and flexion points – are the perceptual pickup points for bimanual coordination. A recent series of experiments conducted in our lab also suggests that the pickup of information at the endpoints of a movement is particularly important for visual coordination. In these experiments [32], participants were instructed to intentionally coordinate with an oscillating visual stimulus that had different phase regions ($0°/180°$, $90°/270°$, $45°/135°$) and phase amounts (80, 120, 160°) of the stimulus' trajectory occluded from view (Fig. 9, top left and bottom). Consistent with information at the endpoints being most important, an analysis of the entrainment variability (SDϕ) revealed that occluding the endpoints (the phase regions at $0°/180°$) resulted in significantly less stable wrist-stimulus coordination compared to when the middle phase

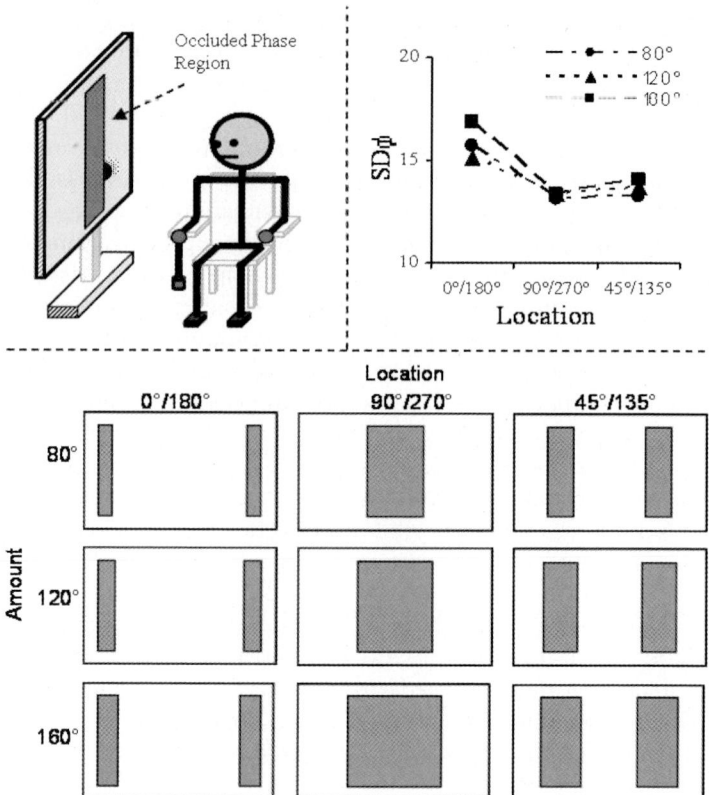

Fig. 9. Experimental setup and stimuli (*left* and *bottom*, respectively) used to study effects of visual occlusion (both location and amount) on intentional visual entrainment. Results (*right*) suggest that the variability of relative phase is only increased by occluding the ends (0°/180°) of the trajectory. Adapted from [32]

regions (90°/270° and 45°/135°) were occluded (Fig. 9, top right). When the middle phase regions were occluded, the stability of the coordination was the same as when none of the stimulus trajectory was occluded. Furthermore, the amount of phase occlusion was found to have no affect on the stability of the visual coordination.

Although the above studies involved only a single individual coordinating with an environmental rhythm (stimulus), the findings can easily be generalized to interpersonal coordination, which is itself a form of environmental coordination (a co-actors movement is a type of environmental rhythm). Identifying the informational and dynamical constraints that influence the stability and emergence of unintentional coordination between people may provide perspective for understanding when and why social and interpersonal interactions succeed or break down.

7 Coordination Dynamics in the Wild II: Social–Psychological Variables

In the interpersonal coordination studies reviewed above, we demonstrated how the stability of dyadic behavioral coordination is influenced by variables of action and perception – influenced by dynamical principles of motor synergies and their constituent variables (e.g., frequency of oscillation, detuning) as well as the kinematic structure of visual information and processes of information pickup and attentional attunement. But given that these action and perception processes occur naturally nested within a social context, one can ask how the stability of coordination dynamics is influenced by variables at the social–psychological scale. Social psychologists have been investigating this question in their studies of "interactional synchrony" in adults [5, 31, 41] as well as in infants [6, 19, 22, 39]. They have found that social variables such as attachment and rapport as well as psychological variables such as learning disabilities, mental illness, and expressivity constrain the degree of interactional synchrony observed in interactions. These investigations have involved naturalistic tasks where the interpersonal coordination is implicit within the interaction. The problem associated with such investigations is the difficulty of measuring the interactors' movements and evaluating their coordination. Consequently, we have performed studies investigating the influence of social variables on interpersonal entrainment using more stereotypic perception and action tasks to circumvent these methodological problems as well as to allow us to study how social variables interact with the dynamics of interpersonal synergies.

Schmidt et al. [72] used intentional antiphase wrist-pendulum coordination to investigate the effects of social competence on interpersonal coordination stability. The coordination task required participants to swing three pairs of pendulums whose length differences comprised three levels of detuning ($\Delta\omega = -0.32$, 0, and 0.32) at two different tempos (slow: 0.65 Hz and fast: 1.5 Hz). Using a shortened version of the Riggio Social Skills Inventory [64], participants were selected to create homogeneous social competence dyads (both participants having high or both participants having low social competence) and heterogeneous dyads (one participant having high and the other having low competence). Socially competent people possess greater skill in social interactions, and are presumably more adept at perceiving the various possibilities for social action, perhaps by being more attuned to the facts of a situation and what types of actions they require [8]. Consequently, an intuitive hypothesis for this investigation was that pairs with higher social competence would have more coordinated interactions.

The social competence variable did affect the coordination dynamics but not as anticipated. The heterogeneous (High–Low) pairs demonstrated significantly fewer breakdowns in coordination than both the homogeneous (High–High and Low–Low) social competence pairs (High–High: 51%, Low–Low: 52%, High–Low: 86% phase-locked trials), indicating a stronger coordination dynamic for the heterogeneous pairs. A marginal interaction between competence and frequency suggested that this result was primarily demonstrated for the faster tempos when coordination is generally less stable [71]. Furthermore, for the trials that were phase-locked, only the High–High pairs did not show the typical phase lag relationship with the detuning variable $\Delta\omega$ (i.e., the shorter pendulum leading in its cycle) that is predicted by (2). This result suggests that these High–High pairs did not use a dynamical strategy for 'solving their coordination problem'. These outcomes can be rationalized by noting that the measure of social competence used correlated with a social control subscale. This correlation suggests that the type of competence characterizing the dyads was that of leadership or dominance. The results indicate that reciprocity (leader–follower) rather than symmetry (leader–leader or follower–follower) of this social competence facilitated the social coordination as well as the appropriating of a dynamical strategy to create the interpersonal coordination. This 'control' hypothesis leads one to expect that the high-competence individuals would be more likely to sustain a stronger maintenance tendency (i.e., a tendency to maintain their own preferred dynamic) than the low-competence individuals. Two results suggest that the high social competence individuals did have a stronger maintenance tendency when paired with a low-competence partner. First, the high-competence participants in these pairs led the low-competence participants in their cycles (by 2°, albeit a non-significant difference), and second, the high-competence participants had lower fluctuations in their periods of oscillation than the low-competence participants ($p < 0.05$). Consequently, the results of this experiment seem to imply that (a) a person's personality characteristics become embodied in their motor movements and, as a result, (b) these personality traits constrain the synchrony of their social interactions.

The above study demonstrates that dispositional properties of individuals can affect the coordination dynamics assembled and these influences can be studied in laboratory experiments. But what about social circumstances? Will the kind of social task a pair of individuals performs affect their coordination dynamics as well? In a recent study performed with social psychologists Lucy Johnston of the University of Canterbury and Kerry Marsh of the University of Connecticut [40], we used the rocking chair unintentional coordination paradigm [61] and the dyadic problem-solving task [62] to study whether the dynamics of entrainment would reflect whether the participants performed a cooperative or a competitive dyadic task. A pair of participants sat in rocking chairs side by side. Cartoon faces were attached to their armrests such that each could only see one cartoon face – that of their partner's. The participants' task was to determine the differences between the two cartoon faces. However, this task was performed in one of two ways to make it either a competitive or a cooperative task: Participants were either jointly rewarded (cooperative condition) or individually rewarded (competitive condition) for the number of differences identified. After the task was performed, all participants were asked to rate how much they liked their partner and how pleasant they found the interaction. Analysis of the results indicated that the cooperative pairs displayed more inphase coordination and marginally greater cross-spectral coherence than the competitive pairs, demonstrating an influence of the social goal on the coordination dynamics. Additionally, regression analyses found that the degrees of liking and perceived pleasantness of the interaction were each correlated with the cross-spectral coherence. These results suggest that the social circumstance in which individuals find themselves will constrain the degree of unintentional entrainment observed and that the stability of this interactional synchrony reflects the experience of psychological connectedness between the participants.

8 Conclusions

What can be said about the processes of interpersonal coordination after 20 years of laboratory research? First, these processes of social coordination need to be understood in terms of the universal logic of stability of natural systems – dynamical processes of self-organization. The work of Kelso and colleagues in the field of coordination dynamics [44] has provided a basis for understanding the dynamics of interpersonal entrainment. Their positing of a dynamical model of interlimb coordination made available a much needed mooring for empirical research investigating the dynamics of behavior. Second, these dynamical processes of interpersonal coordination operate automatically, outside of interactors' awareness and seem to be the basis for unintentional synchrony observed in natural interactions. Third, in order to understand the perception and action processes that sustain dynamical interpersonal synergies, we need to understand not only the dynamical and informational constraints on these processes but also how intrapersonal perception and productions rhythms (e.g., eye tracking and speech) interact with limb movements to create whole-body interpersonal synchronization. Finally, as anticipated by social psychologists many years ago, the psychological properties of a person and the social properties of the environment constrain the dynamics of interpersonal coordination. Indeed, the stability of interpersonally coordinated movements mirrors the stability of mental connectedness experienced in social interactions. Consequently, the dynamical theory of interpersonal coordination appears to be in a position to lay a foundation for a general theory of social perception and action [58].

Acknowledgments

R.C. Schmidt, Department of Psychology, College of the Holy Cross and the Center for the Ecological Study of Perception and Action, University of Connecticut. Michael J. Richardson, Department of Psychology, Colby College and the Center for the Ecological Study of Perception and Action, University of Connecticut. The writing of this paper was supported by a National Science Foundation Grant BCS-0240266 awarded to Richard Schmidt and a National Science Foundation Grant BSC-0240277 awarded to Carol Fowler, Kerry Marsh, and Michael Richardson. The authors wish to thank Kate Curtis, Tracy Espiritu, and Mallory Zeising for help with data collection and analysis.

Correspondence concerning this research can be addressed to Richard C. Schmidt, Department of Psychology, College of the Holy Cross, Box 176A, 1 College St., Worcester, MA 01610 USA, email: rschmidt@holycross.edu.

References

1. Amazeen EL, Amazeen PG, Post AA, Beek PJ (1999) Timing the selection of information during rhythmic catching. J Motor Behav 31:279–289
2. Amazeen PG, Schmidt RC, Turvey MT (1995) Frequency detuning of the phase entrainment dynamics of visually coupled rhythmic movements. Biol Cybern 72:511–518
3. Bang J (2007) Steps towards an ecological approach to thinking. Journal of Anthropological Psychology 18:2–13
4. Baron R (2007) Situating coordination and cooperation between ecological and social psychology. Ecol Psychol 19:179–199
5. Bernieri FJ (1988) Coordinated movement and rapport in teacher-student interactions. J Nonverbal Behav 12:120–138
6. Bernieri FJ, Reznick JS, Rosenthal R (1988) Synchrony, pseudosynchrony, and dissynchrony: measuring the entrainment process in mother-infant interactions. J Pers Soc Psychol 54:243–253
7. Bernieri FJ, Rosenthal R (1991) Interpersonal coordination: behavior matching and interactional synchrony. In: Feldman RS, Rime B (eds) Fundamentals of nonverbal behavior. Studies in emotion & social interaction. Cambridge University Press, New York, pp. 401–432
8. Boudreau LA (1992) Social competence as a determinant of the avoidance of categorization in forming interpersonal impressions. (Doctoral dissertation, University of Connecticut, 1991). Dissertation Abstracts International, 53, 3205B
9. Buck J, Buck E (1976) Synchronous fireflies. Sci Am 234:74–85
10. Byblow WD, Carson RG, Goodman D (1994) Expressions of asymmetries and anchoring in bimanual coordination. Hum Movement Sci 13:3–28
11. Chui K (2005) Temporal patterning of speech and iconic gestures in conversational discourse. J Pragmatics 37:871–887
12. Clark A (1997) Being there: putting brain, body, and world together again. MIT Press, Cambridge, MA
13. Clark HH (1996) Using language. Cambridge University Press, New York
14. Collins JJ (1999) Fishing for function in noise. Nature 402:241–242
15. Collins JJ, Imhoff TT, Grigg P (1997) Noise-mediated enhancements and decrements in human tactile sensation. Phys Rev E 56:923–926

16. Condon WS (1982) Cultural microrhythms. In: Davis M (ed) Interaction rhythms: periodicity in communicative behavior. Human Sciences Press, New York, pp. 53–77

17. Condon WS, Ogston WD (1966) Sound film analysis of normal and pathological behavior patterns. J Nerv Ment Dis 143:338–347

18. Condon WS, Ogston WD (1967) A segmentation of behavior. J Psychiat Res 5:221–235

19. Condon WS, Sander LW (1974) Neonate movement is synchronized with adult speech: interactional participation and language acquisition. Science 183:99–101

20. Copeland J, Moiseff A (2004) Flash precision at the start of synchrony in Photoris frontalis. Integr Comp Biol 44:259–263

21. Csikszentmihalyi M (1990) Flow: the psychology of optimal experience. Harper Collins, New York

22. Feldman R, Eidelman AI (2004) Parent-infant synchrony and the social-emotional development of triplets. Dev Psychol 40:1133–1147

23. Foth E, Graham, D (1983) Influence of loading parallel to the body axis on the walking coordination of an insect. Biol Cybern 48:149–157

24. Franz EA, Zelaznik HN, Swinnen S, Walter C (2001) Spatial conceptual influences on the coordination of bimanual actions: when a dual task becomes a single task. J Motor Behav 33:103–112

25. Furuyama N (2002) Prolegomena of a theory of between-person coordination of speech and gesture. Int J Hum-Comput St 57:347–374

26. Gibson JJ (1962) Observations on active touch. Psychol Rev 69:477–491

27. Gibson JJ (1979) The ecological approach to visual perception. Houghton-Mifflin, Boston

28. Gilmore R (1979) Catastrophe time scales and conventions. Phys Rev A 20:2510–2515

29. Gilmore R (1981) Catastrophe theory for scientists and engineers. Wiley & Sons, New York

30. Goldfield EC, Schmidt RC, Fitzpatrick P (1999) Coordination dynamics of abdomen and chest during infant breathing: a comparison of full-term and preterm infants at 38 weeks postconceptional age. Ecol Psychol 11:209–232

31. Grammer K, Kruck KB, Magnusson MS (1998) The courtship dance: Patterns of nonverbal synchronization in opposite-sex encounters. J Nonverbal Behav 22:3–29

32. Hajnal A, Harrison SJ, Richardson MJ, Schmidt RC (2006) Visual information and environmental coordination. Paper presented at the North American Meeting of the International Society for Ecological Psychology, University of Cincinnati, Cincinnati, OH

33. Haken H (1977/1983) Synergetics: an introduction. Springer-Verlag, Berlin

34. Haken H (1983) Advanced synergetics. Springer-Verlag, Berlin

35. Haken H, Kelso JAS, Bunz H (1985) A theoretical model of phase transitions in human hand movements. Biol Cybern 51:347–356

36. Hanson FE (1978) Comparative studies of firefly pacemakers. Fed Proc 37:2158–2164

37. Henriques DYP, Crawford JD (2002) Role of eye, head, and shoulder geometry in the planning of accurate arm movements. J Neurophysiol 87:1677–1685

38. Hollands MA, Marple-Horvat DE, Henkes S, Rowan AK (1995) Human eye movements during visually guided stepping. J Motor Behav 27:155–163

39. Isabella RA, Belsky J (1991) Interactional synchrony and the origins of infant-mother attachment: a replication study. Child Dev 62:373–384
40. Johnston L, Miles L, Richardson MJ, Schmidt RC, Marsh KL, Yabar Y (2006) Implicit mimicry and synchronization: impact of and on liking. Poster presented at the North American Meeting of the International Society for Ecological Psychology, University of Cincinnati, Cincinnati, OH
41. Julien D, Brault M, Chartrand E, Begin J (2000) Immediacy behaviours and synchrony in satisfied and dissatisfied couples. Can J Beh Sci 32:84–90
42. Kelso JAS (1981) On the oscillatory basis of movement. B Psychonomic Soc 18:63
43. Kelso JAS (1984) Phase transitions and critical behavior in human bimanual coordination. Am J Physiol: Reg, Integr Comp 246:R1000–R1004
44. Kelso JAS (1995) Dynamic patterns. MIT Press, Cambridge, MA
45. Kelso JAS, Delcolle JD, Schöner G (1990) Action-perception as a pattern formation process. In: Jeannerod M (ed) Attention and performance XIII, Vol. 5. Erlbaum, Hillsdale, NJ, pp. 139–169
46. Kelso JAS, Ding M (1994) Fluctuations, intermittency, and controllable chaos in biological coordination. In: Newell KM, Corcos DM (eds) Variability in motor control. Human Kinetics, Champaign, IL, pp. 291–316
47. Kelso JAS, Holt KG, Kugler PN, Turvey MT (1980) On the concept of coordinative structures as dissipative structures: II. Empirical lines of convergence. In: Stelmach GE, Requin J (eds) Tutorials in motor behavior. North-Holland, Amsterdam, pp. 49–70
48. Kelso JAS, Holt KG, Rubin P, Kugler PN (1981) Patterns of human interlimb coordination emerge from the properties of nonlinear limit cycle processes: Theory and data. J Motor Behav 13:226–261
49. Kelso JAS, Scholz JP, Schöner G (1986) Nonequilibrium phase transitions in coordinated biological motion: Critical fluctuations. Phys Lett 118:279–284
50. Kendon A (1970) Movement coordination in social interaction: some examples. Acta Psychol 32:1–25
51. Kugler PN, Kelso JAS, Turvey MT (1980) On the concept of coordinative structures as dissipative structures: I. Theoretical lines of convergence. In: Stelmach GE, Requin J (eds) Tutorials in motor behavior. North-Holland, Amsterdam, pp. 3–47
52. Kugler PN, Kelso JAS, Turvey MT (1982) On the control and coordination of naturally developing systems. In: Kelso JAS, Clark JE (eds) The development of movement control and coordination. Wiley, New York, pp. 5–78
53. Kugler PN, Turvey MT (1987) Information, natural law and the self-assembly of rhythmic movement. Erlbaum, Hillsdale, NJ
54. Large EW, Jones MR (1999) The dynamics of attending: how people track time-varying events. Psychol Rev 106: 119–159
55. Laurent M, Thomson JA (1988) The role of visual information in control of a constrained locomotor task. J Motor Behav 20:17–37
56. Lockman JJ, Hazen NL (1989) Action in a social context. Plenum, New York
57. Lopresti-Goodman SM, Silva PL, Richardson MJ, Schmidt RC (2006) Frequency basin of entrainment for unintentional coordination. Paper presented at the North American Meeting of the International Society for Ecological Psychology, University of Cincinnati, Cincinnati, OH
58. Marsh KL, Richardson MJ, Baron RM, Schmidt RC (2006) Contrasting approaches to perceiving and acting with others. Ecol Psychol 18:1–37

59. Patterson ML, Webb A, Schwartz W (2002) Passing encounters: Patterns of recognition and avoidance in pedestrians. Basic Appl Soc Psych 24:57–66

60. Richardson MJ, Curtis K, Zeising M, Schmidt RC (2006) Visual tracking and unintentional interpersonal coordination. Poster presented at the North American Meeting of the International Society for Ecological Psychology, University of Cincinnati, Cincinnati, OH

61. Richardson MJ, Marsh KL, Isenhower R, Goodman J, Schmidt RC (2006) Rocking together: Dynamics of intentional and unintentional interpersonal coordination. Manuscript submitted for publication

62. Richardson MJ, Marsh KL, Schmidt RC (2005) Effects of visual and verbal information on unintentional interpersonal coordination. J Exp Psychol: Human Percept Perform 31:62–79

63. Richardson MJ, Schmidt RC, Kay BA (2007) Distinguishing the noise and attractor strength of coordinated limb movements using recurrence analysis. Biol Cybern 96:59–78

64. Riggio R (1986) Assessment of basic social skills. J Pers Soc Psychol 51:649–660

65. Riley MA, Turvey MT (2002) Variability and determinism in motor behavior. J Motor Behav 34:99–125

66. Roerdink J, Peper CM, Beek PJ (2005) Effects of correct and transformed visual feedback on rhythmic visuo-motor tracking: tracking performance and visual search behavior. Hum Movement Sci 24:379–402

67. Santvoord AAM, Beek PJ (1994) Phasing and the pickup of optical information in cascade juggling. Ecol Psychol 6:239–263

68. Schmidt RC (1988) Dynamical constraints on the coordination of rhythmic limb movements between two people. Unpublished Ph. D. Thesis, University of Connecticut, Storrs, CT.

69. Schmidt, RC (2007). Scaffolds for social meaning. Ecol Psychol 19:137–151

70. Schmidt RC, Bienvenu M, Fitzpatrick PA, Amazeen PG (1998) A comparison of intra- and interpersonal interlimb coordination: coordination breakdowns and coupling strength. J Exp Psychol: Human Percept Perform 24:884–900

71. Schmidt RC, Carello C, Turvey MT (1990) Phase transitions and critical fluctuations in the visual coordination of rhythmic movements between people. J Exp Psychol: Human Percept Perform 16:227–247

72. Schmidt RC, Christianson N, Carello C, Baron R (1994) Effects of social and physical variables on between-person visual coordination. Ecol Psychol 6:159–183

73. Schmidt RC, O'Brien B (1997) Evaluating the dynamics of unintended interpersonal coordination. Ecol Psychol 9:189–206

74. Schmidt RC, O'Brien B (1998) Modeling interpersonal coordination dynamics: Implications for a dynamical theory of developing systems. In: Molenaar PC, Newell K (eds) Dynamics Systems and Development: beyond the metaphor. Erlbaum, Hillsdale, NJ, pp. 221–240

75. Schmidt RC, Richardson MJ, Arsenault CA, Galantucci B (2007) Visual tracking and entrainment to an environmental rhythm. J Exp Psychol: Human Percept Perform 33:860–870

76. Schmidt RC, Richardson MJ, Curtis K (2005) Effects of perceptual and production rhythms on unintentional synchronization. Poster presented at Progress in Motor Control V: A Multidisciplinary Perspective, Pennsylvania State University, State College, PA, August 2005

77. Schmidt RC, Shaw BK, Turvey MT (1993) Coupling dynamics in interlimb coordination. J Exp Psychol: Human Percept Perform 19(2):397–415
78. Schmidt RC, Turvey MT. (1992) Long-term consistencies in assembling coordinated rhythmic movements. Hum Movement Sci 11:349–376
79. Schmidt RC, Turvey MT (1994) Phase-entrainment dynamics of visually coupled rhythmic movements. Biol Cybern 70:369–376
80. Schmidt RC, Turvey MT (1995) Models of interlimb coordination: equilibria, local analyses, and spectral patterning. J Exp Psychol: Human Percept Perform 21:432–443
81. Schöner G, Haken H, Kelso JAS (1986) A stochastic theory of phase transitions in human hand movement. Biol Cybern 53:247–257
82. Shockley KD, Baker AA, Richardson MJ, Fowler CA (2007) Verbal constraints on interpersonal postural coordination. J Exp Psychol: Human Percept Perform 33:201–208
83. Shockley K, Santana MV, Fowler CA (2003) Mutual interpersonal postural constraints are involved in cooperative conversation. J Exp Psychol: Human Percept Perform 29:326–332
84. Stein PSG (1974) Neural control of interappendage phase during locomotion. Am Zool 14:1003–1016
85. Sternad D, Turvey MT, Schmidt RC (1992) Average phase difference theory and 1:1 phase entrainment in interlimb coordination. Biol Cybern 67:223–231
86. Strogatz SH (1994) Nonlinear dynamic and chaos. Perseus, Cambridge, MA.
87. Strogatz SH (2003) Sync: The emerging science of spontaneous order. Hyperion Press, New York
88. Tickle-Degnen L (2006) Nonverbal behavior and its functions in the ecosystem of rapport. In: Manusov V, Patterson M (eds) The SAGE Handbook of Nonverbal Communication. Sage, Thousand Oaks, CA
89. Turvey MT, Rosenblum LD, Schmidt RC, Kugler PN (1986) Fluctuations and phase symmetry in coordinated rhythmic movement. J Exp Psychol: Human Percept Perform 12:564–583
90. van Donkelaar P (1997) Eye-hand interactions during goal-directed pointing movements. Neuroreport 8:2139–2142
91. von Holst E (1973) The collected papers of Eric von Holst. In: Martin R, Trans. The behavioral physiology of animal and man, Vol. 1. University of Miami Press, Coral Gables, FL (Original work published 1939)
92. Warren WH (1984) Perceiving affordances: visual guidance of stair climbing. J Exp Psychol: Human Percept Perform 10:683–703
93. Warren WH, Yaffe DM (1989) Dynamics of step length adjustment during running: a comment on Patla, Robinson, Samways, Armstrong. J Exp Psychol: Human Percept Perform 15:618–623
94. Wilson M (2002) Six views of embodied cognition. Psychon Bull Rev 9:625–636
95. Yarbus AL (1967) Eye movements and vision. Plenum Press, New York
96. Zanone PG, Kelso JAS (1990) Relative timing from the perspective of dynamic pattern theory: stability and instability. In J. Fagard, P. H. Wolff (eds) The development of timing control and temporal organization in coordinated action. Elsevier, Amsterdam, pp. 69–92

EEG Coordination Dynamics: Neuromarkers of Social Coordination

Emmanuelle Tognoli

Center for Complex Systems and Brain Sciences, Florida Atlantic University, Boca Raton, FL 33431, USA

Abstract. The aim of this chapter is to present a framework of EEG coordination dynamics based on rhythmical entrainment of brain activity. The interest of continuous brain~behavior[1] analysis is underlined and its principles are illustrated with results from a dual-EEG experiment of spontaneous social coordination.

1 Introduction

At the source of coordination dynamics is the will to identify organizing principles in (living) systems transcending both the specifics of their substratum and their levels of description. An example of such research in the wake of Kelso is movement coordination, which has been studied theoretically [1, 17, 24, 25, 29, 34, 35, 41] and empirically, at the behavioral [8, 11, 13, 20, 33, 36, 37, 38, 56, 65] and brain levels [26, 32, 40, 43, 48, 49, 52, 61], not only in single subjects but also in paired individuals [22, 53, 60]. In the present chapter, we present ongoing research on interpersonal coordination coupled by visual information. Brain dynamics was recorded through a specially designed dual-EEG system while vis-à-vis pairs of subject performed and simultaneously observed self-paced finger oscillations. A variety of behaviors emerged from this situation, ranking from independent, to transiently locked, to stable inphase and antiphase synchronous. Our aim was to combine coordination dynamics and EEG to uncover the neural signatures of individual and social behaviors.

In a first movement, we will describe limitations of classical EEG approaches to meet some of the requirements of coordination dynamics, as well as efforts taken to overcome them. In a second movement, we will detail interpersonal coordination dynamics at the behavioral level and isolate homogenous classes of coordination. In a third movement, we will present a description

---

[1] The tilde symbol (~) was introduced by Kelso and Engstrøm to indicate complementary pairs, i.e., "modes of a dynamical system that is capable of moving between boundaries even as it includes them" [42, p. 9].

of brain dynamics in which the neuronal components are delimited but neuronal coordination is not yet apprehended. Ultimately, we will chart strategies to capture interactions between neuronal components across brains, so as to reveal mechanisms through which the social level could emerge.

2 EEG Coordination Dynamics

On one side, we have EEG, a neuroimaging method known for its temporal resolution and commonly used to relate brain and behavior. On the other side, we have coordination dynamics, which seeks to reconcile levels of organization in biological systems and study the temporal evolution of coordination variables derived from the dependency of elements within and/or between the behavior and brain levels. The concord sounds immediate, however, convergence of both fields into an integrated framework is not as straightforward as one would think.

A classical approach to EEG analysis consists of extracting recurring dynamics through averaging of time-locked brain activity in paradigms of repeated events. From the viewpoint of coordination dynamics, qualitative changes in the system's time course are of major interest to understand its dynamical skeleton and associated neural correlates. A first problem is that such impromptus behaviors are difficult to grasp with this classical EEG approach. Their recurrence and temporal definition may not be sufficient to lead to adequate event-related dynamics. Even when such conditions are met, a second problem comes into play: averaged signals are not always faithful representations of continuous brain dynamics. Key to those two problems, the recursive dynamics derived from classical EEG analysis, does not have an inscription in the continuous dimension in which the behavior lives (see Sect. 2.2). Taken together, those limitations lead us to an alternative strategy, i.e., not to use EEG averaging but to develop ways to study both brain and behavior dynamics continuously. A tractable issue with EEG signal-to-noise ratio is to maximize the signal by resorting to rhythmic entrainment of brain dynamics.

2.1 Impromptus of Behaviors

Because of the complexity of the brain and of the continual spatio-temporal overlap of many simultaneous ongoing processes, EEG studies of cognitive or behavioral phenomena usually consists of isolating a small fraction of task-related activity amidst a substantial amount of "background" neural noise. In addition, the signal gathers an even larger amount of instrumental and environmental recording noise. The long-established methodology of EEG analysis to deal with this unfavorable signal-to-noise ratio (SNR) is to average a large number of realizations of the same event so as to cancel-out task-unrelated oscillatory components. This is done both in the time [63, 19, 54, 14] and frequency domains [51, 3, 57, 44], and requires events to be frequent enough (typically in the hundreds) and well defined in their temporal onset (no more than a few milliseconds in jitter or uncertainty).

In parallel, the modus operandi of coordination dynamics is to stimulate qualitative changes in the collective variable. Those transitions clue the investigator toward the identification of the attractors' structure underlying the system. In practice, the collective variable is monitored continuously and its trajectory spends most of its time in one state or another, only briefly undergoing transitions. Such transition events may be too rare and they may not have sufficiently precise temporal onset to be integrated in the classical EEG framework. Reciprocally, if attempting to read the EEG in the continuous dimension of the behavior, one is turned back to the critical issue of noise.

We conclude that EEG investigations of coordination dynamics tract with two distinct time scales: (1) a continuous time scale in which the behavior's collective variable evolves and changes under the constraints of some control parameters, and (2) a recursive time scale in which information-rich variability scattered within the discontinuous windows of brain activity is overwhelmed. Because of the incompatibility between those two time scales, EEG is not directly geared toward the understanding of real-time changes in brain activity that accompany real-time changes in behavior.

2.2 Recursive Time and the Methodology of EEG

One could, however, hope that selective aspects of brain dynamics would be preserved by the process of averaging. Brain activity linked to a precise event (stimulus or response) and consistent across trials would emerge and faithfully reproduce at the recursive time scale what happens over and over again at each discrete event along the continuous time scale of behavioral coordination.

The model of EEG underlying this simple view was once dominant. Its rationale was to extract event-related components (consistent positive or negative deflections phase-locked to the event reflecting cortical activation) which were obscured under a layer of additive noise. Current theories of EEG signal suggest a more complex picture involving at least both event-related activations and changes of underlying populations' phases [2, 46, 64, 16, 27], maybe additional mechanisms. In such a model, potentially large discrepancies exist between the task-related signal present in real-time and the reconstructed dynamics estimated from averaged data (Fig. 1).

The properties of the reconstructed dynamic differ from those of the continuous dynamic. For instance in the spatial domain, maxima in topographical maps can appear at the intersection of two distinct regions which alternate over repetitions of the event (e.g., Cz maximum cumulating alternating temporal left and right temporal maxima). Another example in the temporal domain is the existence of amplitude difference in event-related dynamics which may not express any modulation of the EEG amplitude at the continuous time scale, but rather arise from a more pronounced phase-locking to the

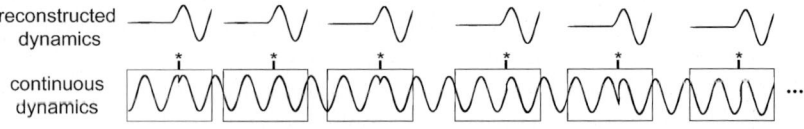

Fig. 1. Two levels of representation of brain dynamics. An event appears recurrently (stars atop second row) during the time course of the behavior and elicits phase resetting of a brain rhythm (lower row), which is hidden by the presence of noise. To identify cortical activity evoked by the event, continuous dynamics of EEG (lower row) is substituted with its event-related dynamics (upper row), the latter consisting of averaged epochs from the former in the time (as here) or frequency domain

event. Those distinct properties limit our ability to infer the continuous brain dynamics from its reconstructed dynamics.

2.3 Maximizing the Signal to Analyze Continuous EEG: Recruitment and Entrainment

In the previous paragraph, we considered three kinds of EEG activities: those which continuous dynamics may be acceptably represented by the event-related dynamics, those whose may be distorted and those which can simply not be handled through averaging. For the former, a possible strategy is to seed duplicates of the event-related dynamics at all appropriate moments and compare continuously the behavior with the reconstructed EEG. For the latter two, grounds for this strategy are lacking and one has to choose an alternative strategy of analyzing the EEG in continuous despite the presence of noise.

This alternative strategy requires to maximize the neural signal. A robust and lasting neural response is indispensable and cannot be obtained with sustained sensory or motor event as they cause the brain to decrease its response through habituation [59, 45]. However, at the interface between spontaneous brain rhythms and discrete evoked responses, some robust EEG activity has been shown to emerge [28]: a rhythmical sensory or motor event will establish persistent brain activity. This is a requirement which is easily met by paradigms of coordination dynamics: while not requiring paradigms of rhythmic behavior [35], the framework of coordination dynamics has a historical predilection for oscillatory systems. With the EEG being extremely inconsistent from moment to moment (e.g., Bullock et al. [9]), the rhythmic strategy attempts to force brain dynamics so that it becomes more stationary due to the periodic event (stimulus or behavior).

We distinguish two types of brain responses elicited by rhythmic events. The first is known as steady state, frequency following response or flicker phenomenon [62, 47, 28] and consists of oscillatory brain activity that resonates at the frequency of the rhythmical event. The advantage of this entrained response is straightforward: at some critical frequencies of stimulation, the amplitude of the brain's oscillatory response to periodic stimuli reaches a maximum [28, 30], i.e., the SNR is increased and task-related activity may rise above the level of background noise. A second type of brain response is the recruitment of neural activities at frequencies other than the events'. As soon as the evoked responses elicited by successive events overlap, continuous task-related neural activity occurs. Robustness is therefore achieved by persistence of a pattern of activity in the temporal domain. In addition, interactions between the events' steady state and brain rhythms at other frequencies are expected, however, such questions are a matter of very recent interest [55, 6, 10, 18], very rarely conducted on periodically forced oscillations (but see [15, 58]). We suggest that due to nonlinear interaction between neural populations, rhythmic events may stabilize robust brain responses in a range of frequencies, establishing the grounds for EEG coordination dynamics based on rhythmic paradigms.

3 Varieties of Behaviors in Spontaneous Social Coordination

In a study of social coordination, we put the preceding principles at work at the behavioral and brain level. Subjects ($n = 16$, allotted to eight pairs) performed repetitive finger movements at their preferred frequency and amplitude and were intermittently exposed to the view of each other in a session consisting of 36 trials. Visual coupling was controlled by a liquid crystal window alternating between opacity and transparency during three periods of 20 s. Participants' finger movements were recorded using light single-axis goniometers affixed to their right index. The task had been developed to provide minimal inducement to the subjects' behavior, and typically resulted in a variety of patterns which seemed to carry various degrees of social valence. The first objective of this research was to classify behavioral patterns obtained during periods of social interaction. Patterns were examined by following continuously the dynamics of the behaviors' coordination variable: the relative phase between the two finger movements.

In a general sense, coordination between two oscillatory components arises as a result of their coupling, i.e., each element leaks its functional state to the other element in a way that influences the other's future state. This coupling can be a physical connection (as is the case in the interactions between neurons) but in the realm of human behavior it is also often supported by an informational connection mediated by sensory systems. In the paradigm of social coordination developed by Kelso and collaborators [53], the primary system connecting the behavior of each subject is visual perception.

Episodes of phase-locking were observed in all pairs of subjects. They favored inphase (both subjects flex their finger in synchrony) and antiphase coordination (one subject flexes his finger while the other extends), although a few pairs sustained more variable locking with a phase-lag which was intermediary between inphase and antiphase. This type of collective behavior is reminiscent of coordination within subjects observed in many studies (see [39] for review).

314   E. Tognoli

Cases with minimal social valence were observed when the relative phase was not deflected by the establishment of the coupling (the slope at which the relative phase drifted remained identical at the onset of the visual contact, see Fig. 2A). This case is described as "social neglect" as it probably requires both subjects to withdraw their attention from their partner's behavior. In contrast, social behaviors were defined when the collective variable was affected at the establishment of visual contact. The consequence was either to repel or to promote coordination. A pattern of behavior which suggested the presence of a coupling despite the failure of synchronization was the drift apart of each oscillator's frequency at the onset of visual contact (Fig. 2B). For 1:1 coordination modes, such behaviors minimize the likeliness of phase-locking and are probably best regarded as instances of social segregation. However, the very existence of this change at the onset of the visual contact testified that social situation affected the subjects' behavior. Its contrary was social integration, encountered when the frequencies of movements produced by each subjects came closer (71% of the trials). Sometimes, this frequency rapprochement did not allow enough proximity to enable phase-locking (13% of the trials). Other times, successful social integration was met, by order of increasing strength: (1) when the oscillations showed occasional phase-locking (33% of the trials; see example Fig. 2C), (2) when the oscillations exhibited

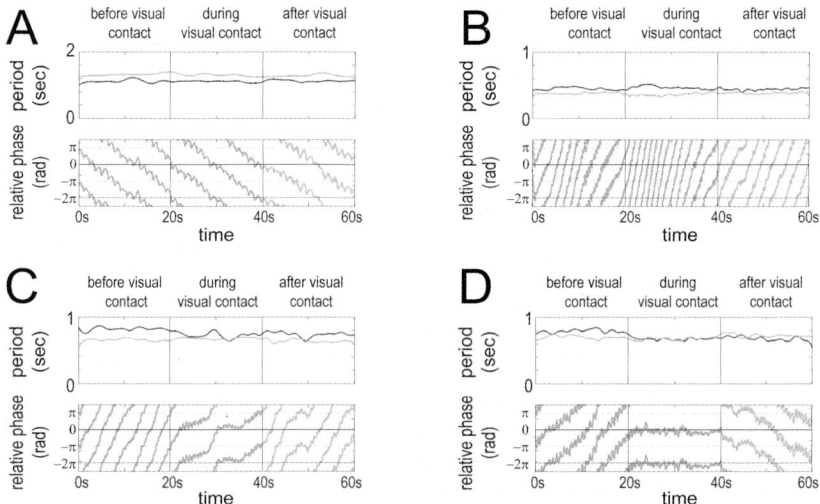

Fig. 2. Representative behaviors of social neglect (A), frequency estrangement (B), transient phase-locking (C) and sustained phase-locking (D). In each case, the upper plot shows the time course of the instantaneous period of movement for each subject and the lower plot shows the relative phase between the two movements. The period of visual contact during which coordination is possible extends from 20 to 40 s in 1-min trials. Social coordination is observed when the frequencies become identical and the relative phase settles

the quick establishment of phase-locking which was sustained until the cessation of visual contact (25% of the trials; see example Fig. 2D), and finally (3), when this phase-locking was sustained so much as to enable its persistence after the visual contact was removed.

4 Personal Brain Dynamics of Interpersonal Behavioral Coordination

Following the classification of behavioral patterns, we attempted to identify neuromarkers of coordinated and uncoordinated social behaviors. We analyzed the subjects' EEG in the spectral domain in the range 7.5–13 Hz. The identification of neuromarkers was addressed in two steps: first, stationary trials were used to detect rhythms induced by different behaviors of social integration and segregation. Second, the identified rhythms were inspected to identify direct brain–behavior relationships, especially during transition from coordination to independent behavior.

Simultaneous acquisition of both subjects' brain activity was achieved with a dual-EEG system, consisting of two 64-electrode montages from the 10 percent system [12], at the exclusion of rows 9 and 10. Reference electrodes were placed over the linked left and right mastoids of each subject and ground electrodes were located on their forehead at location FPz. Impedances were maintained below 10 kΩ. Data acquisition was performed using a Synamp 2 amplifier (Neuroscan, Texas, El Paso). Signals were analog filtered (Butterworth, bandpass from 0.05 Hz–12 dB/octave- to 200 Hz–24 dB/octave-), amplified (gain = 2010) and digitized at 1000 Hz with a 24-bit vertical resolution in the range 950 μV (sensitivity of 0.11nV).

Single-trial EEG spectra were estimated in each condition (before, during and after visual contact) within homogenous classes of coordination behavior (complete phase-locking and complete absence of phase-locking). Transients were discarded at the onset and offset of each 20-s period (3 and 0.6 s, respectively). A tukey window (10%) was applied to the remaining data and EEG spectral amplitudes were derived from a discrete Fourier transform. These spectra best represent brain activity which is sustained over the entire window, under the assumption that underlying brain dynamics is stationary, which is met when the behavior itself is stationary.

Due to the large size of the window over which the spectral analysis was performed (16.4 s, or 16 384 points at a sampling rate of 1000 Hz), the resulting spectral resolution (0.06 Hz) was orders of magnitudes higher than classical behavioral and cognitive EEG studies (usually 1 Hz). It allowed precise differentiation of the EEG rhythms in the alpha range. We applied a colorimetric model to the electrode space to visualize the topographical patterning of the rhythms [60] and identified three rhythms in the range 7.5–13 Hz, located respectively above rolandic, occipital and right centroparietal regions.

When visual coupling was established, two prominent changes in classical EEG rhythms occurred: occipital α and rolandic μ rhythms were depressed as compared to baseline (μ isolated in eight subjects: -20.1%; α isolated in nine subjects: -30.44%). However, changes in those two rhythms were not differentially affected by the pattern of collective behavior. We identify a third rhythm (which we called ϕ) composed of two adjacent spectral components (ϕ_1 and ϕ_2, see Fig. 3A) both appearing above right centroparietal electrodes (Fig. 3B) at frequencies of 10–12 Hz. Their boundaries were identified by plotting spectral difference between the left and right hemispheres. This technique emphasizes ϕ by canceling out large symmetrical components, among which are μ and α (see Fig. 3A).

Both ϕ components were characterized by power increase during visual contact but they were functionally distinct and their reactivity was mutually exclusive. Increase in ϕ_1 always occurred during uncoordinated trials (identified in four subjects: $+12.1\%$; see Fig. 4A). Increase in ϕ_2 always occurred during coordinated trials (identified in five subjects: $+9.6\%$) and never occurred during uncoordinated trials (Fig. 4B). We concluded that ϕ_1 promoted intrinsic behavior and ϕ_2 promoted social behavior.

While the high-resolution spectral analysis provides great details on the spectral band and topography of each rhythm in individual subjects, it is blind to instantaneous changes in brain activity that may accompany behavioral events (see Sect. 2.1). In a second phase, we analyzed the continuous brain dynamics using time–frequency–power plots derived from a continuous wavelet transform (Morlet wavelet). Focusing on the spectral band of ϕ, we attempted to identify its dynamics during trials of sustained or transient coordination. The instantaneous dynamics of ϕ was directly observable in a few subjects,

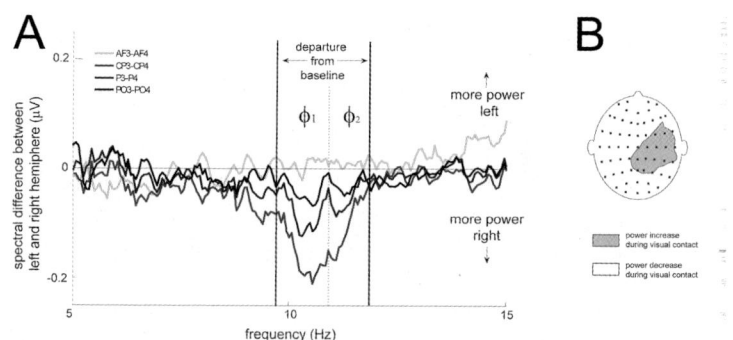

Fig. 3. Spectral (A) and spatial (B) identification of ϕ complex in a representative subject. (A) The two adjacent components of the phi complex are best seen by plotting the spectral difference between pairs of electrodes in the left and right hemispheres, as the symmetrical rhythms (μ and α) cancel out. (B) Representative topographical map in the ϕ band in a subject. The line shows isopotential (same power before and during visual contact), the area shaded in gray shows selective increase in power attributable to ϕ_1. ϕ_2 has a similar topography

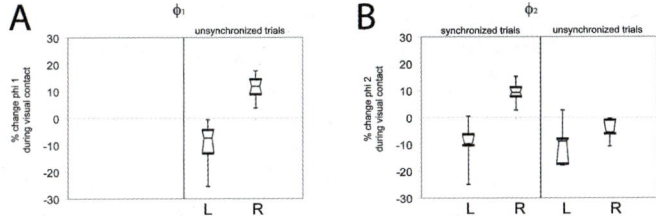

Fig. 4. (A) Box-and-whisker plot of ϕ_1 shows its selective increase in the right hemisphere during unsynchronized trials (ϕ_1 changes during synchronized behaviors not depicted: when ϕ_1 was active in a subject, his/her pair did not produce synchronized behavior). (B) Box-and-whisker plot of ϕ_2 shows its selective increase in the right hemisphere during synchronized trials. (L: left hemisphere – R: right hemisphere)

those whose μ and α bore little spectral power (μ and α decrease during visual contact being potentially compensated by the rise of ϕ in subjects presenting notable spectral density in all three rhythms). During stationary trials of coordination, we observed repeated occurrence of very brief, high-amplitude bursts in ϕ_2 band (number of bursts typically in the tens during the 20-s periods, each lasting only two or three cycles, with an amplitude many fold above the background power in that band). We also identified the disappearance of those bursts during brief periods of loss of coordinative stability between the subjects' movements (Fig. 5). Such observations may

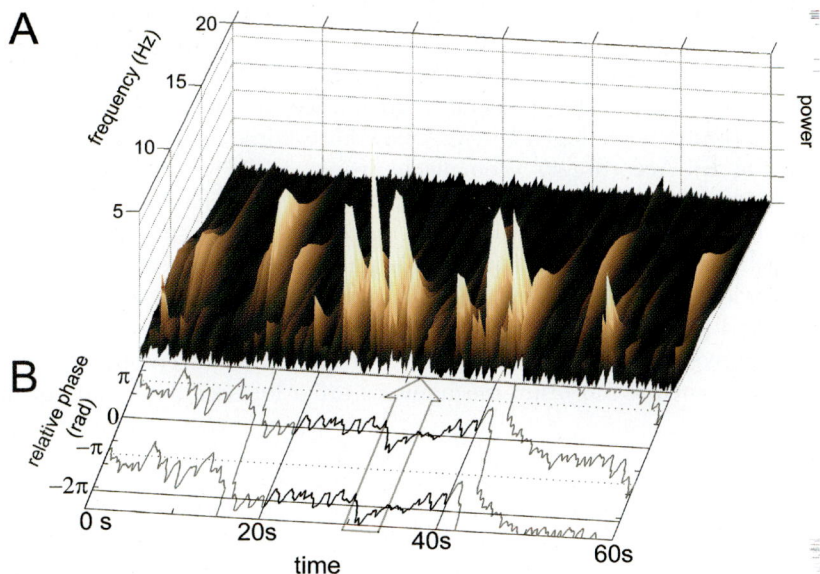

Fig. 5. (A) Time–frequency–power plot showing recurrent bursts of ϕ_2 during visual contact (from 20 to 40 s), with a gap at 32 s corresponding to a brief lapse of coordinative stability, as seen on the behavioral relative phase (gray arrow in B)

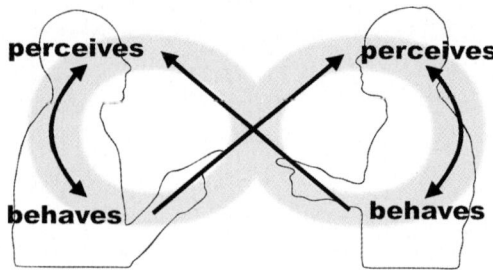

Fig. 6. Action∼perception coupling in pairs of individuals. Between-subjects' action∼perception is an informational coupling whereas within-brain perception∼action is supported by neural connectivity

allow for studying instantaneous brain–behaviors interactions, a strategy that could be furthered with an adequate decomposition of the spectrum into its respective rhythms so as to generalize the study to all subjects exhibiting ϕ rhythm irrespective of their α and μ's spectral power.

5 Brain Dynamics for Coordinated and Uncoordinated Behaviors

While belonging to the field of social neuroscience, the previous results characterized individual brain dynamics. But the coordination between brains remains to be characterized. In situations when humans interact, action∼ perception coupling occurs. It implies dependencies between brain rhythms, which is the next frontier of social neuroscience. In social systems composed of several brains, such dependencies follow hard-wired as well as informational paths (Fig. 6). Accordingly, their mechanisms are likely to be found amongst classical large-scale linear coupling [7] as well as other modes of coordination comprising amplitude/amplitude and amplitude/phase spectral correlation, non-linear coupling between functionally analogous areas of distinct structural and functional characteristics (e.g., the same functional system in different brains) and non-linear coupling between functionally distinct areas (different functional systems in different brains). Conceptual and empirical efforts to characterize such social brain mechanisms are a foreseeable domain of future research.

6 Conclusion

In the first decades of the EEG era, electroencephalographers were spending a lot of time scrutinizing their EEG polygraphs and deriving some instantaneous information about brain activity. A very direct approach of brain∼behavior causality was then available. With the venue of the digital era, this too-qualitative approach went to be replaced with quantitative methods and continuous brain activity got substituted with reconstructed dynamics, especially in the domain of cognitive neuroscience. There, understanding of the brain's instantaneous reactivity to sparse behavioral transitions has been merely lost. We showed that using robust brain responses arising from long-known properties of oscillatory entrainment, it was possible to restore a framework of continuous EEG analysis in the prospect of uncovering brain~behavior co-implications.

Applying this framework of EEG coordination dynamics to the study of interpersonal coordination, we showed the existence of neuromarkers in the 10 Hz range and presented specimens of instantaneous brain~behavior co-occurrence. ϕ_1 was recruited when subjects maintained a behavior which was distinct from their partners'. ϕ_2 was recruited when subjects adopted coordinated behavior. Both social segregation and social integration are probably adaptive facets of human behaviors. During performance of certain specific tasks, it may be advantageous for subjects to adjust their behaviors. For instance, if a pair of subjects is walking while carrying a load, they may need to adopt the same frequency and to coordinate their pace's "phases". There are also tasks during which social segregation, and not integration, may be adaptive.

Social integration is further engaged into higher level social behaviors: overt adjustments has been suggested as a mechanism for social facilitation and empathy [5, 50]; covert adjustments (stimulation of the motor system by perceived action without associated production of a behavior) for action understanding by direct matching of a conspecific's motor behavior into one's motor system [4, 23, 21] and for skill learning [31]. The implication of ϕ in this range of social processes is an open question.

While the identification of the ϕ complex may help to understand a large range of social behavior in a single brain, there is still one step up for the field of Social Neuroscience to which Coordination Dynamics may contribute. Behavioral contexts exert pressure on individual brains to coordinate, realizing some action~perception coupling at the level of "societies of brains". Models of coordination could be extended to the interpersonal level to describe reciprocal effects of brains on each other when subjects are constrained by a common task.

Acknowledgments

My acknowledgments to the one in the light of this book and in the shadow of this chapter. The experimental work was performed in collaboration with J. Lagarde, G.C. de Guzman and J.A.S. Kelso. The suggestions and comments of Armin Fuchs are gratefully acknowledged.

References

1. Assisi CG, Jirsa VK, Kelso JAS (2005) Synchrony and clustering in heterogeneous networks with global coupling and parameter dispersion. Phys Rev Lett 94:018106
2. Başar E (1980) EEG brain dynamics – Relation Between EEG and Brain Evoked Potentials. Elsevier, Amsterdam
3. Başar E, Demiralp T, Schürmann M, Başar-Eroglu C, Ademoglu A. (1999) Oscillatory brain dynamics, wavelet analysis and cognition. Brain Lang 66:146–183
4. Bekkering H, Wohlschläger A (2002) Action perception and imitation: A tutorial. In: Prinz W, Hommel B (eds) Common mechanisms in perception and action (Attention and Performance, 19). Oxford University Press, Oxford, pp. 294–314
5. Billard A (2002) Imitation. In: Arbib MA (ed) Handbook of brain theory and neural networks. MIT Press, Cambridge MA, pp. 566–569
6. Breakspear M, Terry J (2002) Detection and description of nonlinear interdependence in normal multichannel human EEG. Clin Neurophysiol 113: 735–753
7. Bressler SL (1995) Large-scale cortical networks and cognition. Brain Res Rev 20:288–304
8. Buchanan JJ, Kelso, JAS (1999) To switch or not to switch: Recruitment of degrees of freedom stabilizes biological coordination. J Mot Behav 31:126–144
9. Bullock TH, Enright JT, Chong KM (1998) Forays with the additive periodogram applied to the EEG. In: Proceedings of the fifth joint symposium on neural computation, University of California, San Diego 8, pp. 25–28
10. Canolty RT, Edwards E, Dalal SS, Soltani M, Nagarajan MM, Kirsch HE, Berger MS, Barbaro NM, Knight RT (2006) High gamma power is phase-locked to theta oscillations in human neocortex. Science 313(5793):1626–1628
11. Carson RG, Goodman D, Kelso JAS, Elliott D (1994) Intentional switching between patterns of interlimb coordination. J Hum Mov Stud 27:201–218
12. Chatrian GE, Lettich E, Nelson PL (1985) Ten percent electrode system for topographic studies of spontaneous and evoked EEG activity. Am J EEG Technol 25:83–92
13. Chen Y, Ding M, Kelso JAS (1997) Long term memory processes (1/fa type) in human coordination. Phys Rev Lett 79:4501–4504
14. Coles MGH, Rugg MD (1995) Event-related brain potentials: An introduction. In: Rugg MD, Coles MGH (eds) Electrophysiology of mind. Oxford University Press, Oxford, pp. 1–26
15. Cvetkovic D, Cosic I, Djuwari D (2004) The induced rhythmic oscillations of neural activity in the human brain. In: Tilg B (ed) Proceedings of the Second IASTED International Conference on Biomedical Engineering, ACTA, Calgary
16. David O, Harrison L, Friston KJ (2005) Modelling event-related responses in the brain. NeuroImage 25(3):756–770
17. DeGuzman GC, Kelso JAS, Buchanan JJ (1997) The self-organization of trajectory formation: II Theoretical model. Biol Cybern 76:275–284
18. Demiralp T, Bayraktaroglu Z, Lenz D, Junge S, Busch NA, Maess B, Ergen M, Herrmann CS (2007) Gamma amplitudes are coupled to theta phase in human EEG during visual perception. Int J Psychophysiol 64(1):24–30

19. Donchin E (1979) Event-related brain potentials: A tool in the study of human information processing. In: Begleiter H (ed) Evoked potentials and behavior. Plenum Press, New York, pp. 13–75

20. Fink P, Foo P, Jirsa VK, Kelso JAS (2000) Local and global stabilization of coordination by sensory information. Exp Brain Res 134:9–20

21. Fogassi L, Ferrari PF, Gesierich B, Rozzi S, Chersi F, Rizzolatti G (2005) Parietal lobe: from action organization to intention understanding. Science 308: 662–667

22. Foo P, Deguzman GC, Kelso JAS (in press) Intermanual and interpersonal stabilization of unstable systems. J Motor Behav

23. Frith C, Frith U (2005) Theory of mind. Curr Biol 15:R644–R645

24. Fuchs A, Jirsa VK (2001) The HKB Model Revisited: How varying the degree of symmetry controls dynamics. Hum Mov Sci 19:425–449

25. Fuchs A, Jirsa VK, Haken H, Kelso JAS (1996) Extending the HKB-Model of coordinated movement to oscillators with different eigenfrequencies. Biol Cybern 74:21–30

26. Fuchs, A, Jirsa, VK, Kelso, JAS (2000) Theory of the relation between human brain activity (MEG) and hand movements. NeuroImage 11:359–369

27. Fuentemilla L, Marco-Pallares J, Grau C (2006) Modulation of spectral power and of phase resetting of EEG contributes differentially to the generation of auditory event-related potentials. NeuroImage 30(3):909–916

28. Galambos RS, Makeig S, Talmachoff P (1981) A 40 Hz auditory potential recorded from the human scalp. Proc Natl Acad Sci USA 78(4):2643–2647

29. Haken H, Kelso JAS, Bunz H (1985) A theoretical model of phase transitions in human hand movements. Biol Cybern 51:347–356

30. Herrmann CS (2001) Human EEG responses to 1–100 Hz flicker: Resonance phenomena in visual cortex and their potential correlation to cognitive phenomena. Exp Brain Res 137:346–353

31. Heyes C (2001) Causes and consequences of imitation. Trends Cognit Sci 5:253–261

32. Jantzen KJ, Kelso JAS (2007) Neural Coordination Dynamics of Human Sensorimotor Behavior. In: Jirsa VK, McInstosh RA (eds) Handbook on brain connectivity, Springer, Berlin, pp. 421–461

33. Jeka JJ, Kelso JAS (1989) The dynamic pattern approach to coordinated behavior: A tutorial review. In: Wallace SA (ed) Perspectives on the coordination of movement. North Holland Publishers, pp. 3–45

34. Jirsa VK, Fuchs A, Kelso JAS (1998) Connecting cortical and behavioral dynamics: Bimanual coordination. Neural Comput 10:2019–2045

35. Jirsa VK, Kelso JAS (2005) The excitator as a minimal model for the coordination dynamics of discrete and rhythmic movement generation. J Mot Behav 37(1):35–51

36. Kay BA, Kelso JAS, Saltzman EL, Schöner GS (1987) The space-time behavior of single and bimanual movements: Data and model. Journal of Experimental Psychology: Hum Percept Perform 13:178–192

37. Kelso JAS (1981) On the oscillatory basis of movement. Bull Psychon Soc 18:63

38. Kelso JAS (1984) Phase transitions and critical behavior in human bimanual coordination. Am J Physiol Regul Integr Comp 15:R1000–R1004

39. Kelso JAS (1995) Dynamic patterns: The self-organization of brain and behavior. MIT Press, Cambridge, MA

40. Kelso JAS, Bressler SL, Buchanan S, DeGuzman GC, Ding M, Fuchs A., Holroyd T (1991) Cooperative and critical phenomena in the human brain revealed by multiple SQUIDS. In: Duke D, Pritchard W (eds) Measuring Chaos in the Human Brain. World Scientific, New Jersey, pp. 97–112

41. Kelso JAS, DelColle J, Schöner G (1990) Action-perception as a pattern formation process. In: Jeannerod M (ed) Attention and Performance XIII. Erlbaum, Hillsdale, NJ, pp. 139–169

42. Kelso JAS, Engstrøm D (2006) The complementary nature. MIT Press, Cambridge MA

43. Kelso JAS, Fuchs A, Lancaster R, Holroyd T, Cheyne D, Weinberg H (1998) Dynamic cortical activity in the human brain reveals motor equivalence. Nature 392:814–818

44. Kramarenko AV, Tan U (2002) Validity of spectral analysis of evoked potentials in brain research. Int J Neurosci 112:489–499

45. Lu ZL, Sperling G (2003) Measuring sensory memory: Magnetoencephalography habituation and psychophysics. In: Lu ZL, Kaufman L (eds) Magnetic source imaging of the human brain. Lawrence Erlbaum Associates Inc., Mahwah NJ, pp. 319–342

46. Makeig S, Westerfield M, Jung TP, Enghoff S, Townsend J, Courchesne E, Sejnowski TJ (2002) Dynamic brain sources of visual evoked responses. Science 295:690–694

47. Marsh JT, Worden FG (1968) Sound evoked frequency-following responses in the central auditory pathway. Laryngoscope 78:1149–1163

48. Mayville JM, Fuchs A, Kelso JAS (2005) Neuromagnetic motor fields accompanying self-paced rhythmic finger movements of different rates. Exp Brain Res 166:190–199

49. Mayville JM, Jantzen KJ, Fuchs A, Steinberg FL, Kelso JAS (2002) Cortical and subcortical networks underlying syncopated and synchronized coordination revealed using fMRI. Hum Brain Mapp 17:214–229

50. Meltzoff AN, Decety J (2003) What imitation tells us about social cognition: A rapprochement between developmental psychology and cognitive neuroscience. Philos Trans Roy Soc 358:491–500

51. Muthuswamy J, Thakor NV (1998) Spectral analysis methods for neurological signals. J Neurosci Methods 83(1):1–14

52. Nair DG, Purcott K, Fuchs A, Steinberg FL,Kelso JAS (2003) Cortical and cerebellar activity of the human brain during imagined and executed unimanual and bimanual action sequences: A functional MRI study. Cogn Brain Res 15:250–260

53. Oullier O, de Guzman GC, Jantzen KJ, Lagarde JF, Kelso JAS (2005) Spontaneous interpersonal synchronization. In: Peham C, Schöllhorn WI, Verwey W (eds) European workshop on movement sciences: Mechanics-physiology-psychology. Sportverlag, Köln, pp. 34–35

54. Picton TW, Lins OG, Scherg M (1995) The recording and analysis of event-related potentials. In: Boller F et al. (eds) Handbook of neuropsychology: Vol. 10, Event-related brain potentials and cognition. Elsevier, Amsterdam, pp. 3–73

55. Schack B, Vath N, Petsche H, Geissler HG, Moller E (2002) Phase-coupling of theta-gamma EEG rhythms during short-term memory processing. Int J Psychophysiol 44(2):143–163

56. Scholz JP, Kelso JAS (1990) Intentional switching between patterns of bimanual coordination is dependent on the intrinsic dynamics of the patterns. J Motor Behav 22:98–124
57. Tallon-Baudry C, Bertrand O (1999) Oscillatory gamma activity in humans and its role in object representation. Trends Cogn Sci 3:151–162
58. Thomas PJ, Tiesinga PHE, Fellous JM, Sejnowski TJ (2003) Reliability and bifurcation in neurons driven by multiple sinusoids. Neurocomputing 52–54: 955–961
59. Thompson RF, Spencer WA (1966) Habituation: A model phenomenon for the study of neuronal substrates of behavior. Psychol Rev 73:16–43
60. Tognoli E, Lagarde J, De Guzman GC, Kelso JAS (2007) From the cover: The phi complex as a neuromarker of human social coordination. Proceedings of the National Academy of Sciences USA 104: 8190–8195
61. Wallenstein GV, Kelso JAS, Bressler SL (1995) Phase transitions in spatiotemporal patterns of brain activity and behavior. Phys D 20:626–634
62. Walter VJ, Walter WG (1953) The central effects of rhythmic sensory stimulation. Electroen Clin Neurophysiol 1:57–86
63. Walter WG, Cooper R, Aldridge VI, McCallum WC, Winter AL (1964) Contingent negative variation: An electric sign of sensorimotor association and expectancy in the human brain. Nature 203:380–384
64. Yeung N, Bogacz R, Holroyd C, Cohen JD (2004) Detection of synchronized oscillations in the electroencephalogram: An evaluation of methods. Psychophysiology 41:822–832
65. Zanone PG, Kelso JAS (1992) The evolution of behavioral attractors with learning: Nonequilibrium phase transitions. J Exp Psychol Hum Percept Perform 18(2):403–421

Coordination: Neural, Behavioral and Social
Dynamics

J.A. Scott Kelso's Contributions to Our Understanding of Coordination

Armin Fuchs and Viktor K. Jirsa

Center for Complex Systems and Brain Sciences, Department of Physics, Florida Atlantic University, Boca Raton, FL 33431-0991, USA

1 Introduction

"Does old Scotty still make a living from finger wagging?" A question asked by an Irish man who had known Scott Kelso since both were children. The answer: "Yes, and doing quite well, actually" triggered the much tougher question: "What can be studied there for half a life span?" Such was not possible to respond in detail as we were at the airport in Miami and had to catch our flights. But the question remains, in more scientific terms: Why do we study coordination dynamics? Why are not only psychologists and kinesiologists but also theoretical physicists interested in finger wagging? Theorists appreciate laws and first principles, the more fundamental, the better. Coordination dynamics provides such laws. They are the basic laws for a quantitative description of phenomena that are observed when humans interact in a certain way with themselves, with other humans and with their environment.

2 Elementary Coordination Dynamics

The most basic phenomenon in coordination dynamics is easy to demonstrate: When humans move their index fingers in an anti-phase coordination pattern (one finger flexes while the other extends) and the movement frequency is increased, the movement spontaneously switches to in-phase (both fingers flex and extend at the same time) at a certain critical rate. This does not mean that the subjects could not move their fingers faster, say due to biomechanical limitations, in fact they can but only in the in-phase pattern, not in anti-phase. Why is that so? As with most of the 'why' questions the answer is: we don't know. But then Sir Isaac Newton did not know 'why' the apple falls or 'why' the moon moves around the earth, however he had figured out 'how'. So, how does coordination work and how can we describe or model its phenomena quantitatively?

In the early 1980s Scott Kelso met Hermann Haken, a theoretical physi-
cist at the University of Stuttgart, Germany. Their collaboration led to a by
now seminal paper in 1985, where they published what became known as
the Haken–Kelso–Bunz or HKB model [7]. The model was worked out in the
spirit of synergetics [6], a general theory for systems that are far from thermal equilibrium and undergo qualitative changes in their dynamical behavior
(so-called *non-equilibrium phase transitions*) when an external quantity (the
so-called *control parameter*) exceeds a critical threshold. Synergetics further
predicts that even though the systems themselves are complex (in our case
muscles, tendons, bones, joints, controlled by an even more complex system,
the brain), close to transition points they exhibit low-dimensional behavior
and their dynamics on a *macroscopic* scale can be described by a few collective
variables, the so-called *order parameters*. In systems from physics, like fluids
or lasers, these order parameters can be derived from a *mesoscopic* level where
field theoretical equations, in fluid dynamics the Navier–Stokes equations, are
known from basic laws of nature, here the conservation laws of energy and momentum. The strategy for modeling the transition from anti-phase to in-phase
in human movement coordination had to be different as the laws guiding coordination or brain dynamics on the mesoscopic level are not known a priori
and cannot (yet?) be derived from basic principles. Therefore, the approach of
Haken, Kelso and Bunz was top-down rather than bottom-up, i.e., to first find
a description on the macroscopic, order-parameter level and then determine
what kind of lower level dynamics can lead to such a macroscopic behavior.

2.1 The Macroscopic Level: Relative Phase

The first step in a top-down approach for movement coordination consists
of determining one or a few quantities that represent the order parameters
together with a dynamical system for these variables which is consistent with
the experimental observations, namely:

- At slow movement rates subjects can move their finger in either in-phase
  or anti-phase.
- If a movement is initially in anti-phase and the movement rate is increased
  subjects spontaneously switch to in-phase.
- If a movement is initially in in-phase and the movement rate is increased
  or decreased no transitions are observed.

Translated into the language of dynamical systems we can state: the movements of the single fingers are oscillations. Oscillations $x(t)$ are described by
a closed trajectory (a limit cycle) in phase space with an amplitude $r(t)$ and
a phase $\varphi(t)$. The difference between an in-phase and an anti-phase movement of two oscillators is captured by the difference between their phases
$\phi(t) = \varphi_1(t) - \varphi_2(t)$, the *relative phase*. Relative phase is 0 for an in-phase
movement and π or 180° for an anti-phase movement and became the most
important order parameter of coordination dynamics.

Now we can reformulate the experimental findings above for the dynamics of the relative phase ϕ:

- At slow movement rates the dynamics for ϕ has two stable fixed points at $\phi = 0$ and $\phi = \pi$ and the system is bistable.
- At fast rates there is only one stable fixed point at $\phi = 0$ and the system is monostable.
- Relative phase is a cyclic quantity and its dynamics must be periodic modulo 2π, i.e., expressed in terms of sine and cosine functions.

The simplest dynamical system that fulfills all these requirements is given by

$$\dot{\phi} = -a \sin \phi - 2b \sin 2\phi \qquad (1)$$

which can also be derived from a potential function

$$\dot{\phi} = -\frac{\mathrm{d}V}{\mathrm{d}\phi} \quad \text{with } V = -a \cos \phi - b \cos 2\phi \qquad (2)$$

The control parameter in this system is the ratio $k = b/a$ which corresponds to the movement rate. An increase in this ratio reflects a decrease in movement rate and vice versa. The critical value where anti-phase movement is no longer stable is given by $k = b/a = 0.25$. The dynamical properties of (1) are shown in Fig. 1 with the potential function in the top row (a), a phase space plot ($\dot{\phi}$ over ϕ) in the middle row (b) and a bifurcation diagram in the bottom row (c), where solid and open circles indicate branches of stable and unstable fixed points, respectively. A transition occurs when the system is started in anti-phase at values of k greater than 0.25 (corresponding to slow movements) and k is decreased to a value smaller than 0.25 (corresponding to an increase in movement rate beyond its critical value).

2.2 The Mesoscopic Level: Oscillators and Their Coupling

As pointed out before, (1) describes coordination behavior on the macroscopic level of the quite abstract order-parameter relative phase. As shown already in the original HKB paper of 1985 [7] this equation can be derived from a lower level by modeling the oscillator dynamics of the moving fingers. In order to define a relative phase and derive (1), we need two oscillators and a coupling function. What would be a good oscillator to describe human limb movement? The easiest oscillating species, that is linear harmonic oscillators, are not good candidates because they do not have stable limit cycles. If a linear oscillation is perturbed, the system will switch to a new orbit. In contrast, if a human limb movement is perturbed, the oscillation will relax back to its original amplitude. We therefore need nonlinear terms in the oscillator equations and as it turns out the most important ones for our purpose are $x^2 \dot{x}$ (called a van-der-Pol term) and $\dot{x}^3$ (known as the Rayleigh term). Together

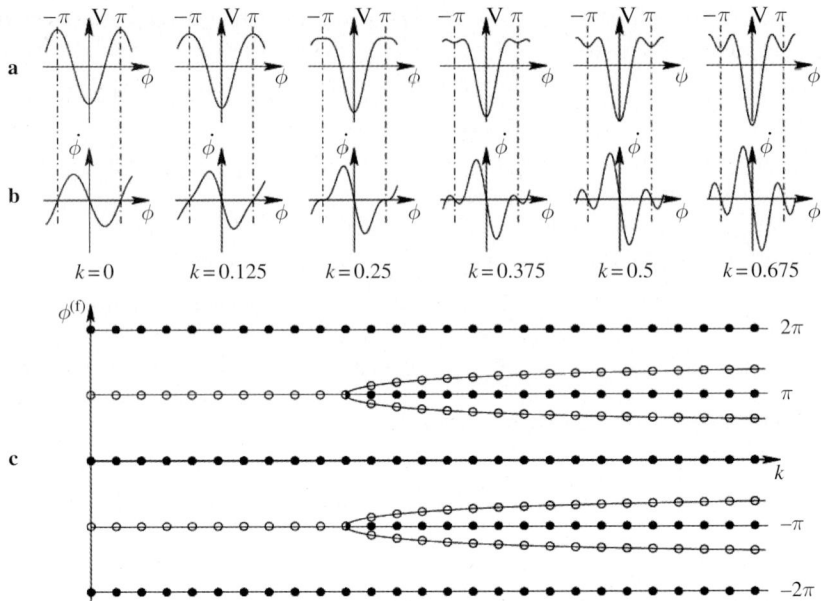

Fig. 1. Dynamical properties of (1) as a function of the control parameter $k = b/a$. Potential function (a), phase space plot (b), and bifurcation diagram (c), where solid and open circles indicate branches of stable and unstable fixed points, respectively

this leads us to a system that has been termed hybrid oscillator and reads explicitly

$$\ddot{x} + \epsilon\dot{x} + \omega^2 x + \gamma x^2\dot{x} + \delta\dot{x}^3 = 0 \tag{3}$$

There are good reasons to pick this specific form. First there is symmetry. In human limb movements the flexion phase is in good approximation a mirror image of the extension phase. This means that the equation which describes such movements must be invariant if we substitute x by $-x$, reflecting a point symmetry with respect to the origin in phase space. This constraint does not allow any quadratic terms because they would violate the required invariance or break the symmetry. The second reason stems from two experimental findings. The amplitude of a moving limb decreases linearly with frequency as has been shown by Kay et al. [13] and the phase portraits of limb movements are almost circular.

To couple two oscillators of the form (3) such that their relative phase follows the dynamics described by (1) is the true challenge. We can think of coupled oscillators as two swinging pendulums connected by a spring. The force exerted by the spring onto the pendulums is then proportional to the difference in their locations $x_1(t) - x_2(t)$. It is easy to show that such a coupling between two hybrid oscillators does not lead to a dynamics for the relative phase of the form (1). Closer investigation reveals that there are several ways that lead to the correct phase relation. The arguably easiest form of a coupling, which was also given in the original HKB paper, consists of a combination of differences in the locations and velocities of the individual components. In this case the complete system of coupled oscillators that leads to the phase relation (1) reads

$$\ddot{x}_1 + \epsilon\dot{x}_1 + \omega^2 x_1 + \gamma x_1^2\dot{x}_1 + \delta\dot{x}_1^3 = (\dot{x}_1 - \dot{x}_2)\{\alpha + \beta(x_1 - x_2)^2\}$$
$$\ddot{x}_2 + \epsilon\dot{x}_2 + \omega^2 x_2 + \gamma x_2^2\dot{x}_2 + \delta\dot{x}_2^3 = (\dot{x}_2 - \dot{x}_1)\{\alpha + \beta(x_2 - x_1)^2\}$$
(4)

The parameters a and b in (1) can now be expressed in terms of parameters in the oscillators and the coupling terms in (4) and read explicitly

$$a = -\alpha - 2\beta r^2 \qquad b = \frac{1}{2}\beta r^2 \quad \text{with} \quad r^2 = \frac{-\epsilon}{\gamma + 3\omega^2\delta}$$
(5)

where r represents the amplitude and ω the frequency of the individual hybrid oscillators.

A numerical simulation of the system of coupled oscillators (4) is shown in Fig. 2. When the system is started in anti-phase and the frequency ω is continuously increased a switch to in-phase occurs at the critical value ω_c (top row). No transition occurs when the oscillators are started in in-phase (bottom row).

It is important to be aware of the fact that there are other ways to introduce a coupling between two hybrid oscillators that leads to the right dynamics for the relative phase. In general, the bottom-up approach from the mesoscopic to the macroscopic level of description of complex systems is unique, the top-down approach is not. However, knowledge of the macroscopic behavior of a system drastically reduces the possible models on the mesoscopic level and also provide important guidance for the design of experiments.

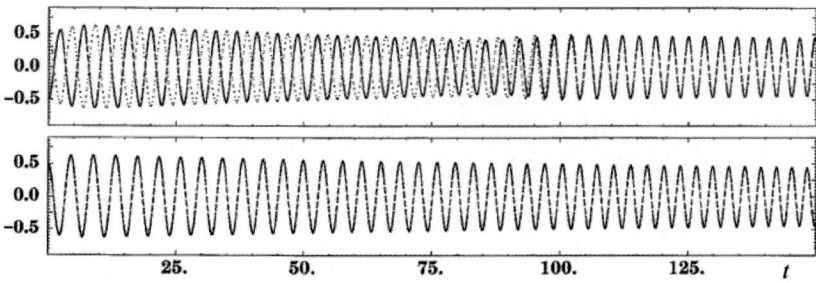

Fig. 2. Numerical simulations of (4) with initial conditions in anti-phase (top) and in-phase (bottom). The frequency is continuously increased from $\omega = 1.4$ on the left to $\omega = 2.1$ on the right. Switching at a critical value of ω_c occurs only in the anti-phase case (other parameters $\epsilon = -0.7$, $\gamma = \delta = 1$, $\alpha = -0.2$, $\beta = 0.5$)

3 Breaking the Symmetry: $\Delta\omega$

The most striking feature of the Haken–Kelso–Bunz model is its validity when we go beyond the simple symmetric cases discussed above. What does the model predict when the two oscillators have different eigenfrequencies ω_1 and ω_2? In human movements such a scenario can be realized by coordinating an arm and a leg, for instance. If the eigenfrequencies are not too different the coupling between the oscillators will force them into 1:1 frequency locking. There are regions in parameter space where the coupled system can perform in-phase or anti-phase oscillations at a common frequency Ω. As before, starting from the system of coupled oscillators that corresponds to (4) but now with eigenfrequencies ω_1 and ω_2, the dynamics of the relative phase can be derived. As it turns out, the phase relation is the same as (1) with an additional constant on the right hand side commonly called the symmetry breaking term $\Delta\omega$, which can be expressed in terms of the two eigenfrequencies of the single oscillators ω_1 and ω_2 and the common frequency Ω of the coupled system

$$\dot{\phi} = \Delta\omega - a \sin\phi - 2b \sin 2\phi \qquad \text{with} \qquad \Delta\omega = \frac{\omega_1^2 - \omega_2^2}{\Omega} \approx \omega_1 - \omega_2 \qquad (6)$$

A finite value for the constant $\Delta\omega$ in (6) leads to qualitative changes in the dynamical behavior of the relative phase. First, the fixed points of (6) are not at $\phi = 0$ and $\phi = \pi$ anymore, but are given by the solution of the transcendental equation

$$\Delta\omega - a \sin\phi - 2b \sin 2\phi = \Delta\omega - \sin\phi \{a - 4b \cos\phi\} = 0 \qquad (7)$$

For small values of $\Delta\omega$ these fixed points are shifted proportional to $\Delta\omega$ and can be written as

$$\phi^{(0)} = \frac{\Delta\omega}{a + 4b} \qquad \text{and} \qquad \phi^{(\pi)} = \pi - \frac{\Delta\omega}{a - 4b} \qquad (8)$$

The different eigenfrequencies not only lead to a shift of the fixed points, but also break the cyclic symmetry, i.e., the points $\phi = 0$ and $\phi = 2\pi$ are not the same anymore. The transitions that occur when the movement rate exceeds its critical value now have a preferred direction toward either 0 or 2π depending on the sign of $\Delta\omega$.

The potential as a function of k and $\Delta\omega$ is shown in Fig. 3. The symmetry breaking term leads to an additional slope in the HKB potential which destabilizes the fixed points at smaller k values as compared to the symmetric case. With decreasing k and increasing $\Delta\omega$ first the fixed points at $\phi = \pm\pi$ disappear and finally also the fixed point at $\phi = 0$, which corresponds to in-phase movement, becomes unstable. At this point no fixed phase relation between the two oscillators exists anymore and a phenomenon known as *phase wrapping* occurs. Details of the behavior of relative phase are shown in Fig. 4 where ϕ is plotted as a function of time. $\Delta\omega$ is kept at 1.5 and k is decreased from

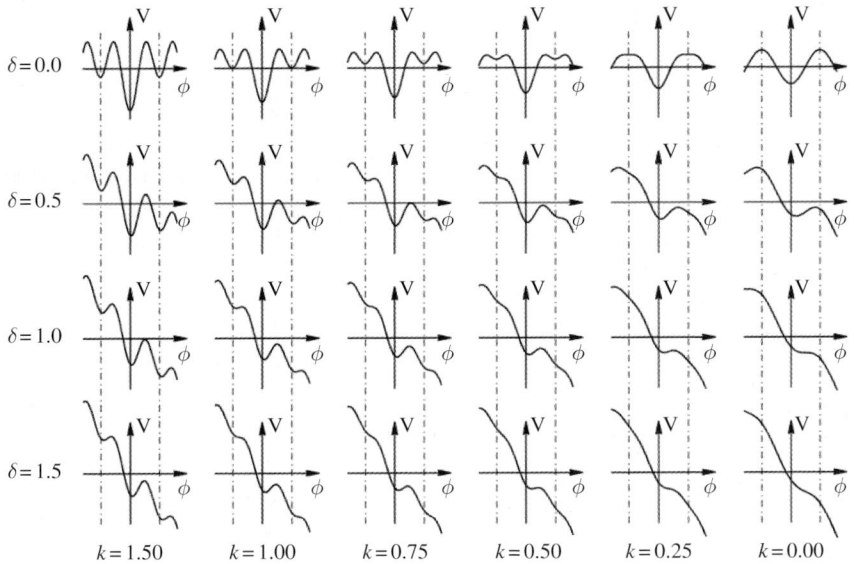

Fig. 3. The potential as a function of $k = b/a$ and $\delta = \Delta\omega$

an initial value of 1.5 by 0.1 at each of the vertical lines. First the systems settles in anti-phase but with decreasing k the fixed points drift away from $\phi = \pm\pi$. At a first critical value $k_\text{c} = 1$ the anti-phase movement becomes unstable and a switch to in-phase takes place. The new stable relative phase has shifted away from $\phi = 0$ and again shows a drift. As k decreases further a second critical value is reached where the in-phase movement becomes unstable and the relative phase starts wrapping. However, reminiscence of the in-phase fixed point can still be seen in this parameter region in form of a shallower slope around $\phi = 0$. As the movement rate increases further (with k decreasing) the curve develops more and more toward a straight line.

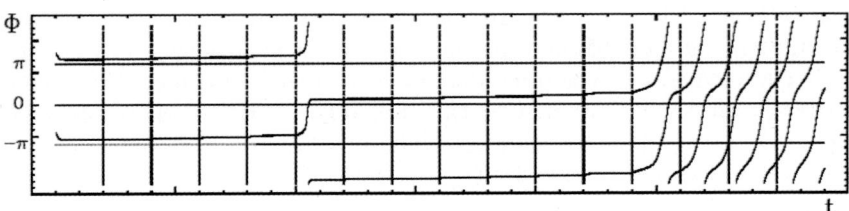

Fig. 4. Fixed point drift with decreasing k for a finite value of $\Delta\omega$. A switch from anti-phase to in-phase takes place at a first critical k_c and beyond a second critical value phase wrapping occurs

4 Beyond the HKB Model

Beyond the classic bimanual rhythmic coordination paradigm, Scott Kelso and his colleagues explored various extensions of this task. Or better: the bimanual coordination paradigm was used as an experimental window to address issues in learning, sensorimotor coordination, attention and many other areas. Here we wish to illustrate two of these extensions.

4.1 On Sensorimotor Coordination

Experimental movement paradigms are very rarely independent of environmental influences. If they are, they are referred to as 'self-paced'. In most cases though, we deal with 'paced' movements, i.e., an external stimulus is delivered to the subject. This stimulus prescribes a pace at which the subjects should perform the finger movements. But in reality the stimulus does much more: it changes the stability and the variability of the relative phase of the movement; it also changes the variability of the target point (called anchoring) and the variability of the movement amplitude. In short, it couples the environment to the perception–action system. Kelso and colleagues investigated the role of environmental information in the dynamics of bimanual coordination. A central finding that emerged from this study was that external information may serve to stabilize states that would otherwise have switched to more stable modes of coordination. Jirsa et al. [8] accounted for this effect by coupling the external information parametrically to a set of limit cycle oscillators. The main idea elucidated by the above studies is that perception and action, environmental information and the dynamics of movement, are inextricably linked. This linkage, or coupling, has been shown to be of parametric nature (mathematically speaking: multiplicative coupling). In the following, we summarize the properties of this type of coupling and point to some of its implications.

The HKB model does not account for the presence of the metronome and in its original form can be described as a model for the intrinsic dynamics of the system. Recent experiments by Fink et al. [4] and Byblow et al. [2] have established that the metronome can modify the trajectories of movement. Further, Fink et al. demonstrated that the presence of the metronome not only causes local changes in the trajectory of movement but also introduces global effects to the dynamics, such as a shift of the critical frequency at which a phase transition from the anti-phase to the in-phase mode of coordination takes place. In light of these experiments it becomes crucial to include the metronome into (4). The HKB model correctly reproduces a number of experimentally observed phenomena including phase transitions and hysteresis [7], critical fluctuations [16] and critical slowing down [21] among others. Hence, it is prudent to retain the HKB model at the core of other models that include environmental influences in a description of the dynamics of unimanual and bimanual coordination. For the latter, this was achieved by

Schöner et al. [22] and Jirsa et al. [8]. Schöner et al. used an additive linear driving term to describe the effect of the metronome. The environmental information was introduced as an additional force acting on the order-parameter dynamics attracting ϕ to the phase of the metronome. In contrast, to account for the results of Fink et al.'s recent experiments, Jirsa et al. used a parametric driving term to describe the effect of the metronome. In the limit of negligible coupling to the metronome both models reduce to the original HKB equations. Using equations of motion proposed by Jirsa et al., the coupling functions (right hand side of (4)) become

$$
\begin{aligned}
(\dot{x}_1 - \dot{x}_2)\{\alpha + \beta(x_1 - x_2)^2\} + \epsilon x_1 \cos \Omega t \\
(\dot{x}_2 - \dot{x}_1)\{\alpha + \beta(x_1 - x_2)^2\} + \epsilon x_2 \cos \Omega t
\end{aligned}
\tag{9}
$$

where the first terms denote the HKB coupling, Ω is the frequency at which the metronome is presented and ϵ is the strength of the parametric coupling.

In contrast to a linear driving or coupling, the multiplicative coupling can operate at multiple frequency ratios. In other words, such coupling allows for the stabilization of movement–stimulus frequency ratios 1:1, 1:2, 3:2 and others. This is essential to allow for a maximum of flexibility for environment–subject (or agent) coupling and has been followed up by various other researchers [3, 14]. The multifrequency coupling regimes are illustrated in Fig. 5. In order to see frequency and phase locking in the experiment it is necessary that the Arnol'd tongue structures are wide enough so that the system does not fall into a qualitatively different solution due to small perturbations that are present in any biological system. The two broadest Arnol'd tongues correspond to the 1:1 and 1:2 modes of coordination and there are no other stable coordination modes in between.

Beyond flexibility in the frequency domain, multiplicative coupling also provides for a differential stabilization in the time domain. This is clearly illustrated in Fig. 6, where phase flows obtained from computational simulations of the HKB model under parametric stimulation are shown. In particular, on the left, a situation is shown where only one stimulus per movement period is provided (single metronome). Here the driving frequency is at the center of the 1:1 Arnol'd tongue (1.2 Hz). The points of peak flexion and extension are marked in the figure. Clearly, the trajectory shows lower variability at peak flexion in comparison to peak extension. This effect can be attributed to the presence of a stimulus at peak flexion which is absent at the opposite reversal point. On the right, the figure shows the effect of the stimulus on the trajectories of motion for the double metronome condition at a frequency at the center of the 1:2 Arnol'd tongue. Here, unlike in the previous case on the left of the figure, the trajectories are symmetric in phase space. In the double metronome condition, the environmental stimulus occurs at both peak flexion and peak extension. Hence, the trajectories at the two reversal points show similar variability. Figure 7 shows the corresponding situation for single and double metronome for real human data. Clearly, the same differences in the

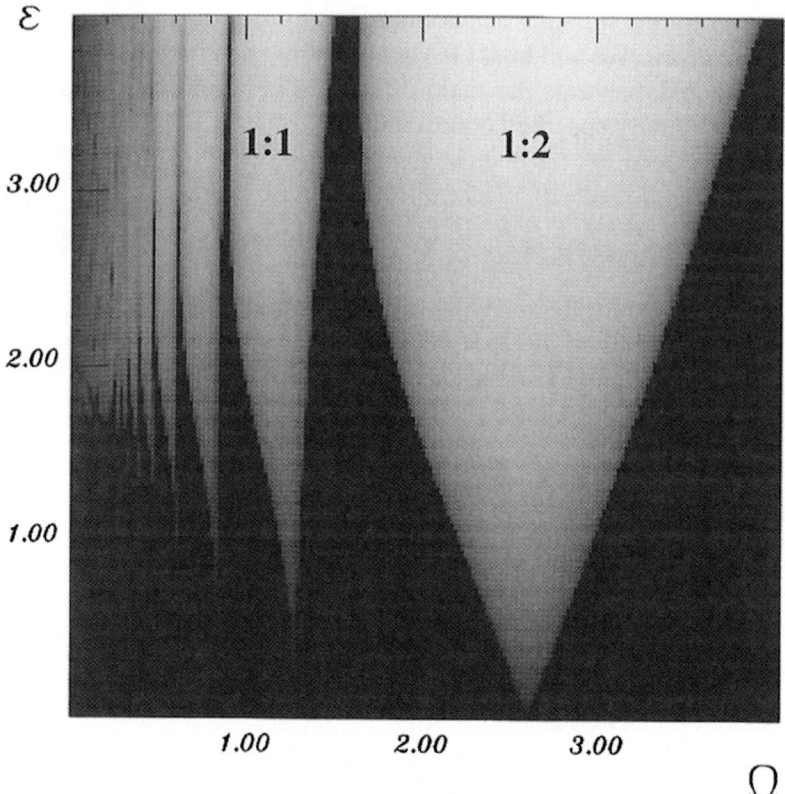

Fig. 5. Stability regions of parametric coupling are referred to as Arnol'd tongues and are plotted in light gray

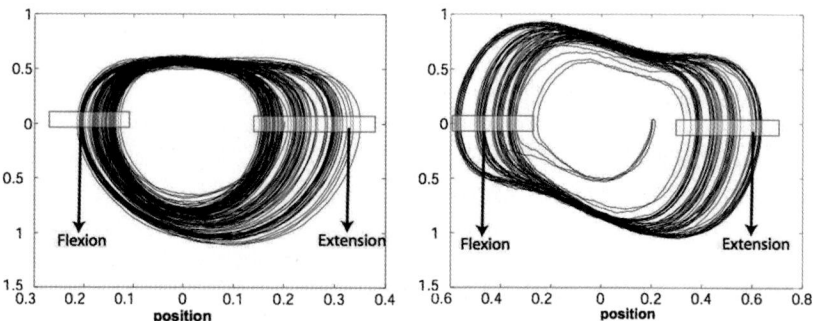

Fig. 6. Phase space trajectories when one stimulus (left) or two stimuli (right) are provided per movement cycle. Data are obtained from simulations

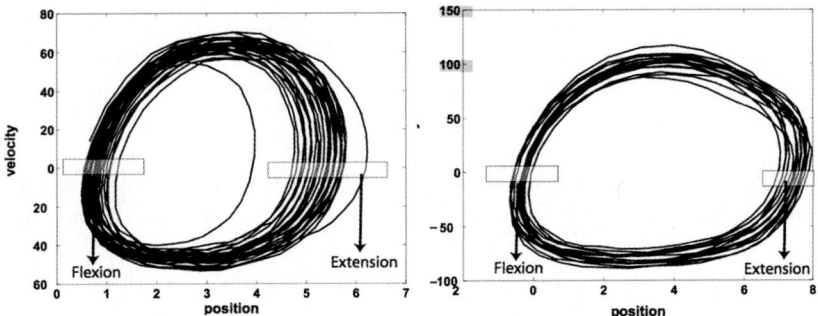

Fig. 7. Experimentally obtained phase space trajectories when one stimulus (left) or two stimuli (right) are provided per movement cycle

variability of the phase flows are observed, indicating differential stability in the time domain.

Despite the effect of *local* stabilization in the time domain, or better, in phase space, the following *global* consequence of multiplicative coupling is even more intriguing. It turns out that bimanual coordination under the double metronome stimulation is more stable than in the single metronome condition. This is reflected by a larger critical transition frequency and reduced variance of the relative phase between the left and right hands (and not between hand and stimulus!). We consider the coordination pattern of both hands (for instance, the in-phase or anti-phase pattern) as a global characteristic, since various information on both effector movements is needed, whereas a point (and its trajectory variability) in the phase space of one effector (such as maximum flexion of the left hand) does not require this information. Hence, can we naively infer that additional information locally in the phase space stabilizes the global coordination pattern? Probably not. However, it seems safe to say that there exist local manipulations of phase space trajectories, which allow to control certain aspects of the entire movement system. Since a phase space trajectory is an entity which characterizes the dynamics of the movement, one could rephrase the former statement such as "if you know when and where to stimulate the end effector, then you control the whole system". The difficult part is the "where" and "when" for complex movements. For bimanual rhythmic coordination, Kelso and colleagues answered that question.

4.2 On Non-rhythmic Coordination

Often the question has been raised (repeatedly also by the current authors themselves), why are bimanual rhythmic movements interesting? The answer is that they are not, at least not from the perspective of real world applications. Per se these movements are quite limited. However, the bimanual rhythmic coordination paradigm has served as a beautiful entry point to probe the functioning of the nervous system. Still, this point taken, what do we learn about coordination in real world problems? In 2005, Jirsa and Kelso [12] extended the applicability of dynamic system theory to a wider range of movements, including discrete and continuous movements. The power of their approach lies in the generality of phase flows. All time-continuous and deterministic dynamics must be captured by the phase flows in phase space. Such is not just another model, but actually a theorem underlying the temporal evolution of dynamical systems. Furthermore, the topology of the phase flows qualifies as a candidate invariant as shown by various theorems in the theory of dynamical systems.[1] Hence, two dynamical systems (or in this context: movement systems) are different if, and only if, their phase space topologies are different. If the topologies are not different, then a transformation exists that allows to map one system upon the other. Hence, the systems are not really different. This insight provides us with a beautiful approach to classify human movements on the phenomenological level, i.e., without making reference to the underlying neural substrate. However, this shall not be the focus of our discussion here (instead, see Huys et al. (this volume) for a discussion of phase space topologies in the context of timing and Sternad (this volume) for a discussion of the discrete vs. rhythmic movement debate). Rather we assume the existence of phase flows (and leave it to the cited authors to dwell on these issues) and ask the question, how does the HKB paradigm contribute to the discrete–rhythmic movement debate? Or equivalently, does the HKB coupling extend to non-rhythmic movements and hence coordination? Kelso and colleagues [11, 12] went forward and addressed this question theoretically. First of all, rhythmic movements are limited to closed (in the ideal case circular) structures in phase space as illustrated in Fig. 8. The in-phase movement then corresponds to a motion where the coordinates (position u and velocity v) of both effectors (1 and 2) coincide (see lower row in Fig. 8) in phase space indicated here by x and y on the axes. The anti-phase movement corresponds to the situation when the two effectors assume the maximum distance in the allotted space (that is the circle) as shown in the upper row of Fig. 8.

Since these two states are both stable (flows convergent to the particular state) below a critical frequency, then, when following the line along the circle and separating the two effectors, (u, v), there must be a point where the flow switches its direction. This point on the circle identifies the maximum divergence in the coordination of the two effectors. If the HKB coupling is removed, the entire circle consists of neutral points only, that is the flow along the circle is zero, or put differently once again, any phase relation is allowed. Hence, Kelso and colleagues put forward that the nature and function of a coupling

---

[1] This statement is precise for two degrees of freedom. In higher dimensions, one may identify cases where the situation is less clear (such as low- or high-dimensional deterministic chaos). However, it is equally unclear and questionable if these special cases will ever be of any relevance for human movements.

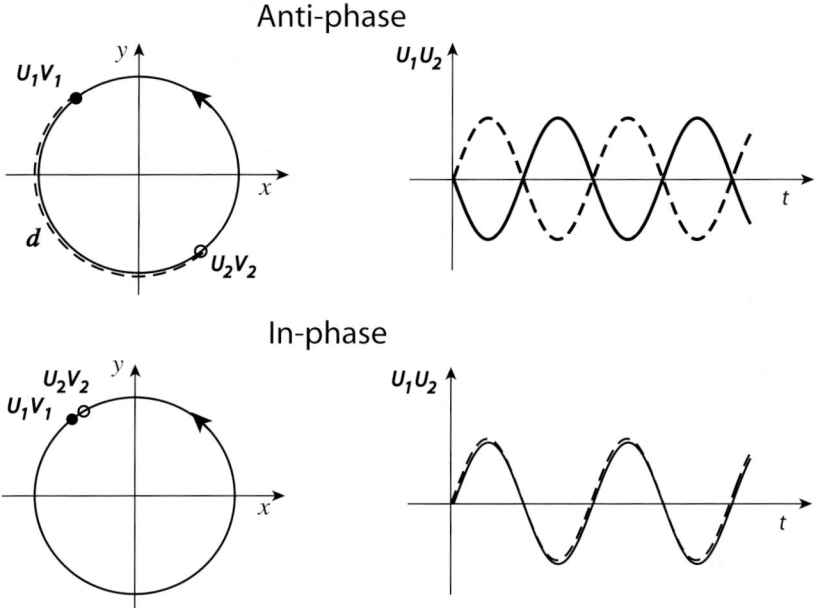

Fig. 8. Euclidean distance d of two limit cycle oscillators is maximal for anti-phase motion as a consequence of divergent dynamics and minimal for in-phase motion due to convergent dynamics

is to alter the existing flows in phase space in a meaningful manner [12]. This may be true for cognitive systems [11], but in particular for the movement system. If true, then such shall be true for arbitrary movements also (assuming that the nervous system does not adapt its couplings when switching from one movement type to another). Following this line of thought, Kelso and colleagues [12] implemented the HKB coupling in numerical simulations of arbitrary movements. The latter are characterized by not being constrained to the circle as in rhythmic movements (Fig. 8), but may explore a much larger phase space. The findings of these simulations are summarized in Fig. 9.

Essentially the convergence and divergence hypothesis translates to acceleration and deceleration phenomena of individual effectors in arbitrary but coordinated movement tasks. More specifically, when two discrete movements are executed bimanually (first with the one effector, then with the other) following two stimuli, then the movement time of the effectors will be influenced by the HKB coupling. Up to a certain maximal inter-stimulus interval (ISI) the two movement trajectories will converge in phase space. Beyond the critical ISI, the two movement trajectories will diverge. Translating this effect of convergence and divergence to more common measures in movement sciences, we obtain the difference in movement time as illustrated in Fig. 9. This effect is a clear prediction – based upon the HKB coupling – within the

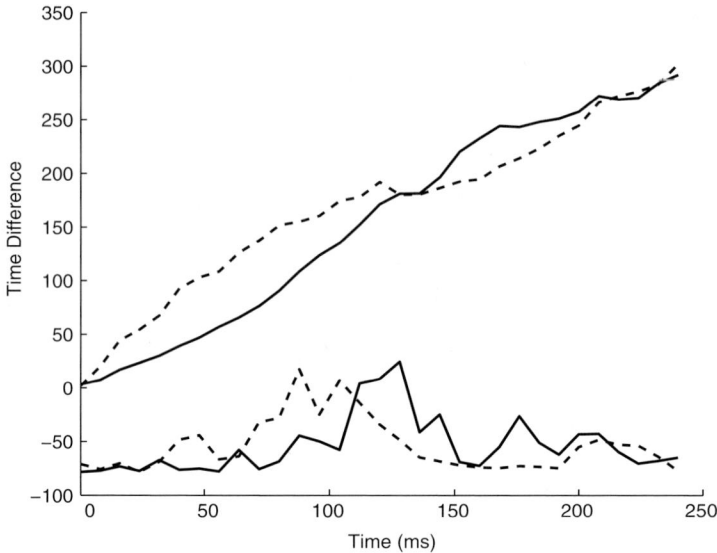

Fig. 9. Mean time difference of the two effectors (upper graphs) and its variance (lower graphs) are plotted as a function of the inter-stimulus interval. The time units are in ms. The dashed lines refer to the uncoupled situation, the solid lines to the coupled situation. The variance is not plotted on the same scale as the time difference graphs. The maximum variance is about $800 \, \text{ms}^2$ at an inter-stimulus interval of approximately 130 ms

domain of discrete movements and remains to be tested experimentally. In any case, the development of the convergence–divergence hypothesis of couplings illustrates beautifully how the HKB paradigm accomplishes in enriching other domains within movement sciences, which are at first sight beyond its reach.

5 Neural Correlates of Coordination

Starting in the early 1990s a series of experiments have been conducted using noninvasive brain imaging techniques to identify the regions and networks in the brain that are activated when humans perform coordination tasks. Especially, the syncopation–synchronization paradigm (see below) has been studied using electroencephalography (EEG), magnetoencephalography (MEG) and functional magnetic resonance imaging (fMRI). These technologies are sensitive to different quantities and provide complementary aspects about information processing inside the brain. EEG and MEG reflect the electric neural activity with a high temporal resolution of milliseconds, whereas fMRI is a measure of the metabolism of cell clusters but provides true three-dimensional information on a spatial scale of millimeters.

5.1 MEG

In several experiments MEG was used to record the magnetic field originating from neural activity while subjects performed coordination tasks [5, 17]. We will restrict ourselves here to results involving the *syncopation–synchronization paradigm* [15]. In this experimental setup a single limb (finger) is moved such that peak flexion occurs between the beats of an external metronome, i.e., in a syncopated fashion. When the metronome frequency is increased a critical value is reached where the subjects switch spontaneously to a movement where peak flexion occurs on the beat, i.e., in synchronization with the metronome.

While subjects performed the task, both their finger movements and neural activity were recorded. The movement profile was detected as pressure changes in a small air cushion connected to a transducer outside the magnetically shielded room that housed the MEG device. For brain recordings a fullhead magnetometer with about 150 SQuID (Superconducting Quantum Interference Device) sensors was used to measure the radial component of the magnetic field originating from tiny currents reflecting neural activity inside the brain.

The main findings from these experiments include:

1. At low coordination rates the strongest signal power is found bilaterally in sensors over auditory and sensorimotor cortices. At high movement rates, where subjects can only synchronize, the activity over the ipsilateral hemisphere with respect to the finger movements disappears (Fig. 10).
2. In channels over the contralateral sensorimotor areas the spectrum component switches from the coordination frequency to its first harmonic [17].
3. The phase of the Fourier component corresponding to the coordination frequency in these channels undergoes a shift by 180° at the point where the movement undergoes its transition from anti-phase to in-phase.
4. The spatial pattern of the dominating mode from a principal component analysis corresponds to a pattern that reflects auditory activation for syncopation at low movement rates, whereas after the switch to synchronization the dominating pattern reflects sensorimotor activity.

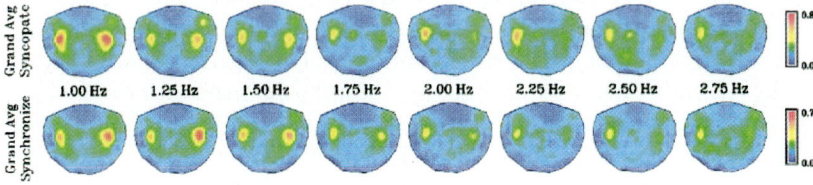

Fig. 10. Global signal power changes from a bilateral pattern at low coordination rates to a pattern with only a single maximum over the contralateral hemisphere at fast rates. There is no significant difference between the conditions (note that at high rates the subjects have switched to synchronization)

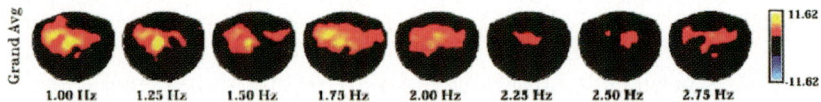

Fig. 11. Spatial distribution of activity differences synchronization minus syncopation in the β-band for different movement rates. All differences are positive, indicating a stronger β-desynchronization during syncopation. Interestingly, even after the switch to synchronization has occurred in the syncopation condition there is still a significant difference to β-activity when the initial coordination pattern was synchronization

5. The power in the β-band (15–30 Hz) is significantly larger during synchronization compared to syncopation, or in other words, β-desynchronization is larger during syncopation than synchronization (Fig. 11).

5.2 fMRI

In recent years functional MRI has become widely available and the brain imaging technology of choice for many neuroscientists. The fMRI signal is based on the change in the magnetic properties of the hemoglobin molecules, the oxygen carrier in the blood. When oxygen is released to a cell oxyhemoglobin, which is diamagnetic, is transformed into deoxyhemoglobin, a paramagnetic molecule. The two forms interact differently with the spins in the regions where the oxygen release takes place, which leads to a detectable change in the magnetic resonance signal intensity. Therefore, fMRI is a not a direct measure of neural activity but of metabolism. Nevertheless, it provides a spatially high resolution real volume measure of the brain regions that are active when certain tasks are performed. This information complements the knowledge gained from technologies with high temporal resolution like MEG and EEG. Moreover, fMRI allows for detecting sub-cortical activity that cannot be picked up by the electrophysiological technologies.

Straightforward comparison of brain regions that are active when subjects performed syncopated and synchronized timing patterns revealed both similarities and differences [18]. When compared to rest, both coordination types trigged a larger fMRI signal in auditory cortices, the contralateral sensorimotor areas, the premotor and supplemental motor areas and the (primarily ipsilateral) cerebellum as shown in Fig. 12. During syncopated movement, nevertheless, not only was activity in these brain regions increased as compared to synchronization, but also an additional network of areas including the basal ganglia, and prefrontal and temporal association cortices became active. Interestingly, there was not a single region where the activity during syncopation was smaller than in synchronization. When compared to a control condition where the subjects performed self-paced movements at about the same rate activation in the cerebellum turned out to be increased for syncopation but decreased for synchronization.

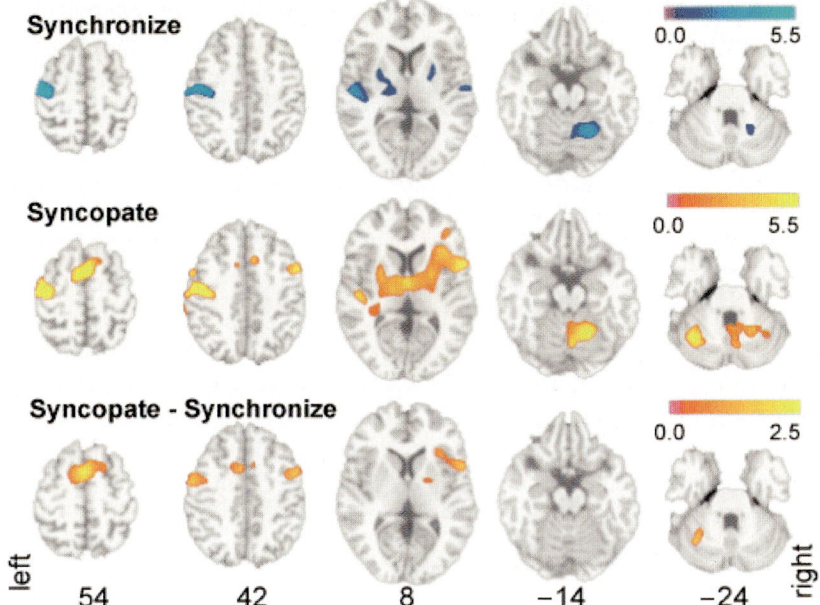

Fig. 12. Active brain areas during synchronization (first row), syncopation (second row) as well as the difference in activation between the two conditions (third row)

Taken together these results led to the conclusion that the synchronization timing pattern can be carried out relatively automatically as a sequence, whereas the syncopation pattern requires planning, initiating, monitoring and execution of each movement cycle individually.

5.3 Modeling of Neural Activity

The experimental findings from the MEG experiments summarized above were used by Jirsa and Haken [9, 10] to formulate a model of the underlying neural dynamics on a mesoscopic scale of the so-called neural field. Starting from the work of Wilson and Cowan [23, 24] and the wave equation approach by Nunez [19, 20], both dating back to the early 1970s, they derived an integral equation for the neural field $\psi(\mathbf{x}, t)$ on a cortical surface Γ which reads explicitly

$$\psi(\mathbf{x}, t) = a \int_{\Gamma} d\mathbf{x}' f(\mathbf{x}, \mathbf{x}') \, S\{\rho \, \psi(\mathbf{x}', t - \frac{|\mathbf{x} - \mathbf{x}'|}{v}) + p(\mathbf{x}', t - \frac{|\mathbf{x} - \mathbf{x}'|}{v})\}$$

(10)

Here $f(\mathbf{x}, \mathbf{x}')$ represents the coupling between the locations $\mathbf{x}$ and $\mathbf{x}'$, p is an external input at location $\mathbf{x}'$ that affects the field at location $\mathbf{x}$ with a delay given by the distance $|\mathbf{x} - \mathbf{x}'|$ divided by the propagation velocity v. $S\{X\}$ is a sigmoidal function, and a and ρ are constants that represent synaptic weights and fiber density, respectively.

The form (10) is quite general in the sense that first the dimension of the cortical surface is not set and second the connectivity function $f(\mathbf{x}, \mathbf{x}')$ is not explicitly defined, allowing both homogeneous connections between a location and its neighbors as well as heterogeneous connections realized as fiber bundles in the cortical white matter connecting distant brain regions. In the original work it was assumed that the cortical "surface" is one-dimensional that there are no long-range or heterogeneous connections in the system and that the short-range homogenous connectivity between locations falls off exponentially with distance

$$f(x, x') = \frac{1}{2\sigma}e^{|x-x'|\sigma} \tag{11}$$

The latter is a good approximation for the short-range connectivity that is found experimentally in mammals [1].

Under these assumptions (10) can be written in form of a partial differential equation which is much easier to deal with than the retarded integral equation (10)

$$\ddot{\psi}(x, t) + (\omega_0^2 - v^2\frac{\partial^2}{\partial x^2})\psi(x, t) + 2\omega_0\,\dot{\psi}(x, t)$$

$$= a(\omega_0^2 + \omega_0\frac{\partial}{\partial t})S\{\rho\,\psi(x, t) + p(x, t)\} \quad \text{with} \quad \omega_0 = \frac{v}{\sigma} \tag{12}$$

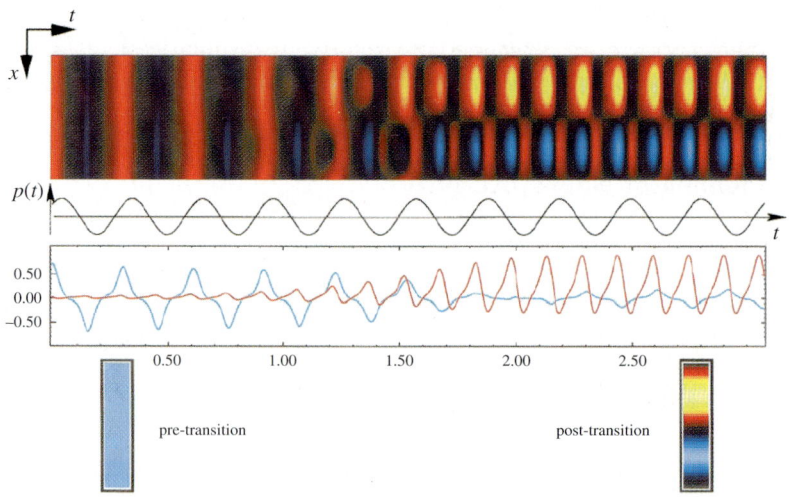

Fig. 13. Numerical simulation of (12). Top row: Space–time plot of the spatiotemporal activity. The dynamics undergoes a transition from a spatially homogenous oscillation pattern to an oscillation with a pattern that has a maximum and a minimum in space when the frequency of the external driving (second row) exceeds a critical value. Third row: Temporal amplitudes of the two dominating spatial patters (shown in the fourth row)

Results from a numerical simulation of (12) are shown in Fig. 13. The top row represents the spatiotemporal pattern of the one-dimensional neural sheet in a space–time plot. A qualitative change in the dynamical behavior from an oscillation of a spatially constant pattern to a pattern that has a maximum and minimum at each point in time is evident. The second row shows the external signal (metronome), driving the system with a constantly increasing frequency. The third row shows the amplitudes of a projection of the spatiotemporal signal onto the two patterns in the bottom row, also showing the transition in the dominating spatial pattern at the critical driving frequency. This simulation demonstrates that even with such a one-dimensional model the main experimental findings from the MEG experiments can be reproduced. Work on more realistic models using a two-dimensional surface and also heterogeneous long-range connections between distant locations is currently under way.

6 Conclusion

From its beginning more than a quarter century ago coordination dynamics has come a long way. Not only has coordination behavior been modeled quantitatively, it is one of the few cases in the life sciences where such has been achieved for a system as complex as the human body. But the phenomena, in particular the transition phenomena that are studied in coordination dynamics, have also been used as a probe into the most complex system known to us, the human brain.

In this chapter we have summarized a small portion of the many contributions Scott Kelso has made to advance the science of coordination dynamics. We have done so from our point of view, which is the perspective of two theoretical physicists. What makes coordination dynamics fascinating to our species is the insight that it can be put on a solid quantitative foundation and that the same basic laws govern on all the levels covered in this volume: behavioral, neural or social dynamics.

In any case, the Irish man would be astounded of where Scott's finger wagging has led us to.

Acknowledgments

We thank Kelly Jantzen for providing Fig. 12. Work supported by NINDS grant NS48229 to J.A. Scott Kelso, and Brain NRG JSM22002082 and ATIP (CNRS) to Viktor K. Jirsa.

References

1. Abeles M (1991) Corticonics. Cambridge University Press, Cambridge
2. Byblow WD, Carson RG, Goodman D (1994) Expressions of asymmetries and anchoring in bimanual coordination. Hum Mov Sci 13: 3–28

3. Daffertshofer A, Huys R, Beek PJ (2004) Dynamical coupling between locomotion and respiration. Biol Cybern 90:157–164
4. Fink PW, Foo P, Jirsa VK, Kelso JAS (2000) Local and global stabilization of coordination by sensory information. Exp Brain Res 134:9–20
5. Fuchs A, Jirsa VK, Kelso JAS (2000) Theory of the Relation between human brain activity (MEG) and hand movements. Neuroimage 11:359–369
6. Haken H, (1983) Synergetics: An Introduction. Springer, Berlin
7. Haken H, Kelso JAS, Bunz H (1985) A theoretical model of phase transitions in human hand movements. Biol Cybern 51:347–356
8. Jirsa VK, Fink P, Foo P, Kelso JAS (2000) Parametric stabilization of bimanual coordination: A theoretical model. J Biol Phys 1:85–112
9. Jirsa VK, Haken H (1996) Field theory of electromagnetic brain activity. Phys Rev Lett 77:980–963
10. Jirsa VK, Haken H (1997) A derivation of a macroscopic field theory of the brain from the quasi-microscopic neural dynamics. Physica D 99:503–526
11. Jirsa VK, Kelso JAS (2004) Integration and segregation of behavioral and perceptual function. In: Jirsa VK, Kelso JAS (eds) Coordination dynamics: Issues and trends. Springer, Berlin pp. 243–259
12. Jirsa VK, Kelso JAS (2005) The excitator as a minimal model for the coordination dynamics of discrete and rhythmic movement generation. J Motor Behav 37:35–51
13. Kay BA, Kelso JAS, Saltzman EL, Schöner GS (1987) The space-time behavior of single and bimanual movements: Data and model. J Exp Psychol Hum. Percept. Perform. 13:178–192
14. Kay BA, Warren WH Jr (1998) A Dynamical model of coupling between posture and gait. In: Rosenbaum DA, Collyer CA (eds) Timing of behavior. MIT Press, Cambridge pp. 293–322
15. Kelso JAS, DelColle JD, Schöner GS (1990) Action-perception as a pattern formation process. In: Jeannerod M (ed) Attention and performance. Erlbaum, Hillsdale, Vol. 13, pp. 139–169
16. Kelso JAS, Scholz JP, Schöner G (1986) Nonequilibrium phase transitions in coordinated biological motion: Critical fluctuations. Phys Lett A 118:279–284
17. Mayville JM, Fuchs A, Ding M, Cheyne D, Deecke L, Kelso JAS (2001) Event related changes in neuromagnetic activity associated with syncopation and synchronization timing tasks. Hum Brain Mapp 14:65–80
18. Mayville JM, Jantzen KJ, Fuchs A, Steinberg FL, Kelso JAS (2003) Cortical and subcortical networks underlying syncopated and synchronized coordination revealed using fMRI. Hum Brain Mapp 17:214–229
19. Nunez PL (1972) The brain wave equation: A model for the EEG. Math Biosci 21:279–297
20. Nunez PL (1981) Electric fields of the brain. Oxford University Press, Oxford
21. Scholz J, Kelso JAS, Schöner G (1987) Nonequilibrium phase transitions in coordinated biological motion: Critical slowing down and switching time. Phys Lett A 123:390–394
22. Schöner G, Kelso JAS (1988) A synergetic theory of environmentally-specified and learned patterns of movement coordination: II. Component oscillator dynamics. Biol Cybern 58:81–89
23. Wilson HR, Cowan JD (1972) Excitatory and inhibitory interactions in localized interactions of model neurons. Biophys J 12:1–24
24. Wilson HR, Cowan JD (1973) A mathematical theory of the functional dynamics of cortical and thalamic nervous tissue. Kybernetik 13:55–80

Index

Understanding Complex Systems

Printing: Krips bv, Meppel, The Netherlands
Binding: Stürtz, Würzburg, Germany